Paul Rasmussen

ADAPTIVE MANAGEMENT
OF RENEWABLE RESOURCES

BIOLOGICAL RESOURCE MANAGEMENT

A Series of Primers on the Conservation and Exploitation of Natural and Cultivated Ecosystems

Wayne M. Getz, Series Editor
University of California, Berkeley

Adaptive Management of Renewable Resources, by Carl Walters

Building Models for Wildlife Management, by A. M. Starfield and A. L. Bleloch

Mathematical Programming for Economic Analysis in Agriculture, by Peter B. R. Hazell and Roger D. Norton

Range Economics, by John P. Workman

ADAPTIVE MANAGEMENT OF RENEWABLE RESOURCES

CARL WALTERS

Institute of Animal Resource Ecology
University of British Columbia, Canada

MACMILLAN PUBLISHING COMPANY
NEW YORK

Collier Macmillan Publishers
LONDON

Macmillan Publishing Company
866 Third Avenue, New York, NY 10022

Collier Macmillan Canada, Inc.

Printed in the United States of America

1 2 3 4 5 6 7 8 9 10   6 7 8 9 0 1 2 3 4 5

Library of Congress Cataloging-in-Publication Data

Walters, Carl J., 1944–
  Adaptive management of renewable resources.

  (Biological resource management)
  1. Wildlife management. 2. Fishery management.
I. Title. II. Series.
SK355.W35 1986   333.95′15   86-2762
ISBN 0-02-947970-3

Contents

Preface

This book is about ways of dealing with uncertainty in the management of renewable resources, such as fisheries and wildlife. My basic theme is that management should be viewed as an adaptive process: we learn about the potentials of natural populations to sustain harvesting mainly through experience with management itself, rather than through basic research or the development of general ecological theory. The need for an adaptive view of management has become increasingly obvious over the last two decades, as management has turned more often to quantitative model building as a tool for prediction of responses to alternative harvesting policies. The model building has not been particularly successful, and it keeps drawing attention to key uncertainties that are not being resolved through normal techniques of scientific investigation. We keep running up against questions that only hard experience can answer, and a basic issue becomes whether to use management policies that will deliberately enhance that experience. Such policies would represent a radical departure from traditional prescriptions about how to deal with uncertainty, namely to proceed with great caution or to act as though there were no uncertainty in hopes that mistakes and opportunities will automatically reveal themselves.

My major conclusion is that actively adaptive, probing, deliberately experimental policies should indeed be a basic part of renewable resource management. The design of such policies involves three essential ingredients: mathematical modeling to pinpoint uncertainties and generate alternative hypotheses, statistical analysis to determine how uncertainties are likely to propagate over time in relation to policy choices, and formal optimization combined with game playing to seek better probing choices. In pursuing these ingredients, I have been led to examine a variety of concepts and methods, such as adaptive parameter estimation for dynamic models, that have not been widely applied to renewable resource problems. Thus, I hope the book will serve two purposes: to provide general motivation for deliberately treating management as an adaptive process, and to bring together a collection of tools for adaptive policy analysis that have previously been scattered through literature sources ranging from engineering control theory to resource economics.

The book is intended for two audiences. The introductory and concluding chapters, and chapter introductions throughout, are aimed primarily at practicing resource managers and administrators who want to get some feeling for basic issues and concepts. The later sections in Chapters 4–10 look more deeply at various mathematical tools for analysis; these sections are aimed at analysts (modelers, statisticians, stock assessment experts) who are concerned with the practice of policy design. For the technical sections, I assume the reader is familiar with introductory calculus, matrix algebra, and introductory statistics, including regression analysis. Fourth-year undergraduates and graudate students in resource ecology and economics generally have these prerequisites, and I have included problem sets to make the book more usable as an advanced text. The problems for each chapter are graded, so the first few require no special background while the later ones assume at least some skill at microcomputer programming.

It would be nice to claim that the book presents a coherent theory for management under uncertainty, with appropriate recipes for all circumstances. Unfortunately, I rather doubt that such a theory can be developed, even in principle; an essential feature of dealing adaptively with uncertainty is to reject recipes and rituals, in favor of a search for better processes to promote imagination and learning. We can now describe with some rigor how particular types of adaptive processes (such as sequential parameter estimation given a fixed dynamic model structure) are likely to perform, but only under very restrictive assumptions. It is useful to examine such processes, if for no other reason than to promote frustration and a search for better assumptions. But the really key processes are those by which we search for better assumptions upon which to base rigorous analysis, and here I can offer only experience with a few techniques such as modeling workshops.

In trying to get adaptive management ideas across to various resource agencies, I have become acutely aware that management is done *by* people, as well as for people. We all have limited backgrounds, interests, and abilities to assimilate new ideas; these limits are inevitably carried into the work place, so that decision making about renewable resources is anything but the coldly rational process usually assumed in introductory and theoretical texts. I feel that this point is crucial for students to understand, so I return to it repeatedly with comments and suggests about the need to develop simple and understandable models, to communicate analyses vividly in terms of tricks like microcomputer games and verbal summaries, and to recognize that there are fundamental conflicts of interest that no analysis can resolve. My preoccupation with communication is reflected at places in a chatty or anecdotal writing style and in oversimplification of some technical presentations; the choice was deliberate, and I make no apologies to readers who would prefer a more precise, academic style. On the other hand, I also make no apology for extensive use of mathematical models and notation; only a fool would dare approach the study of dynamic resource systems without these tools.

This book was written mainly under the support of the International Institute for Applied Systems Analysis, Laxenburg, Austria. There Dr. C. S. Holling provided me essential support and protection to write. He saw the value of forming a small research team, the Adaptive Resources Project (ARP), through which many of these ideas were developed. For stimulating discussions and assistance with the mathematical development, I am especially grateful to ARP members Joe Koonce, Anatoly Yashin, John Casti, Valeri Federov, and Mike Staley. Our project secretary, Shirley Wilson, handled the project administration and organization with a competence that gained us much time for research. For patience with my abominable writing, I thank typists Ann Tedards, Susan Riley, and Bonnie Riley. For pushing me to complete the book, special thanks to Bob Duis and his editorial staff, especially Valerie Jones. I have also received much support and valuable advice from colleagues at the University of British Columbia, especially Ray Hilborn, Don Ludwig, and Donna Chin. For much patience on too many long evenings and weekends, thank you Sandra, Daniel, and William. Finally, no thanks would of course be too much for Ralf Yorque.

The International Institute for Applied Systems Analysis

The International Institute for Applied Systems Analysis is a nongovernmental research institution, bringing together scientists from around the world to work on problems of common concern. Situated in Laxenburg, Austria, IIASA was founded in October 1972 by the academies of science and equivalent organizations of twelve countries. Its founders gave IIASA a unique position outside national, disciplinary, and institutional boundaries so that it might take the broadest possible view in pursuing its objectives:

To promote international cooperation in solving problems arising from social, economic, technological, and environmental change

To create a network of institutions in the national member organization countries and elsewhere for joint scientific research

To develop and formalize systems analysis and the sciences contributing to it, and promote the use of analytical techniques needed to evaluate and address complex problems

To inform policy advisors and decision makers about the potential application of the Institute's work to such problems

The Institute now has national member organizations in the following countries:

Austria
The Austrian Academy of Sciences

Bulgaria
The National Committee for Applied Systems Analysis and Management

Canada
The Canadian Committee for IIASA

Czechoslovakia
The Committee for IIASA of the Czechoslovak Socialist Republic

Federal Republic of Germany
Association for the Advancement of IIASA

Finland
The Finnish Committee for IIASA

France
The French Association for the Development of Systems Analysis

German Democratic Republic
The Academy of Sciences of the German Democratic Republic

Hungary
The Hungarian Committee for Applied Systems Analysis

Italy
The National Research Council

Japan
The Japan Committee for IIASA

Netherlands
The Foundation IIASA—Netherlands

Poland
The Polish Academy of Sciences

Sweden
The Swedish Council for Planning and Coordination of Research

Union of Soviet Socialist Republics
The Academy of Sciences of the Union of Soviet Socialist Republics

United States of America
The American Academy of Arts and Sciences

Chapter 1

Introduction

The first step to knowledge
is the confession to ignorance.

Weinburg (1975)

Man has proved remarkably adept at developing harvests from potentially renewable natural resources, such as fish, wildlife, and forests. But he
has shown considerably less skill in devising schemes for sustaining the harvests over long periods of time. Until the early part of this century, most
resource developments of the past few hundred years proceeded more like
mining operations, with a boom followed by stock depletion and collapse;
either no thought was given to the long term, or the resources were considered so abundant as to be inexhaustible. Then there was a dramatic shift
in viewpoint, with the emergence of theories about the limits of sustainable
harvests, development of monitoring systems that demonstrated the
deterioration of some resources, and the organization of public conservation
movements that brought political pressure on governments to act as regulators of harvesting activity.

Particularly with the growth of government involvement in management, there developed a strong demand for basic research and university
training programs. By the 1950s it had become common to claim that
fisheries, forestry, and wildlife management had developed into truly
scientific disciplines, with well defined paradigms for research and practical
action. There was, of course, considerable fragmentation into schools of
thinking that centered on different factors and investigative approaches (for
example, some wildlife biologists emphasized problems of "habitat," while
others were concerned more with population dynamics), but the general attitude was optimistic: the details would sort themselves out in due time, given
diligent research and longer experience.

But by the late 1960s some workers were beginning to question seriously whether the resource sciences were making any real progress. There continued to be many gross instances of apparent mismanagement, such as the collapse of the Peruvian anchoveta fishery and the realization that some North American forests were being slowly depleted. Weaknesses in theory and practice were particularly highlighted by the emergence of the environmental movement, with its critical scrutiny of government policies and demand for predictions in the form of environmental impact assessments.

It appears now that there were at least two fundamental flaws in the early development of the renewable resource sciences. The first flaw has been obvious to scientists from other disciplines, particularly economics, for many years: research and management have concentrated primarily on biological/ecological and technical harvesting issues, with only token consideration to the socioeconomic dynamics that are never completely controlled by management activities. This imbalance of concern appears even in reviews that purport to raise wider concerns; for example, Gulland (1981) in a recent paper on operations research in fishery management, stated that "biological models lie at the heart of fishery management." Considering that most resource scientists are trained as ecologists, it is particularly surprising to see such attitudes: harvesting systems are very much predator–prey associations (man–ecosystem), with all the potentials of such systems for unexpected dynamic response when viewed in a fragmentary way; attention has also been often focused only on the prey, thus ignoring problems that develop because the predators do not sit still either.

The second fundamental flaw in the development of natural resource science is equally serious, and provided the central motivation for this book. This flaw concerns the strategic question of how we should proceed to develop better understanding of managed system responses and potentials in a world of great uncertainty, limited research resources, and continuing pressure for more intense exploitation. The traditional dogma has it that the answer is to invest more in basic research, especially in ecology, while very cautiously regulating harvests so as not to destroy potentials before they are understood; "better" understanding is usually taken to be synonymous with "more detailed" analysis of the components of production processes. When I phrase the traditional view this way, at least two questions should immediately leap into the reader's mind: how is anyone going to put the component pieces together, if ever they are all understood? If the systems to be understood are managed conservatively, how can we possibly make all observations and experiments necessary to predict how they will behave when the conservative policies are replaced by more optimum regimes as *extrapolated* from the component understanding? My basic contention is that these questions cannot be answered affirmatively, implying that we must seek a fundamentally different approach to scientific management. I will argue that one possibility is to treat management as an adaptive learning process, where management activities themselves are viewed as the primary tools for experimentation.

Before proceeding to sketch out the basic issues of adaptive management, and thereby provide an outline for this book, let me examine more carefully the two questions raised in the last paragraph. Unless my negative contention about these questions is clearly understood, the reader may see no point in trying to work through the rather arcane machinery of analysis that forms the bulk of the text.

Does Understanding Accumulate?

Consider first the question of how to assemble bits and pieces of scientific understanding into an overall framework for management. Most students, at least of biology, are taught that knowledge and understanding accumulate, perhaps by fits and starts, toward a complete picture of nature. I first became suspicious about this view in reading Thomas Kuhn's *Structure of Scientific Revolutions* (1962), which argues that science proceeds through occasional sharp changes of view (paradigm shifts), often involving simplification and discarding of amassed results rather than accumulation. Then the point was brought home very forcefully to me one evening during an otherwise blurry barroom conversation with W.E. Ricker, the noted fisheries analyst. In reminiscing about the history of Pacific salmon management, Ricker lamented that the current generation (1978) of management biologists seemed to use less information in their decision making than had their predecessors 20 years earlier, in spite of much greater expenditures for inventories and research over those 20 years. The biologists in question fully agreed, but argued that conditions had changed so much that the older concepts, data, and management problems were simply no longer relevant. In effect, they were saying that there is no single, structurally stable system out there in nature to be understood. I cannot agree completely with that argument, because one can always argue that they are just emphasizing different aspects of a larger system. But how large is large enough to remain interesting?

Even if managed systems do not keep slipping away and changing under us, there remains the problem of how to use accumulated data effectively. Some would argue that this is not a problem, and that the human mind is quite capable of intuitively grasping and making valid inferences with complicated relationships. A more honest, humble, and realistic proposal has been to develop mathematical models that somehow integrate the complexities in a systematic way, and then to use these models as "deductive engines" for prediction. Such models are now widely used in resource management agencies (sometimes partly disguised through termi-nology like "stand table" in forestry) but their impact on actual policy choice is difficult to assess. Indeed, I have been a strong proponent of computer simulation modeling, especially since cheap and friendly microcomputers have made it much easier for people without a strong mathematical training to become involved in model development.

But much modeling experience suggests even deeper problems with the notion of accumulating experience. In the late 1960s, C.S. Holling and I began to develop a process, now known as adaptive environmental assess-ment (Holling, 1978), for more effectively constructing and testing simula-tion models for natural resource management. Our goals were to compress the time required for model development to an absolute minimum, and to involve a wide variety of key actors (disciplinary scientists, managers, policy people) in modeling for policy purposes. Using the process, we involved a very large number of scientists and developed literally dozens of rather com-plicated simulations, for cases ranging from Pacific salmon management and forest insect problems in Canada, to environmental/economic issues of ski area development in the Austrian Alps. In all these cases, major uncertain-ties about economic/ecological processes were apparent, and the processes involved showed two characteristics: (1) their effects were only clearly evi-dent on large spatial and/or temporal scales, and (2) they were not the sub-ject of intensive research investment.

Let me put this observation more vividly: we keep encountering key processes, identified as necessary causal ingredients for prediction by many scientists, that have effects that are only clearly visible over large areas and/or long time periods. These processes are therefore either very expen-sive to study experimentally, or do not offer the speedy rewards that scien-tists need to keep publishing, get research grants, and so forth. A few exam-ples are worth mentioning: large-scale dispersal of insect pests; changes in recruitment rates of fish with varying densities of parental spawners; and changes in the attack rates on prey by large vertebrate predators (like wolves) with changes in prey density.

It is a sad but understandable fact that most scientists base their research programs not on broad analyses of uncertainties, but instead on the investigative tools (nets, etc.) and analytical methods that they learned in university or find popular among colleagues. This means that some ecological/economic research paths are deeply trodden, while others remain untouched. For example, most population dynamics work in fisheries centers on the use of well defined catch sampling programs and statistical analysis procedures, aimed at estimating a few parameters of growth, mor-tality, and average recruitment rate that some modelers during the 1950s (Ricker, Beverton–Holt, etc.) prescribed as necessary to yield assessment. Of course there are shining exceptions, but these are rare enough to make one wonder whether the systems we study might not be changing faster than our understanding of them.

It has been argued that the large-scale processes that cause so much grief for modelers should not be studied directly in the first place, and instead should be logically decomposed into detailed subprocesses that are more amenable to experimentation. So far as I can tell, those who make this recommendation must never have actually tried it, or they would realize that: (1) the number of details grows explosively as a process is decomposed into finer details; (2) a model is still required eventually, to put all the details back together, and small errors in this model can have large cumulative effects; and (3) predictions from the detailed analysis must ultimately be tested by reference back to, and experiments upon, the behavior of the overall process of original interest. Also, you can be sure that there will be at least one of the detailed subprocesses that does not lend itself to existing experimental technique and resources, and the whole logical house of cards will not stand without this detail.

Dangers of Extrapolation

We have come now to my second basic question about the traditional scientific approach: how can we guarantee to conduct all observations and experiments necessary to extrapolate an optimum management regime, when the system of concern is perhaps maintained away from that regime by conservative management policies? More simply, how can you know that something will work until you try it? In the natural resource sciences, we are not dealing with engineered systems about which correct predictions should be possible, at least in principle, because the system components are deliberately chosen with supposedly known characteristics. Even in these systems, the regular occurrence of nasty surprises should be a warning to anyone who would claim that we can understand how a complicated system will act without actually testing it.

A Curved Example

Scientists who harbor some hope of successful extrapolation might consider the example of Figure 1.1, which plots harvestable stock of a major Pacific salmon system (Fraser River sockeye) resulting from allowing different numbers of adults into their spawning rivers four years earlier. There are obviously many processes at work in determining how many of the eggs laid by the spawners will hatch, survive their freshwater rearing period, then go to sea and survive to return as harvestable adults. (I spell this out a bit to evoke a sense in the biologist reader's mind of just how complex the whole survival question is.) The key feature of these data is how they appear to demonstrate that the system has been overexploited, at least since accurate records have been kept: over the range of recent experience, harvestable stock appears to be about linearly proportional to spawners. This means that spawners should be allowed to increase, until an appropriate point of diminishing returns is reached.

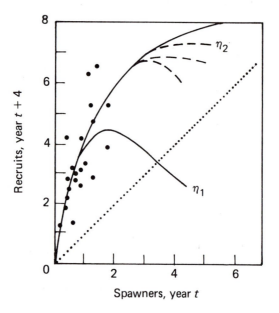

Figure 1.1. Relationship between number of sockeye salmon allowed to spawn in the Fraser River, BC, and the number of resulting offspring measured as recruits to the fishery four years later. Data are for 1939–73, omitting every fourth (cycle) year beginning in 1942. The curves η_1 and η_2 are alternative extrapolations of response to increased spawning stock. η_2 predicts higher yields if more fish were allowed to spawn. *Source:* Walters and Hilborn (1976); see also Walters (1977), Holling (1978).

Now, Figure 1.1 shows just two curves extrapolating the pattern of response to such an increase; infinitely many such curves could be drawn, all consistent with historical experience but predicting different responses to management aimed at increasing spawners. How might we resolve which of these curves of average response is correct, and thereby determine the optimum number of spawners to allow? The scientific answer to this question seems so obvious as to be hardly worth stating: let more fish spawn, and see what happens. But this answer is not at all what the traditional approach would prescribe, and in fact was strongly opposed by some biologists when we proposed it several years ago (Walters and Hilborn, 1976). The basic objections were: (1) the fisherman would protest loudly (more spawners means less immediate catch), and would lose whatever confidence they might have had that the managers knew what they were doing in the first place; and (2) the uncertainty can be resolved without disturbing the existing management balance, by conducting more basic research on the factors that limit production (cause the curves to bend over at high spawning densities). The first of these objections is more serious than it may initially appear, and I will return to it and related issues in Chapter 2.

Reflect for a moment on the complexity of the salmon life cycle (and it is simple by fisheries standards), and consider studying the various limiting factors that could operate at each stage in this cycle, *without deliberately manipulating the number of fish entering each stage.* We could, of course, do local experiments on small groups of fish in limited areas, without appreciably affecting the overall system. But what about those factors, such as disease transmission, that are not apparent at all except when large populations/areas are studied? (Disease outbreaks on spawning areas are a real concern with Fraser River sockeye.) What about the effects of overall densities of spawners on straying rates of fish to colonize new spawning areas and old areas that were depleted many years ago? The list of these experimentally difficult questions is practically endless, and would make any prediction based on small-area studies very dubious indeed! So, even if a conservative harvest regime were followed for a long time so that all the feasible local experiments could be conducted, there would still be fundamental uncertainties about the effect of increasing overall spawning numbers. In short, the waiting would not solve the problem, and considerable opportunities for increased harvest might be lost over the waiting period.

The fact that adaptive learning through management "experiments" may proceed much more quickly than through conservative management and basic research has been noticed by some practicing managers for many years, and has helped fuel an unhealthy split and mutual contempt between managers and researchers in many agencies. This split makes the valuable basic advances that do occur much more difficult to put into practice, and isolates researchers from the wealth of experimental opportunities afforded by whole-system manipulations by managers.

It is implicit in the above discussion that every managed resource system is somehow unique, with at least some quantitative characteristics that cannot be inferred from experience with other (replicate) cases. This assumption is obviously not correct in many situations, particularly in forestry and agriculture, where large and relatively homogeneous systems can be subdivided into many spatial units, with some proportion of these units devoted to experimentation that can be directly applied to the other units. Indeed, one of the wisest resource management decisions in American history was embodied in the Homestead Act of 1865, which systematically reserved a vast grid of farmland areas for agricultural experimentation. Testing of innovative schemes on these areas has probably been one of the main factors responsible for the dramatic success of American agricultural development. Of course, even in these situations there are large-scale processes such as wind erosion (producing dust bowls) and plant disease epidemics that could not reasonably be anticipated on the basis of local research activities.

Fisheries and wildlife workers have been slow in taking advantage of existing opportunities for replicated experimentation on a fairly large scale. A very delightful and surprising paper by Bilton et al. (1982) illustrates this well. They studied the effect of body size and date of release into the wild on the survival rates of juvenile coho salmon produced in one of the many salmon hatcheries that line the Pacific coast of North America. By varying these two hatchery "operating parameters," they showed that there is a rather sharply defined optimum combination to shoot for, involving smaller sizes and later releases than have generally been used in production hatcheries. The surprising thing about this study is not that it found such an optimum, but that it was done more than 20 years after large-scale investment began in salmon production hatcheries. The earlier record of experimentation and even standard operating results from these very costly hatcheries is so spotty and pathetic that one must wonder how the investments were ever justified in the first place.

Issues of Adaptive Management: A Preview

The various questions raised above are disturbing, and they have certainly not been resolved by my brief discussion. My hope has been only to raise doubt in the reader's mind about some very basic (and usually unquestioned) notions of how we should proceed in developing better resource management systems. The remainder of this text will explore an approach that has come to be called "adaptive management." This approach begins with the central tenet that management involves a continual learning process that cannot conveniently be separated into functions like "research" and "ongoing regulatory activities," and probably never converges to a state of blissful equilibrium involving full knowledge and optimum productivity.

The business of designing adaptive management strategies appears to involve four basic issues:

(1)   bounding of management problems in terms of explicit and hidden objectives, practical constraints on action, and the breadth of factors considered in policy analysis;

(2)   representation of existing understanding of managed systems in terms of more explicit models of dynamic behavior, that spell out assumptions and predictions clearly enough so that errors can be detected and used as a basis for further learning;

(3)   representation of uncertainty and its propagation through time in relation to management actions, using statistical measures and imaginative identification of alternative hypotheses (models) that are consistent with experience but might point toward opportunities for improved productivity;

(4)   design of balanced policies that provide for continuing resource production while simultaneously probing for better understanding and untested opportunity.

The following chapters will look into some details about each of these issues, using a mixture of theoretical arguments and relatively simple case examples. The emphasis will not be on presenting recipes for successful practice, but rather on stimulating critical thinking about the issues.

Chapter 2 examines the very treacherous matter of deciding what management is about. In systems analysis this is sometimes called "bounding the problem," and most, if not all, resource policy analyses go astray right at this starting point. It is very easy to build a lovely fairy castle of policy based on some incorrect presumption about what management should or can do, then to be shattered upon presenting it by hearing just two words: "who cares?"

Chapters 3–5 are concerned with representation of understanding through mathematical models. Chapter 3 reviews the so-called adaptive environmental assessment (AEA) process mentioned earlier, which tries to make model building a more effective adaptive process for the people involved in it. Chapter 4 looks at some of the models that have been widely used in fisheries analysis, and at some of the biological and economic processes that these models have failed to address adequately. Chapter 5 is concerned with the very exciting possibility that many resource problems can be well represented with very simple and understandable models resulting from "compression" of the quite complex models that usually emerge early in the AEA process. Simple models are more attractive to decision makers, and are usually all that can be identified or realized in the face of the very noisy data available.

Chapters 6 and 7 look at the sometimes discouraging problems of embracing uncertainty by comparing models with experience in the form of time series data from exploited resources. Chapter 6 discusses uncertainty in general, then reviews the Bayesian idea of assigning probabilities to alternative models and the recent systems theoretic idea of "realization," in which the analyst seeks to find all possible dynamic structures that are consistent with the data available. Chapter 7 borrows some ideas from the theory of stochastic control, to model the sequential learning process involved in resource development and continued probing of responses; a particular concern in this chapter is with statistical methods of "tracking" parameter changes over time in relation to unmodeled processes such as environmental change.

Chapters 8–11 take up the matter of finding optimal, or at least relatively good, harvest regimes in the face of great uncertainty. Chapter 8 reviews some basic ideas about stochastic optimization, with emphasis on the notion of designing feedback policies to cope with unpredictable variation. Chapter 9 deals with the design of actively adaptive management policies, in which there is a deliberate attempt to find some optimal balance between conservative, usually stabilizing, harvest regimes versus the disruptive probing necessary to gain better understanding of long-term potentials. Chapter 10 looks at the design of actively adaptive policies for replicated systems, where a number of spatial subunits (stocks, areas) are managed together and may be informative about one another. Finally, Chapter 11 examines the design of adaptive policies for complex and ambiguous situations where formal optimization is not practical because of the number of variables involved.

Problems

1.1. Consider the uncertain production relationship in Figure 1.1. Identify at least three factors that might make average recruitment lower (dome-shaped curve) at higher spawning stock densities. Indicate why "local" (laboratory, field pilot scale) experimental data about these factors would not provide a reliable basis for predicting responses of the whole system.

1.2. Again in relation to Figure 1.1, identify two major reasons why a return to historical escapement levels might not result in an immediate return to historical recruitment rates.

1.3. Table 1.1 shows population size and harvest data for whitetail deer introduced into a large fenced area, the George Reserve (McCullough, 1979). Assuming that population size next year equals population this year plus net production minus harvest, estimate net production for each year as population next year minus population this

Table 1.1. Population size and harvest of whitetail deer in the George Reserve. Estimates from **Figure** 3.1 of McCullough (1979).

Year	Prehunt population	Harvest	Year	Prehunt population	Harvest
1927	16	0	1949	132	35
1928	30	0	1950	130	58
1929	50	0	1951	87	14
1930	80	0	1952	121	43
1931	140	0	1953	127	45
1932	160	10	1954	129	40
1933	220	96	1955	158	84
1934	155	19	1956	112	58
1935	215	41	1957	91	22
1936	170	100[a]	1958	107	52
1937	145	36	1959	96	35
1938	140	37	1960	101	55
1939	143	46	1961	77	13
1940	130	51	1962	112	53
1941	130	43	1963	108	30
1942	120	33	1964	118	41
1943	110	30	1965	122	44
1944	106	22	1966	121	47
1945	104	11	1967	125	63
1946	160	52	1968	103	50
1947	130	30	1969	94	43
1948	103	20			

[a] Value of 80 assumed in Chapter 5 estimation.

year plus harvest. Plot your estimates as a function of population this year. Does this plot leave you with any doubt about whether the population has recently been held near the level that would maximize average net production?

1.4. A controversial aspect of big game management in Canada and Alaska concerns the validity of wolf control (predator removal) as a means of increasing ungulate productivity and stock sizes. A typical wolf pack needs to kill around 100 moose per year to maintain itself (survive, reproduce, etc.), and under "normal" conditions a pack will defend a territory large enough to contain at least 1000 moose; under such conditions the wolves are apparently quite selective about taking mainly weaker moose that would likely have died anyway. But what might happen in situations where the moose population has been reduced by hunting, habitat loss, hard winters, or other factors? Identify several specific alternative hypotheses about the effect of wolves in such situations, and suggest experiments that might help resolve the uncertainty.

1.5.   Develop a computer program to test how many observations would be required to tell which curve in Figure 1.1 is correct, given different escapement levels. The program should (1) plot the historical data and curves η_1, η_2; (2) secretly (randomly) select one of the curves, then generate simulated data from it at any escapement level you select; and (3) pause after generating and plotting each data point, so you can decide visually whether the correct curve is apparent. For the dome-shaped model, use the equation

$$R_t = s_{t-1} \exp(1.96 - 0.44s_{t-1} + w_t)$$

where R = recruitment, s = escapement, and w_t is normally distributed with mean zero and standard deviation 0.3. For the asymptotic model, use

$$R_t = \frac{s_{t-1}e^{w_t}}{0.1237 + 0.1025s_{t-1}}$$

with w_t as above. On average, how long does it take you to be confident about the correct curve when the escapement is 1.5? 2.0? 3.0? Be sure to conduct at least five trials (called Monte Carlo simulations) at each escapement level, since the number of data points required for clear discrimination is itself a random variable. How many mistakes did you make by picking the wrong model after too few observations were available? Based on this experience, what do you think about the policy recommendation of "incremental experimentation" on the Fraser River, which would involve slowly increasing escapements toward 2 million over a 10–20-cycle period?

Chapter 2

Objectives, Constraints, and Problem Bounding

I, the Onceler, felt sad
as I watched them all go,
BUT ...
Business is business!
And business must grow
regardless of crummies in tummies, you know

Dr Seuss, *The Lorax*

It hardly bears repeating that resource policy analysis cannot proceed intelligently without at least some articulation of the objectives of management, the constraints placed on managers to reflect broader resource allocations and implied social objectives, and the scope of factors considered worthy of study. Most textbook discussions have concentrated on debates about obvious objectives, such as maximizing sustainable yield or economic efficiency of harvesting, and I will review these discussions briefly below. However, actual management practice rarely proceeds in accordance with simple objectives; after all, decision makers are people who, like the rest of us, are guided partly by motives that are often not so lofty and are not spelled out clearly. These implied objectives are usually expressed in terms of risk-averse behavior and resistance to change in general, and so can (and do) make it extremely difficult to implement adaptive policies that call for variation and change as essential to learning. So if the reader is otherwise excited by the idea of adaptive management, he or she should understand right at the outset that there are formidable obstacles to practical implementation.

The discussion will proceed in five steps. First, I will provide some background to the need for management, by reviewing the usual course of events in unmanaged systems; two extreme patterns are possible, depending on the basic structure of resource ownership. Second, I will review some of the obvious management objectives that have developed in response to the general perception that unmanaged systems do not behave optimally. Third, I will look into the delicate matter of how management institutions actually behave, in view of implied objectives and the fact that there are always conflicting actors who bargain to better their positions. Fourth, I will argue that institutional factors lead not to productive equilibrium, but rather to a rhythm that occasionally produces crises and adaptive opportunity, so the timing of policy design and implementation is crucial. Finally, I will turn to the difficulties of bounding problems for analysis, to draw the disturbing conclusion that there are no natural boundaries; the very definition of a problem is itself an adaptive process.

Behavior of Unregulated Systems

The quotation at the beginning of this chapter comes from a delightful children's book about the "conflict" between ecology and economics; it presents a model of development that I have noticed is also held by most university undergraduates, namely that unregulated development driven by the greed of resource exploiters always results in the total destruction of resources. In this model, government regulation is always necessary if there is any concern for the rights and opportunities of future generations.

In reality the need for management is rarely so clear-cut, and depends on at least two basic factors. The first is resource tenure or ownership; even in the absence of public intervention, resources held in "private ownership" tend to be husbanded by the owners. The second is the economic behavior of resource users even when the resource is held "in the commons" (public ownership); this behavior can result in harvesting cycles or "bionomic equilibrium" (Clark, 1976) where it does not pay the resource users to destroy the resource completely.

Tenure and resource husbandry

In the early development of North America, it was established as a legal principle that fishery and wildlife resources be held in public ownership, for the use and enjoyment of all. This choice was understandable in an environment of plenty, and where the creatures were usually seen to move about over large areas. However, a very different situation has developed in parts of Europe, and this situation gives important clues regarding basic human attitudes and values about long-term resource maintenance.

Wildlife and freshwater fish are remarkably abundant in Central Europe (the Germanies, Austria, Czechoslovakia, Hungary), especially considering the human population density. There is a complex system of resource tenure and ownership, with harvesting rights and management responsibility held largely by individuals, families, or game cooperatives in a mosaic that has evolved over many years. Individuals gain access to harvest through various arrangements, ranging from long-term leases to daily licenses, and the costs are astronomical by North American standards. In 1982, a day of Austrian trout fishing cost me about $30; a friend's 130-hectare deer hunting lease cost him about $8000 per year. These charges do not even entitle the harvester to keep what he takes; the kill may be retained by tenured owners for sale through the ordinary meat market.

Such high charges for hunting and fishing have led some economists to conclude that resources are maintained in Central Europe simply because it pays to do so in the short term. This conclusion is both naive and misleading; it does not explain, for example, why people like my hunter friend are willing not only to spend a lot of money every year, but also to put much time and effort into husbandry activities like habitat improvement and winter feeding. Some of those activities will not bear fruit until long after my friend and the people from whom he leases are gone, yet he pursues them joyfully. There is a concern for the long-term future that can hardly be explained by today's balance sheets.

The term "resource husbandry" is often applied to situations where the users, without external regulations, show great concern for maintaining the resource in spite of other short-term economic opportunities. The attitudes and priorities of husbandry seem to reappear wherever the resource user can sequester fairly stable rights or tenure, suggesting that these attitudes reflect rather basic human values that transcend local cultural circumstances. Attitudes very similar to those of my hunter friend are regularly expressed by American environmentalists, by Canadian big-game guides who hold relatively exclusive territories for trophy hunting, and by marine commercial fishermen when they lament about the competitive conditions of large fleets pursuing a shared resource.

Ray Hilborn and I accidentally stumbled on a vivid demonstration of husbandry attitudes by Canadian Pacific salmon fishermen, who normally operate under viciously competitive circumstances involving large numbers of boats fishing short weekly openings in small areas off the British Columbian coast. We were conducting short courses for the Canadian government, intended to expose commercial fishermen to some of the management principles and difficulties faced by the government's biologists; it was hoped we could help defuse some of the government/industry conflict for which the salmon fishery had become notorious. We decided it might be helpful to have a simple simulation game on a microcomputer, where a fisherman playing the game would have to manage a model salmon population over time by choosing catches and spawning escapements each year. In effect we made the game player a sole owner of the resource, and he could do with it as he wished. Several dozen fishermen have played this game, and we never saw a single instance where the player deliberately chose to take a quick killing and get out! All the players adopted a "timeless" attitude, probing the model's responses to try and find a good operating level (spawning stock) for maintaining high long-term productivity. After playing they continued to complain bitterly about the government, but only about how it went about regulating them. The basic objective of husbandry was never questioned.

These fishermen were not a culturally homogeneous group who might have been educated from childhood to accept a conservation ethic. They ranged from native Indians, who very explicitly claim such an ethic as part of their tradition, to some very hard-nosed businessmen of various European descents. Their educational backgrounds ranged from third-grade illiteracy to a masters degree in economics. Some had grown up with fishing and had relatively small financial investments in the business, while others had recently purchased quarter million dollar seine fishing vessels and were obviously out to make a fortune.

There is an extensive literature on resource husbandry, most notably in this century by writers like Aldo Leopold and Roderick Haig-Brown. Rather than dwell on it further, for the remainder of this book I will take the demand for long-term management as a basic given for policy design. Having a long-term view is an obvious condition for adaptive management; there is no point in learning more about something you intend to destroy shortly.

Behavior of common property systems

Most really grim instances of resource mismanagement (from the husbandry point of view) have occurred in situations where no one claimed ownership of the basic resource, and where there was relatively open access for new harvesters to enter the game. In such situations it makes no sense to the individual harvester to engage in conservation practices, since the benefit of his actions may be immediately taken by someone else. It is little wonder that North American resource users have often been branded as shortsighted and greedy; it would usually be a waste for them to behave otherwise.

Let us sketch out the typical pattern of development that occurs in a common property resource that is initially unexploited, and where the users do not agree over time to accept some collective or external restraint. This pattern has been well documented in many texts; for a good discussion of it in terms of dynamic models, see Clark (1976). I find it easiest to visualize the processes involved by thinking of the pattern as having four basic stages, the last of which may be lost under certain conditions: (1) discovery/initiation of development; (2) bandwagon growth; (3) the fallback; and (4) evolutionary development. Each of these stages is discussed in the following paragraphs.

The discovery/initiation phase of development is poorly understood and curiously unpredictable. It involves much more than a general perception that there is an abundance of something worth harvesting. Some resources have been passed over for many years, even after their abundance was noted, and it seemed quite clear that there were potential markets for the products. The initial development appears to require special risk-taking individuals, driven either by exploratory curiosity or by desperation about the disappearance of opportunities elsewhere. Particularly in the development of commercially sold resources, the risks can appear to form a very formidable barrier: will the harvesting equipment work as planned? Can a product marketing network be developed? Will the resource collapse naturally, sooner than expected? Will the rapid entry of other harvesters deplete the resource so quickly that even initial investments will not be repaid? These questions may not loom so large in the minds of the government or large corporation agents who initiate many modern resource developments, but they remain as brakes to slow the first harvesting steps.

Once it is reasonably clear that the resource is abundant enough to support more users and is profitable to pursue, development enters the bandwagon growth phase. There is rapid diffusion of information among potential resource users, sources of capital financing are offered (if appropriate), and there is rapid learning about how the resource is distributed and can be captured efficiently. There may also be rapid learning, of the avoidance type, by the organisms. These learning changes are critical from the point of view of adaptive management, since they greatly complicate the interpretation of data acquired from resource harvesters, during a period of great disturbance that would otherwise represent a valuable chance to learn more about response potentials of the resource.

A number of factors may cause potential harvesters to overestimate how well they would do by entering the game, as the bandwagon phase proceeds. First, the information they receive on profitability is generally a bit dated, and is often a biased sample emphasizing those already in the game who have done well. Second, it may take time, several years even, from the point of the decision to enter until capital equipment (boats, etc.) are ready to use; the entering harvester must make some forecast, which usually means assuming that conditions will remain as they have been for at least a few more years. Third, examination of harvest data may give a serious overestimate of the renewal potential of the resource; the initial development usually "mines out" some unproductive stock components (older organisms, marginal substocks that take long periods to build up, etc.; see Ricker, 1973b).

Considering the above factors, the third or fallback phase is practically inevitable. Rather suddenly, the harvesters find themselves in an environment of substantially reduced resource abundance and strong competition from other harvesters. As economists put it, the profits are dissipated, so no new entry decisions are made. However, there may still be new entrants, due to misinformation and to the lags from decision to action, so the squeeze becomes even tighter. The outcome is, of course, predictable: catches fall, many harvesters cannot meet costs or feel they could do better somewhere else, and harvesting effort also declines. Sometimes the decline is exaggerated by some natural catastrophe or environmental change that would have done less damage to the unharvested stock; perhaps the most dramatic example in this century was the collapse of the Peruvian anchoveta fishery (Figure 2.1).

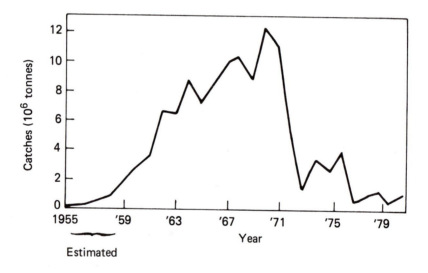

Figure 2.1. Development of the Peruvian anchoveta fishery. The sharp collapse in 1972–73 was apparently associated with a major oceanographic change known as El Niño. For an excellent analysis, see Glantz and Thompson (1981). *Source:* Glantz (1983).

Under certain conditions the fallback is followed by an evolutionary development period, which is characterized by relatively stable or cyclical yields, and by occasional periods of rapid technological change in the harvesting industry as new techniques are discovered and diffuse among harvesters. In biological terms, it is a period of "punctuated" evolution, with periods of near equilibrium interspersed with episodes of rapid change (Caddy, 1983). These episodes may be driven by harvesters trying to gain comparative advantage over one another, but unfortunately this generally involves getting better at finding the resource. In the long term, technological improvements may drive the resource stock so low that it collapses to a point of commercial, or even biological, extinction.

Three conditions must be met if there is to be a period of bionomic equilibrium or evolutionary development after the fallback, in the absence of management. First, the resource must become effectively more difficult (costly) to find as its abundance declines. This means it must not be spatially clumped in some very accessible manner; the passenger pigeon was apparently hunted to extinction in North America because the last few flocks and individuals kept returning to the same trees, where they were easily shot. Second, there must not exist some alternative resource that can continue to support high harvesting effort after abundance has declined. This was a basic problem with blue whales in the Antarctic; they would not have been driven so near extinction if the whaling fleets had depended on them alone. The fleets kept going, and killing blue whales incidentally, because they could still take fin, then sei, then minke, and other whales. The third condition is that the stock should not exhibit some production difficulty when densities are low, such as trouble in finding mates (so-called Allee effect), or increased natural mortality rates due to predators that would ordinarily take only a small fraction of the stock but remain efficient hunters when the stock is low (so-called depensatory mortality effects).

A great deal of concern has been voiced by resource biologists about the third of those conditions, probably just because it is a biological condition that their training prepares them to think about. There are plenty of examples of the first two conditions causing at least commercial extinctions, but surprisingly little empirical evidence that the third condition is a common danger.

The key point to take away from this brief description is that resource development in the commons involves a number of fundamental forces or processes, both economic and biological, that do not just disappear when management is attempted. Profitability, risk, recent experience, the competitive scramble, and technological evolution remain as variables influencing the behavior of harvesters. Biological factors such as erosion of stock structure through loss of less productive subunits result in a changing pattern of productivity over time. On both sides of the ecological/economic interaction, there are forces at work to prevent the long-term maintenance of any happy equilibrium or balance that may be achieved, and these forces come back to haunt the manager who tries to view his problem from any narrow perspective or disciplinary emphasis.

Explicit Management Objectives

In practical situations resource management plans may be developed from detailed scientific analyses, but they are ultimately the subject of essentially political debate and action. In such debates it has proven essential to define and focus argument on relatively simple measures or standards of performance. There is nothing wrong with structuring debates in this way; indeed it may result in clearer illumination of broader concerns and tactical requirements for action than would any deliberate attempt to systematize debate to cover all issues. The following review of simple objectives is presented in a roughly historical sequence, to show how explicit objectives have become more sophisticated over time.

Sustainable yield

One simple management objective has been just to maintain the status quo, preventing resource deterioration and decline of yields from whatever average has been recently achieved. More precisely, the management authority maintains a running account of recent performance, usually in the form of graphs plotting production trends over time, and tries to prevent these trends from turning downward. Management is viewed as a holding action against the forces of resource depletion.

This objective is obviously foolish from any long-term perspective, since action guided by it will prevent both development of potential and recovery from historical depletion. Yet many management agencies in North America, particularly those concerned with wildlife and sport fisheries, have acted as though it were a sufficient basis for wise decisions. When asked why they do not try to do better, proponents from these agencies have responded with arguments like: (1) ecology is so complex that there is no such thing as a relationship between stock and sustainable yield, so any deviation from the current ("tried and true") regime might result in some unexpected disaster; (2) movement to a more productive regime would involve allowing higher immediate yields, and these would result in unrealistically high expectations about the long term; or (3) movement to a more productive regime would involve lower immediate yields, which are politically difficult to sell. We see in these arguments some of the implicit management objectives that will be discussed in the next section.

Maximum sustainable yield

It has long been recognized that there is likely to exist some relationship between resource stock or population size, and the sustainable rate of harvest. Very low stocks are likely to produce only small sustainable yields, if for no other reason than that not enough organisms are around to do the reproducing and growing that constitutes potential production. Very high stocks are also likely to be unproductive, due to competitive interactions that reduce the performance of the average individual; also a high stock might perhaps deplete the other ecological resources upon which it depends. It has been argued that any natural population that we see today *must* exhibit some pattern of excess production related to stock size, at least on the long-term average; populations without this characteristic should have already gone extinct due to inevitable natural variations.

So, the argument goes, there should exist some intermediate stock or production base that produces the greatest excess above maintenance requirements; this excess is the maximum sustainable yield (MSY). Rather elaborate models have been developed to make inferences about precisely what stock size should produce MSY, and about management tactics (such as regulating the size of organisms harvested) to maximize particular measures, such as total weight of organisms harvested. We will discuss some of these models in Chapters 4 and 5.

Some authors have tried to view MSY as a rigidly deterministic, equilibrium concept, and have argued that it is silly even to compute such a simple number to describe the potential of populations that are subject to various effectively random disturbances. They note further that it would be impossible to maintain the MSY, as a fixed quota, for any length of time. Even the slightest downward disturbance in stock size would result in lower production, which would not balance the MSY quota, so stock size would decline further. There would ensue a vicious circle of more and rapid declines. The decline of the Pacific halibut (Figure 2.2) may have been partly due to such a vicious circle, where the initial disturbance was an increase in incidental catches of halibut in other fisheries (Deriso, 1983).

Another problem with MSY is Ricker's (1963) "big effects from small causes." The exploitation rate that produces MSY is likely to be near a cliff edge, with slightly higher rates driving the stock toward extinction. The exact location of the cliff edge can never be known exactly, and it is a real problem to decide how conservative the management should be.

In response to problems of deterministic assessment, it has been suggested that managers regulate harvests according to a "feedback policy" that varies the yield over time in some relation to changing stock size. As we shall see in Chapter 8, it is not difficult to construct such a policy if the

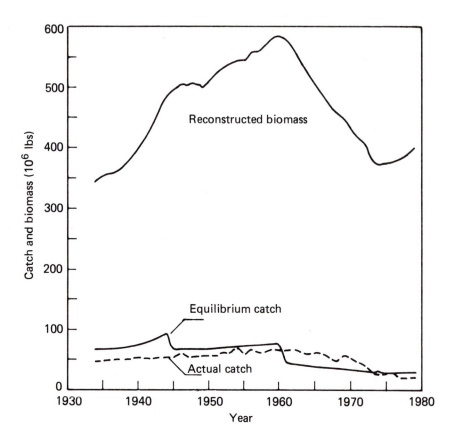

Figure 2.2. A reconstruction of historical changes in stock biomass, equilibrium (sustainable) catch, and actual catch for the halibut fishery of the northeast Pacific Ocean. Notice that the reduction in actual catches was initiated around 1970, a full 10 years after a marked stock decline had begun. *Source:* Deriso (1983).

objective is a stochastic analogue of MSY, namely to maximize the average yield over a long period. Then the optimal feedback policy to use is ridiculously simple. The manager should set a particular base stock or escapement computed from studying production rates at different stock sizes, then each year harvest the excess above this base. Unfortunately, such base stock or fixed escapement policies require annual assessments of stock size (which can be very costly), and result in high variability of harvests from year to year. They also prevent the variation in stock size that is needed to make statistically valid, adaptive assessments of the stock–production relationship, which is likely to be changing over time.

Risk-averse utility

Many resource managers show strong aversion to any policies or developments that involve risk. As noted above, managed systems are often viewed as delicately balanced, with any change inviting disaster. In an effort to counter such rigid and intuitive views, while providing some rational way of dealing with the reasonable objective of avoiding unnecessary risks, it has been suggested that management should attempt to maximize what statistical decision theorists call a "risk-averse utility function" (Raiffa, 1968; Keeney, 1977).

A utility function is just a way of measuring the uncertain outcomes of alternative "gambles," on a single value scale. I can explain the idea most simply with an example, a situation my brother-in-law faced a few years ago. Through some lucky circumstances, he was involved in the discovery of a potentially very rich gold mine. Shortly after the discovery he was approached by a large company, which offered to buy the claim for $250 000. So he was faced with a hard choice: take the certain quarter of a million, or gamble that by keeping the mine he could get much more. He felt the mine could produce anything (after development costs) between zero and about $2 million, and his "expected value" (outcomes weighted by odds) was about $1 million. Well, he chose to take the sure quarter million; his "utility" for this was higher than for the gamble with an expected value four times higher. He claims he would not have been able to decide if the sure offer had been only $100 000, so we say he had equal utility for the choices "100 000 for sure" and "gamble on a million." The utility measure defined (and assessed) this way can also be used to compare various gambles against one another, and to define a general pattern of risk aversion in the decision maker's mind. If my brother-in-law had required a million for sure before he would forgo the gamble, we would call him "risk neutral." Had he required an even larger sure payment, we would call him "risk prone."

In resource management it is tempting to construct utility functions to measure perception of policy gambles involving harvest, so as to help select the best possible (utility-maximizing) gamble. Such assessments require determining the utility functions of the decision makers, and also doing an honest job of admitting and clarifying uncertainties and placing odds on alternative outcomes. Thus the assessments are in the general scientific spirit of making management as open and objective as possible. There might be serious technical problems about placing odds on outcomes, but these odds are somehow perceived and used anyway in making intuitive decisions, so they should at least be aired for debate.

Unfortunately, there is a pretty fundamental difficulty: whose utility function should guide the decision? Ray Hilborn once assessed the utility functions for catches of Canadian government salmon managers and some commercial fishermen. He found the government biologists to be uniformly very risk-averse, as were the younger commercial fishermen he interviewed. These young fishermen had new families and mortgages to worry about, and were concerned to maintain a steady, even if lower, income. But there were some risk-neutral individuals, and even a few (mostly older) risk takers who saw fishing as a grand gambling game; they claimed to highly value occasional chances at the big haul. Why should the public, the basic owners of the resource, act to support any of these very personal attitudes? The biologists were certainly not vested with any special understanding of public values and needs. Indeed, since the public holds and somehow benefits from many resources, perhaps it would be best served by treating each in a risk-neutral way.

Risk aversion is an important concern in adaptive management. Consider the choice, for example, between some conservative, status quo harvesting policy versus a more daring experimental policy with uncertain outcomes. Suppose we are confident that the conservative policy will continue to produce average harvests of 10 units per year, while available data indicate a 75% chance that the experimental policy will result in yields of 20 units per year, but a 25% chance that the average will be reduced to 5 per year. Especially when many such policies and outcomes are to be compared, using some formal computational procedure (optimization algorithm), we are forced to find some objective way of ranking or ordering the choices. If we ignore risk aversion, and weight each possible outcome by its probability of occurrence, we would assign the experimental policy an expected value of $(0.75)(20) + (0.25)(5) = 16.25$; this calculation obviously favors the experiment. However, for a risk-averse manager, the statistical expectation will have little meaning; the 25% chance of low returns will weigh heavily on his mind.

Economic efficiency

Most resource management in this century has centered mainly on the husbandry objectives outlined above, with the implicit assumption that regulation and enhancement of biological harvests will also lead to economic well being. This assumption is incorrect, and like many shortsighted and expedient choices it can lead to precisely the longer-term maladies that it seeks to prevent. Recall again that management rarely proceeds as a simple "command and control" process with management agencies fully in charge; instead there is a bargaining process with at least some power vested in the resource harvesters. When their economic interests are jeopardized, the harvesters are usually in a position to prevent, retard, or delay needed husbandry actions.

Consider what happens when a lid or quota is placed on harvest from a developing (or recovering) common property resource. This lid does not remove the economic incentives (profits, high wages) that attract new harvesters and investment in better equipment; indeed, it may have the opposite effect by providing potential investors at least some assurance that a biological fallback is not imminent. So economic development proceeds, and the limited yield is shared among progressively more harvesters. Ultimately, profits are dissipated and wages fall exactly as in the unregulated situation, until no new investors are attracted. Far more labor and capital than are needed are now employed in pursuing the resource, but even further changes are to come. In order to achieve harvest limitation, the management authority employs tactics such as closed seasons and restricted harvest areas. Harvesters, seeking ways to better their individual takes, start responding to the regulations with various technological innovations (such as faster boats to reach the fishing areas first). To survive, other harvesters are forced to copy these innovations, and progressively more stringent regulation is required to stay in step with the developing technology. New technologies often involve economies of scale through larger harvesting units, so the industry becomes more capital intensive with larger debts to be serviced (and therefore stronger insistence on high harvests in the short term). So, in the long term, harvesting industries tend to develop toward configurations of technologies and income levels that are highly vulnerable to unexpected events and natural variations; such industries become powerful lobbying forces against management actions that are intended to cope with the variations.

A general hypothesis in economics is that maximization of a society's welfare requires at least that the basic factors of production (labor, capital, resources) be "efficiently employed," i.e., shuffled around through market incentives or central planning until each productive activity uses no more of each factor than necessary. According to this hypothesis, governments as resource owners should restrict entry to resource harvesting, even if there is no concern about long-term bargaining relationships, to that level of harvesting effort where adding another unit would add more to harvesting costs than to revenues. For a more precise discussion of the conditions for economic optimum, see Clark (1976).

We may quarrel about the welfare maximization hypothesis, questioning everything from the economists' welfare measures to the assumptions they like to make about things like perfect markets. But the key point is that lack of economic management is bad for almost everyone concerned with a resource; it pays to take at least some steps toward economic efficiency. There have been some very clever proposals about how to accomplish this. One approach is to place a direct tax on the harvest (but not on resource entry; high "license fees" often make matters worse in the long term), so the public both takes its cut of the profits directly and makes new entry unprofitable before the industry gets so large. Another is to assign individual (or vessel) quotas, usually with some tenure or ownership and transfer rights so more efficient harvesters can buy them up, with each quota large enough to assure reasonable wages and return to capital investment. Considering the dangers of quota regulation in relation to generally uncertain overall resource production, an alternative scheme is to allocate transferable shares as fixed proportions of whatever variable total harvest is allowed.

Discounted value

Concern with economic regulation has brought attention to a very fundamental issue in resource management, namely how to value harvest deferred to the future relative to what could be taken today. Surely we should not claim that a ton of codfish harvest taken 100 years from now is worth as much to society as a ton taken next year; what is our responsibility to try and ensure that the later ton is available, especially considering the gross uncertainties involved? No one may even want codfish 100 years from now, and even better substitutes may be available. Also, we can take the money from that ton of codfish next year, and invest it in other productive activities whose value grows with time; should we save the codfish if its growth potential is lower?

Economists approach these questions with two notions: option value and discounting (Krutilla and Fisher, 1975). Option value is defined as the amount of money that we would be willing to bid now, just to ensure the availability of a resource to future generations. Its determination is essentially a political and social matter, difficult to do in practice (how do you elicit correct responses from people about what they would be willing to pay?). Discounting is a more practical approach, and is based on an argument of rational consistency in comparing any two time periods in the future: if we value a unit of harvest deferred until next year at 90% of its value if taken this year (a 10% discount rate), should we not also assume that decision makers 5 or 10 or 20 years from now will place the same relative value on the next year's harvest? If so, we should value a unit harvest two years from now as 0.9^2 or 81% of its value now, three years from now as 0.9^3 or 73% of its present value, and so forth. The total future value, say V, can then be viewed as an infinite sum of annual values v_t, weighted by discounting:

$$V = \sum_{t=0}^{\infty} \delta^t v_t$$

where δ is the discount factor (0.9 in the above example). This formulation of value has the advantage of not assuming an arbitrary cutoff point or time horizon, as is sometimes done in making calculations about future opportunities.

If it is agreed to measure resource value in terms of discounted incremental values, it becomes a critical problem to choose an appropriate discount rate or at least to show that the ranking of any proposed policy options is the same for all reasonable rate choices. Some economists have argued that this is a nonproblem, that resources should be treated like any other investment opportunity and valued with current lending rates. Others argue for very low "social discount rates." The basic issues at stake are the ethical responsibility of governments today to future generations, and whether to be optimistic that people in the future will find ways to take care of themselves no matter what we do today. Obviously these issues cannot be resolved completely: values are not universally shared, and the future will remain uncertain.

Clark (1973) published the disturbing conclusion (for resource managers) that it is optimal to let a renewable resource be driven toward extinction, if the maximum renewal rate of the resource is less than the discount rate used in valuing it. It may not pay the harvesters to drive the resource to extinction, but government should not intervene to prevent a low bionomic equilibrium. This conclusion came at a time of great controversy about the world's whale stocks, whose maximum production rates of 3–15% are well below the interest rates that were then prevalent in investment decision making, and there were heated debates about it. Although it is difficult now to cut through the confusion of issues in that debate (i.e., whether it is ethical to kill whales at all), it appears that consensus was reached that the International Whaling Commission should base its decisions on very low discount rates. Other agencies and commissions will probably follow this model, and I will assume low discount rates in most places where the question arises in this book.

Economic stability

Resource harvest is fed into marketing systems, or at least the harvesting activity supports some secondary industry (guides, tourist resorts, etc.). These induced activities may be impossible to sustain without some trust among the actors involved, in the form of reasonably stable employment, supplies, orders, reservations, and so forth. What bank, for example, would give a fish processing plant or fishing resort the revolving loans it needs to meet operating expenses when income is delayed or seasonal, if there is no assurance that fishing will even be allowed next year? When you think about the complex web of transactions and consensus required to maintain resource-based economic activity, especially in systems that have been functioning for many years, it is easy to understand why about 90% of the discussion in government/industry forums (commissions, hearings, user committees, etc.) is usually about issues of short-term stability, and 10% about the "broader" objectives outlined above. Yet there has been very little theoretical or empirical research about the importance of short-term stability; I suspect this reflects a lack of experience by most researchers of the practicalities of business decision making.

Prevention of rapid change can be in direct conflict with other management objectives. I noted earlier that maximization of average harvest is achieved by a "fixed escapement" policy, which causes large variations in catches; this conclusion also applies when catches are discounted into the future, and when the objectives involve simple measures of economic efficiency, such as total long-term profit. For stocks that are already seriously depleted, fixed escapement and related policies imply that harvesting should be discontinued immediately and completely, in order to minimize the recovery time to productive stock levels. The adaptive probing policies to be discussed later in this book call for strong variations in harvests around the level that appears best on the basis of historical data.

For stocks that exhibit strong short-term variations, there have been a few attempts to quantify the trade-off between maximizing long-term yield versus maintaining some constancy over time. Figure 2.3 shows an example (Walters, 1975) for sockeye salmon of the Skeena River in British Columbia. Here I did a series of optimizations, each with a progressively higher target (or average) catch, and at each target tried to find the feedback policy that would simply minimize the variance (mean squared deviation from average) of catches around the target over a long time horizon. The resulting curve of mean versus variance is called a *Pareto frontier*, and it shows the best achievable combinations of these two measures. The results show that it might be possible to almost completely eliminate variation in catches, but only by accepting a considerable (30%) reduction in the average. This conclusion is hopeful; most fish stocks are not as variable as the Skeena sockeye. However, objective analysis cannot show where on the Pareto frontier of Figure 2.3 it would actually be best to operate; that is a matter for compromise between government and industry.

Other mechanisms and policies can act as buffers against variations in natural production, if it is considered important not to stabilize harvests directly. In market economies, prices for major resource commodities are inversely related to supplies; this tends to dampen variations in income, at least to the immediate resource harvesters. Better technologies for product storage (canning, etc.) allow processing industries at least some flexibility to stabilize delivery of harvests to markets, although these industries must then maintain inefficiently large capital facilities if they are to take advantage of bumper crops. Government may take an active role to stabilize incomes,

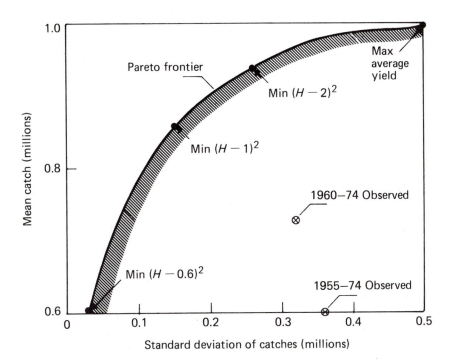

Figure 2.3. Best possible trade-off between mean and standard deviation of catches from a salmon population. Each point along the dotted "frontier" represents the lowest standard deviation that can be achieved by following a feedback policy that holds the mean as shown on the vertical axis. For Skeena River sockeye, the historical combinations achieved were far below the optimum. *Source:* Walters (1975).

especially for harvesters, through various insurance programs. Usually these programs, such as unemployment insurance, serve the whole society; it would be possible to design more focused resource insurance schemes to tax the harvesters in good years and pay them back in bad ones. Indeed, feedback saving/spending schedules could be computed cheaply and separately for each individual, to reflect his own time preferences and risk aversion.

Economic and cultural opportunity

Some resources appear to be managed as glorified welfare programs, with far more people employed in management and harvesting than would appear to be economically efficient. Canadian fisheries are an example;

government expenditures for management, industrial subsidies, and unemployment insurance probably exceed total resource revenues (quantities times wholesale prices) in some areas and years, although exact accounting is impossible because of shared programs and so forth. Such policies are justified in the short term as politically expedient (buying votes, etc.) and socially merciful in view of the human suffering that would be caused by dislocating and retraining people for other employment. But might there not be long-term values as well, especially in an era where the labor requirements for many economic activities are falling due to changing technologies?

A valid objective for government policy might be simply to help maximize the diversity of choices and opportunities for people to seek employment and lifestyles that suit their personal preferences. Resource harvesting usually involves rather unique cultural situations and lifestyles, with dimensions ranging from outdoor work to the challenges of scramble competition to the very special community life that develops in places like isolated fishing villages. It has been argued that the attractiveness of these situations is one of the reasons why resource industries become overdeveloped, with some new entrants knowingly accepting lower wages and profits than they could achieve elsewhere. In effect, the wage differentials measure how much these entrants are willing to pay to enjoy such lifestyles; the problem, of course, is that by taking a part of the limited pie, the new entrants force already established harvesters also to accept less. Perhaps the answer to this problem is to develop systems of legal rights, based on precedence of entry and general welfare concerns, analogous to those used in the management of water resources.

Implicit Objectives

The above discussion was intended to lead the reader rather gently from a consideration of obvious theoretical objectives through to thinking about some of the deeper social issues that should concern resource managers. Now we shall step back a little, and look at how management institutions function as collections of people with limited rationalities and less lofty personal objectives that color their responses to threat and opportunity.

In a brilliant study of how public decisions are made, Allison (1971) notes that analysts have used three models to interpret how management institutions "behave" in practice. Simplest is the *rational actor* model, in which the institution is seen as behaving like a single individual, somehow making rational choices using information and explicit objectives. Allison notes that this model often does not work; actual decisions bear little relationship to stated objectives. More realistic is the *standard operating procedures* model, which argues that most government actions are the result of people who try to follow standardized operating procedures and rituals. The idea is that thinking in general, and risky decision making in particular, makes most people very uncomfortable; so they try whenever possible to fall back on "tried and true" measures, and generally try to minimize the amount of hard work required to get on with the job. Also, following standard procedures frees people to spend time working on personal objectives, such as moving up in the government hierarchy, and to shift blame for bad outcomes onto "the system."

Allison's third model, of *push-and-pull among power factions*, is perhaps even more relevant to understanding how resource decisions are made. The notion is that government agencies and their constituents are divided into power blocks or factions, each with rather narrow objectives. Decisions "emerge" from direct bargaining among factions, hidden power plays, differential access to higher authorities, and so forth. The key point is that lines of authority and influence are seldom so clear-cut as you might think from looking at organization charts, committee systems, and commission agendas.

It is easy enough to see the power faction model in terms of broad government/industry/public factions, but it is equally important to understand that there are usually factions *within* government agencies as well. In large agencies especially, people are charged with specialized tasks. It can become impossible for individuals to maintain a balanced view of how important their tasks are to the management system as a whole, and they often feel a need to press task objectives strongly just to survive in the organization and maintain personal esteem. My favorite example, from personal and family experience, is the people who run government fish hatcheries. It is virtually impossible to work around a hatchery for very long, with its very tangible outputs produced through a lot of drudgery like pond cleaning and crises like disease outbreaks, without developing a very strong personal commitment and feeling that it must all be worthwhile. So when broader management analyses indicate (as is often the case) that the hatchery fish are not surviving to be caught or are surviving well but causing problems with wild production through competition or attracting too many fishermen, it is little wonder that hatchery people often become viciously defensive and try to build their power bases through simplistic arguments and direct appeals to the public for support.

Management agencies are supposedly organized as hierarchies of responsibility, with people at the top concerned about broad strategic issues and people at the bottom about day-to-day operational tactics. But this intended organization usually breaks down to some degree in practice. Power factions, in pursuing limited objectives, do not respond precisely to command/control decisions from above. People at top decision levels do not concentrate solely on the difficult and ambiguous strategic questions; they often take refuge instead in worrying about (and interfering directly with) the more understandable details of routine operations.

So throw together some hatchery managers, law enforcement officers, ecological researchers, welfare economists, statisticians, policy planners, resource biologists, administrative personnel, and perhaps quite a few others. Call this a management agency. Now "interface" it somehow with its constituents, ranging from politicians worrying about the next election, to concerned conservationists, to careful business entrepreneurs, to "cowboys" out to take the biggest catch this year. Be sure to throw in a few characters with complex motives, like an operator of sport fishing charters who loudly opposes fishing regulations that would make it easier to catch fish without his help. Finally, consider the resource itself, a complex ecological system that is too expensive to monitor thoroughly, changes unpredictably in response to environmental factors, and generally offers all sorts of conflicting signals that are open to every interpretation from imminent disaster to grand opportunity. There you have the modern management situation. It is little wonder that progress appears to be almost nonexistent, that only major crises seem to elicit concerted response, and that resource managers are often branded as cynics with little concern for resource husbandry.

Rhythms of Crisis and Opportunity

The last paragraph was not meant to be completely discouraging, and it contained what I believe is a critical phrase: "crises seem to elicit concerted response." The behavior of managed systems appears to follow a rhythm, or pattern, of punctuated evolution that has much in common with the evolutionary development of unmanaged resources: there are time windows when substantial policy change and adaptation is possible, interspersed with longer periods of normal operation when changes are actively resisted or ignored (Caddy, 1983). It appears that whole economies display such rhythms, as evidenced by Marchetti's (1980) analysis of major technological innovations over time.

Cycles of opportunity for change are practically inevitable in large management systems, when various management responses are delayed through insistence on standard operating procedures and where each move involves bargaining among conflicting interests or factions. Most actions will follow Lindblom's (1959) prescription of "incrementalism," making small improvements without taking large risks. The trouble is, of course, that no procedure or policy is perfect when dealing with a large, open system; even if the system's environment is not changing, policies contain the seeds of their own destruction. Holling (1980) cites the examples of insecticide spraying to control spruce budworm outbreaks in eastern Canadian forests, and forest fire control in the western United States. These policies were initially very successful, but set in motion an accumulation of fuel (for fires and insects) that made control costs grow rapidly and the disasters more severe when they did occur. Examples already mentioned from fisheries include the eventual collapse of fish stocks managed under fixed-quota harvests, and the decline of wild stock production following the introduction of hatcheries intended to supplement that production.

It appears to be a general feature of policy failure that the deterioration starts out slowly (while the policy works almost as planned), then accelerates as the system state moves further from its intended level when the policy was designed. In principle it should be possible to design feedback policies where state changes are monitored to prevent entering such "domains" of rapid change, but in practice the key variables are either not recognized and considered part of the problem, or they are measured inaccurately enough so there is continuing excuse for inaction until the changes become too large to ignore. In short, Lindblom's presumption of incremental changes and responses simply does not work.

So the rapid changes induce a sense of crisis among the actors involved in policy formulation, leading to a period of willingness to reexamine basic objectives and assumptions. "Systems views" are solicited, and there is likely to be good funding for us modelers. What happens next appears to depend somewhat on luck. If the debates and analysis stumble upon some really good new approach, it may be quickly adopted and blessed with a chance to become standard operating procedure. More often, the most pressing problems seem to solve themselves; the people most hurt by the crisis give up and go away, and the system settles into a new apparent "steady state" (period of tolerably slow change). Unfortunately, the new steady states resulting from inaction tend to be progressively less productive relative to long-term potential, with the resource stock eroded and generally higher costs of production. This deterioration is often masked (and left open to debate) by well intentioned decisions during the crisis to introduce new resource monitoring systems; rarely is there sufficient attention to make sure that the old and new data sets can be precisely compared.

Government employees often complain that crises develop all too frequently, so they are forced into "firefighting" new problems every few days or weeks, and have no time for long-term planning. They rightly develop a feeling of unease that all the tactical "band-aids" are hiding the patient's real illnesses, or are making them worse. Their concerns are justified, but what they are usually seeing are the seeds of disaster rather than the deeper culmination of effects that represents major crisis and opportunity for strategic change. Such opportunity is likely to come only once every 10–30 years. For example, in the history of Pacific salmon management in Canada, the following periods stand out:

1913:   Natural disaster reduces Fraser River sockeye, fishery develops for other stocks.

1925:   New fishing techniques, overcapitalization, high exploitation rates coastwide.

1947:   Widespread stock depletion, initiation of systematic catch and escapement monitoring.

1969:   Collapse of herring fishery, rapid growth in various government programs.

1980:   General feeling of economic crisis and failure of regulatory systems, initiation of Royal Commission on Fishery Policy.

There were, of course, other developments between these very fuzzy dates; for example, a major salmon enhancement (habitat improvement, hatcheries, etc.) program was initiated in 1974, and it helped contribute to the sense of crisis that emerged around 1980.

One should not be discouraged that major opportunities for change do not come very often—perhaps only once or twice in a manager's working lifetime. There is plenty to be done at other times, and it may even be possible to deliberately engineer changes in the period and depth of the rhythm. Especially with messy data sets that are open to alternative interpretation, and with a diverse community of economic interests containing some disgruntled people all the time, there is always the chance to form vocal coalitions of unease. This apparently happened in the salmon history outlined above; Pearse's (1982) analysis of biological and economic statistics does not show all that much change in the late 1970s. Obviously, crisis engineering cannot be recommended as a regular management tool; it is too easy for fabricated cases to be shattered, and with them the credibility needed for ongoing management. A wiser strategy is to be alert to real crisis development, and then to try to manage its course so as to turn it more quickly into adaptive opportunity.

Bounding Problems for Analysis

It should be obvious from the discussion so far in this chapter that there are no "natural" boundaries for defining renewable resource systems or the limits of management responsibility in dealing with them. Once one admits that it is not enough to focus on biological resource husbandry, the domain of potential concerns becomes a matter of practicality and continuing adaptation. Suppose, for example, you have been asked to examine future management options for the Peruvian anchoveta fishery. You might begin with a look at its biological history of performance, especially as related to the oceanographic changes (El Niño) that apparently led to its collapse

(Figure 2.1). But you would quickly become concerned with why the stock has not been allowed a better chance to recover, and this would involve analysis of the political and economic environment within which decisions are made. Since the fishery is a major component of the Peruvian economy, perhaps you should be worried about how managing the fishery would affect overall measures of economic performance, such as employment and investment in other industries. That would lead you in turn to issues of international trade, such as Peru's balance of payments and world demand for the anchoveta as an industrial product. Anchovetas have been used mostly for fertilizer, so you might worry about world demand for fertilizers in general, and in particular about substitutes that may enter the market. In the limit, what seemed at first to be a tidy biological problem turns out to have even global dimensions and opportunities for policy intervention (Glantz and Thompson, 1981). Frightened by this scenario of broadening economic concerns, you might choose the opposite tack of delving more deeply into the biology and oceanography. Need I even point out that you would, if you proceed systematically, expose an equally frightening chain of causality leading down into biological details and outward across the Pacific Ocean?

It is easy to become caught either in a case of "paralysis through analysis," or the equally foolish extreme of rejecting analysis entirely and falling back on the problem bounds set through earlier experience or codified in someone's textbook. It should be possible to *improve upon* either extreme, without ever pretending that there exists a single "best" model and course of action. In this view, analysis should proceed by deliberately looking at the system from several vantage points, each differing in four key bounding dimensions:

(1)   *breadth* of factors considered;
(2)   *depth* of analysis into detail;
(3)   *spatial scale* of variables considered;
(4)   *time scale* or horizon for prediction.

The classical "theory of fishing" mostly prescribes analysis at only one point in these four coordinates: (1) breadth—a single biological species; (2) depth —average rates of growth, mortality, and recruitment; (3) spatial scale—the so-called "unit stock" of interbreeding individuals with similar movement patterns; and (4) time scale—a few fish generations, or sidestepped by assuming equilibrium. In retrospect, many historical fisheries situations (such as the anchoveta) could have been better understood by broader analysis (including economic factors), perhaps even less concern for biological details, by looking at larger spatial scales in relation to factors like movement of fishermen from stock to stock, and by deliberately examining regulatory options and effects on several time scales. In other cases, probing into details might have revealed dangerous pitfalls or interactions that would become important later.

The approach of deliberately looking at a system from several perspectives (different problem boundaries), hoping to learn something from each, is related to a common debate that I consider rather trivial and deceptive. That is the argument about "top-down" versus "bottom-up" approaches to analysis. According to top-down proponents, you should begin by looking at the system very broadly and simply, as a few subsystems, then begin decomposing each subsystem until an acceptable or ideal level of detailed description is achieved. The bottom-up view is that you should begin, instead, with the system's most primitive or elementary units (individual fish, boats, etc.), and reconstruct (predict) aggregate behavior of direct management interest by analysis of the interactions of the elementary units. My objection to looking at problems according to either of these recipes is simple enough. Neither has a "natural" starting point in the first place, and by pretending that one exists you are quite likely to enter a long process of analysis leading directly away from the relationships that should be of concern. You can only avoid this in the top-down approach by beginning always with a global perspective, which would usually be a waste of time, and in the bottom-up approach by always starting with molecules, which is equally silly. Also, human beings generally do not think very systematically, especially in cases where imaginative solutions may be needed, and are likely to leave out something important when following any particular recipe. That key element of imagination and creativity can most readily be stimulated by deliberately and repeatedly jolting your perspective to look at things from different angles.

One way to make the business of problem bounding a little less painful and ambiguous is to think carefully at the outset about what products the analysis should produce. It has been taken for granted by too many analysts that the ultimate goal should be detailed and quantitative predictions about the future of the system. But in practice such predictions are seldom examined very carefully or taken seriously by the actors involved in decision making. Indeed, data and predictions are often used very selectively to back up narrow positions and even deliberately to promote confusion in what have been called "battles of models." Detailed analysis may be necessary to establish credibility or to explore particular tactical options, but the key product should usually be a small set of strong (robust) qualitative arguments and conclusions that can be understood and debated by actors without quantitative skill. Each step in the analysis (and each proposal for data gathering, for that matter) should be first examined in terms of its likely contribution to qualitative arguments.

(1) Identification of key indicators

(2) Articulation of variables and processes that directly affect indicators

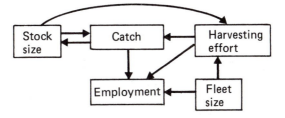

(3) Elaboration of further causal pathways and model boundaries

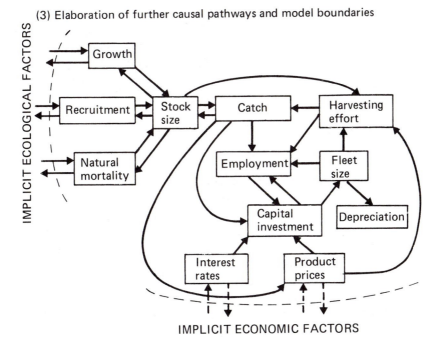

Figure 2.4. Problem boundaries are often identified by working outward from a few key performance indicators. Broader and more detailed concerns are added at each step by thinking about factors that influence each variable already identified, until boundaries are reached where it is felt that further elaboration would be impractical or unnecessary.

Experienced model builders often approach problem bounding by "working outward" from a few key indicator variables, as illustrated in Figure 2.4. For example, early discussions might indicate that it is critical to make at least some predictions about total catch and employment. A next step is to identify major factors that influence or determine the key indicators, and that are likely to change over time in relation to policy actions and uncontrolled natural dynamics. Catch is obviously influenced by available stock size and harvesting effort, and these variables are in turn interdependent (arrows in Figure 2.4), since changes in stock size will influence the willingness of harvesters to exert effort. Then, in a series of further steps, the variables and interactions (dependences) identified at each previous step are examined more closely, with a view to identifying (1) component factors that may be important (for example, stock size is influenced by growth, recruitment, and natural mortality, as well as catch); and (2) boundaries for the analysis, beyond which further elaboration would be impractical. In this approach, the "boundary" is defined by making a series of explicit decisions to treat factors as constant or related to factors already identified, while admitting that, in fact, there are other influences at work. In Figure 2.4, a decision to treat interest rates and prices as constant, or related only to catch, is made as an explicit choice *not* to look into the larger economic system where availability of financing and product demands are determined.

The Hilborn Plan

Let me close this chapter with a brief case example of policy analysis in fisheries. This example illustrates some difficulties and pitfalls of analysis, but also the potential for uncovering new policies that sidestep apparently unresolvable conflicts in objectives.

For several years, a small research group at the University of British Columbia had been looking into various problems of Pacific salmon management in Canada. In cooperation with government biologists, we examined a rather staggering mass of data ranging from historical population trends to fishing fleet behavior to the details of fish movements and regulations in local fishing areas. We constructed literally dozens of management models, from simple stock–recruitment curves to a giant coastal simulation that traced how over 100 stocks move through the various fisheries each year.

Then in 1981, Peter Pearse was appointed to lead a Royal Commission on the status and potential of the Canadian Pacific fisheries. Royal Commissions bring many actors together in a format of public hearings and written briefs, and the Commissioner produces a report with recommendations for government action. We were asked to prepare an analysis of the biological potential of the stocks, using our experience plus a sequence of assessment workshops involving government scientists and managers, and to present key results to Commission audiences in the rather innovative form of microcomputer simulation games that could be used by the audiences to explore alternative policy options. By the time any of these games were ready, the Commission's hearings had already fostered some very lively debate about the future of the resource.

Our earlier analyses had indicated two basic paths that future production might follow (Figure 2.5), both rather bleak from some points of view. One extreme path involved "bite the bullet" rehabilitation of overexploited natural stocks to more productive levels as quickly as possible, by not fishing at all in various places (see above discussion on MSY objective). An alternative we called SEP or "the American plan," to stress a pattern that had already developed in Washington and Oregon. This option would involve maintaining and trying to increase harvest through massive investment in enhancement (hatcheries, etc.), and it was just beginning in Canada with a $300 million enhancement program. The basic trouble with this plan is that the enhanced stocks are mixed with wild fish in most fishing areas, yet can withstand much higher exploitation rates. So to reap the benefits of enhancement, it appeared necessary to allow high exploitation rates that would cause further decline in wild stocks. The end result would be a fool's bargain, with the same total production concentrated in relatively few engineered systems, operated at least partly at public expense.

We could see no way around a hard choice between these two options, until Dr Mike Healey commented during an assessment workshop that we should be looking at broader policy options such as catch quotas. We had avoided discussion of quotas because of their obvious dangers, noted earlier in this chapter, and had concentrated instead on an agonizingly detailed analysis of optimum spawning stocks and exploitation rates. Then Ray Hilborn made what I can only call a brilliant intuitive leap, seeing a new option that all of us had missed (Figure 2.5).

Hilborn's idea is almost ridiculously simple. It comes from noting that the exploitation rate, which must be reduced if wild stock recovery is to occur, is basically just the ratio of catch to stock size. We usually assume that the ratio can be reduced only by reducing catch. But in the salmon case, enhancement can be used to increase the denominator, stock size; in fact, this direction is politically much preferable to reducing catches immediately. Hilborn's idea is to hold catches at constant quotas for a while as enhancement comes on line, so initially the exploitation rate must fall. This starts to allow wild stock recoveries, which in turn contribute to driving the rate down. The end result is like eating your cake and having it too: both wild stock and enhancement production are increased, and stable economic returns are maintained during the transition period. To hold the quotas in the face of increasing abundance of fish, it would be necessary to

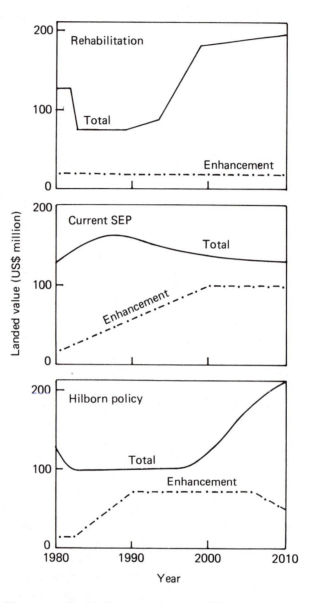

Figure 2.5. Three scenarios for future development of the salmon fisheries of British Columbia. The rehabilitation scenario involves reduced catches immediately to allow stock recoveries. The current salmonid enhancement program (SEP) scenario involves continued development of artificial production systems, and replacement of natural populations. The Hilborn policy scenario involves a combination of artificial production and fixed catch quota until the mid-1990s. *Source:* Walters et al. (1982).

progressively decrease fishing effort (days open, net sets per fisherman, etc.) and thus, possibly, employment in the industry, but that reduction would probably be necessary under the other plans as well.

Hilborn's plan is an excellent example of "counterintuitive behavior" in dynamic systems; two apparently dangerous policy instruments when considered separately (quotas and large-scale enhancement) can in combination produce improvements measured by a whole variety of management objectives. We followed up the initial idea with much careful quantitative modeling, to look for dangerous side effects and to determine more precisely what quotas and enhancement outputs would be needed to give various time patterns of development. But this was easy after seeing the basic need to look at the combined effects of the policy instruments, rather than the incremental effects of each.

So modeling and analysis led us initially to a point of frustration, which it then took a step of creativity and imagination to break. But even the point of frustration might not have been reached if we had tried to sidestep a lot of careful, quantitative analysis. The creative step then led to another round of analysis, and eventually to the simple policy arguments of the previous paragraph. It is these arguments that were finally debated in the Pearse Commission.

Problems

2.1. Criticize the proposition that "the objective of management should be to maintain the population at the level where productivity is highest, so as to provide the maximum sustainable yield." What must be meant here by "productivity," and how does the proposition read if you insert this meaning instead of the single word? Is there likely to be such a well defined level? What investments might be necessary to achieve and maintain any particular population level?

2.2. Try to determine your own degree of risk aversion regarding uncertainty about future incomes, by identifying what level y of sure income you would barely accept if the alternative choice were a 50:50 gamble with possible outcomes 0 and X dollars. (Be sure to estimate y for several values of X, including some X values well above what you ever expect to earn.) You are risk averse if your y values come out far below $X/2$, and risk neutral if they are near X. Then comment on the following questions: Have you been (or can you be) completely honest with yourself about how you would react if the choices were real? If you are some day in a government policymaking or advisory position, should you advocate public policy choices that "feel comfortable" in terms of your own personal risk aversion?

2.3.   Suppose you are asked to advise about how to manage a population
        that has been depleted through overharvesting or a natural disaster,
        and will likely remain low unless deliberate steps are taken to reduce
        harvest rates. Suppose it has been estimated that the best stock size
        (for long-term production) is roughly double the current stock size,
        and that the population will increase by $10 - X$ percent per year
        toward that best level, where X is the annual percent harvest rate
        allowed over the recovery period. Show how the choice of X will
        affect the length of the recovery period.

2.4.   Suppose the harvest value in each year for problem 2.3 is presumed
        to be $v_t = X_t N_t \lambda^t$, where $\lambda = 0.98$, and the total resource value is
        measured by $V = \Sigma_{t=1}^{100} v_t$. Suppose the population is to be held at N_t
        $= 2N_0$ after it reaches this level, by setting $X = 10\%$ per year. Show
        how the total value V will be affected by the choice X_t of harvest rate
        during the recovery period. What happens to your conclusion if $\lambda =$
        0.92? $\lambda = 0.85$?

2.5.   Consider the stochastic recruitment models presented for Fraser River
        sockeye salmon in problem 1.5. For each model, use Monte Carlo
        simulations to estimate the mean and variance of catches over the next
        50 cycles (generations) for three possible "feedback policies" for set-
        ting C_t, the annual catch:

        (1) $C_t = R_t = 1.0$ (but $C_t = 0$ if $R_t < 1.0$). Here 1.0 is a "fixed
               escapement" target.
        (2) $C_t = 0.75 \, R_t$. Here 0.75 is a "fixed harvest rate" target.
        (3) $C_t = 2.5$. Here 2.5 is a "fixed quota" independent of N_t.

        For all tests, let $s_t = R_t - C_t$. Which policy would you prefer if you
        were a commercial fisherman?

2.6.   Using the simulation developed for problem 2.5, try to find a feed-
        back policy function $C(R_t)$ that will give *higher* variance of catches
        over 50 cycles than the fixed escapement policy. Then try to find a
        policy function that will keep the catch as near to 4 million as possi-
        ble, deviating from this target only when the stock size gets very low.

Chapter 3

A Process for Model Building

Modeling is much too important to be left to modelers.

F.E.A. Wood (1978)

Chapter 2 emphasized the complex environment of renewable resource decision making, and the need for both careful quantitative analysis and some "imaginative synthesis" to occasionally cut through the complexities. This chapter turns to a process, called adaptive environmental assessment (AEA), that uses the construction of dynamic models as an intellectual *device* to help people clarify issues, communicate effectively about shared concerns, and explore objectively the consequences of alternative policy options. Details of the process have been described elsewhere (Holling, 1978; ESSA, 1982), so this chapter will provide only an overview with emphasis on how the process is employed to promote adaptive policy development.

Why Bother?

Let me begin with a brief discussion about why it is important for resource analysts to engage in quantitative model building in the first place. This issue has been discussed at length in dozens of treatises from practically all scientific disciplines, and at least one (general systems theory) is deeply preoccupied with it. Yet among resource scientists, and particularly biologists, just mentioning the word "model" can still be an invitation to long and heated debate, or even immediate personal rejection. All sorts of myths about what modeling can and cannot accomplish are promoted on the one hand by people who seem threatened by it, and on the other by inexperienced proponents who try too hard to defend it. Let me review a few of these myths, and show that there is really nothing to be frightened or defensive about.

Perhaps the worst myth is what I call the *all or nothing* stance. According to this myth, if models are to be useful they must be capable of detailed and/or precise quantitative predictions. It is not made clear what is meant by "useful," but the implicit presumption is either that scientific hypothesis testing hinges critically on small deviations from predictions, or that selection among policy options requires accurate prediction. If these presumptions were true, it would indeed be folly to engage in renewable resource modeling. The old adage about computers, "garbage in, garbage out," could be (and is) interpreted literally and applied to all resource models, if for no other reason than the lack of natural bounds for resource systems (Chapter 2). One can always find some "boundary assumption" about the constancy of larger systems surrounding the resource or unimportance of some details that is almost certain to be incorrect. So some people argue that if you cannot do it perfectly, do not do it at all. This argument should carry about as much weight in modeling as it does in other human affairs.

Another myth is that *modeling is a substitute for experience*. I know few model builders who would make this claim, which goes back to the Baconian notion that the behavior of natural systems can be deduced *a priori* from basic principles. But the claim is often made that models can be used as "laboratory worlds" to "test" the possible effects of policy options that are too big or expensive to study experimentally. Here I think modelers have been at fault for not choosing their words more carefully; there can, of course, be no test (in the scientific sense) when experimentation or observation is impossible. A more precise statement of the "laboratory world" notion requires some cumbersome verbiage, and goes something like this: we must still make policy choices even when experimentation is impossible, and choice is always based on some sort of inference about alternative outcomes; since inference (i.e., prediction) is unavoidable, we should make the assumptions underlying it (i.e., the laboratory world) as clear as possible, if for no other reason than to avoid mistakes of reasoning (hidden assumptions, incorrect deductions). In other words, modeling in some general sense is unavoidable, so do it openly.

It is really an empirical question about whether explicit modeling (as opposed to intuitive inference) really helps to avoid bad reasoning. Few would doubt this in fields like physics, but some historians have promoted the myth that "complicated" sciences like biology have proceeded productively without resorting to the "mental crutches" of modeling. So they must deny that the models of the mathematician Thomas Malthus helped Charles Darwin crystallize his ideas about natural selection, that Gregor Mendel and his successors did not benefit from thinking about and extending his genetic models, and that an army of field ecologists has not gone forth to study why Gause's competitive exclusion principle (deduced from a trivial model) has not prevented the incredible richness that we see in natural communities. One may argue in each of these cases that the models were too simplistic and should have been discarded or modified long before they were, that they were woven into dogmas that retarded scientific progress; but this is an indictment of scientists, not of the models!

The value of modeling in fields like biology has not been to make precise predictions, but rather to provide clear caricatures of nature against which to test and expand experience. It seems to be a very fundamental human need or requirement to construct such caricatures as a basis for learning. Moskowitz (1978) noted that even the learning of language by children seems to involve a modeling process. His argument is that children form explicit hypotheses (models) about how to say words, then use feedback from parents to correct these models. This argument has a counterintuitive consequence supported by some experimental evidence: "baby talk" by parents can promote rather than retard learning, by giving children feedback that is closer to their initial models (and so makes it easier for them to modify these models).

Modeling involves two fundamental phases of thinking that alternate with one another in a sort of adaptive dance. There is an inductive, creative, synthetic, constructive phase when we try to decide what and how to include in the caricature of reality. Then comes a deductive, more mechanical phase when we use mathematical analysis and simulation, employing the caricature as a "deductive engine." We then compare the deductions with expectations and, if inconsistencies are revealed, the dance continues with another round of induction. Learning is involved in both these phases; we may be equally surprised by the gaps in understanding that attempts at synthesis usually reveal, and by the predictions themselves. Some workers have stressed one or the other phase, claiming that we learn either mostly from the discipline of thinking constructively, or from rigorous deductive analysis. So we find people content to develop flow charts and "conceptual frameworks," and others preoccupied with mathematical tools for optimization, sensitivity analysis, and so forth. These extreme stances have contributed a lot to the confusion.

When you reject the extreme stances and recognize modeling as a very human way of groping for understanding, it should be obvious who will benefit most from it: *those who engage in it directly.* A great deal of money has been wasted by government agencies on contracts to model builders, in the hope that grand predictive models will be produced and then used by the agency. The modelers certainly learned a lot from these efforts, and have produced many lovely (and largely ignored) reports detailing formulations, predictions, and uncertainties. A few of the models have seen some use, but mostly as interactive computer games that are not taken seriously, or as generators of thick printouts to impress audiences who will never read them.

How Adaptive Environmental Assessment Works

AEA was developed as a way of getting people involved in modeling as a learning process, rather than as something you hire a specialist to do. The basic idea in AEA is to bring people with a mix of knowledge and talents together for brief periods of intense interaction in "modeling workshops." The usual workshop involves:

(1)   *a modeling team* with some experience in the details of mathematical formulation and computer simulation;

(2)   *research scientists* with various disciplinary backgrounds and specializations;

(3)   *resource managers* with experience in the nuts and bolts of monitoring and regulation, and a feeling for the history of the system;

(4)   *policy analysts/decision makers* with some broad responsibility for defining management objectives and options.

It is made an explicit objective of each workshop to develop and test (deduce predictions, run on computer) a quantitative model of the management problem during the time available. In early sessions, this model is usually developed as a computer simulation, representing various components of the managed system in some detail. Later sessions may involve "compressing" the model to eliminate unnecessary details and to provide a clearer, simpler basis for developing qualitative arguments about policy options.

That it is possible to build rather complicated and realistic models during short workshops was discovered by accident, around 1970. Our group at the University of British Columbia was asked to give a three-day seminar on systems analysis to research project leaders from the International Biological Program (IBP), which consisted of several large (and at that time, unique) interdisciplinary research teams, each attempting to study a "whole ecosystem." These leaders had been encouraged to include model builders in their teams and to synthesize the disciplinary results into ecosystem models, and most of them were rightly suspicious of the idea. To give them an inside view of modeling potentials and pitfalls, and to stress the observation that it is mostly the modelers who learn from modeling, we decided it might be possible to actually have them construct and run a little ecosystem model during the seminar. So we put together a team of graduate students to help with the programming, and Ray Hilborn wrote a skeleton program to make it easier to enter model relationships and plot results. That seminar was three days of sleepless nightmare for the modeling team; the research leaders enthusiastically elaborated a far more complicated ecosystem model (of a hypothetical lake) than we could program for them, our mini-computer broke down several times, and the only simulation results that we finally did get were utter nonsense. We feared that the seminar had precisely the opposite effect than intended, and that the participants would leave with even deeper doubts about the value of model building.

Figure 3.1. The construction of quantitative models is an important step in problem analysis, but the models should not be seen as final products. Here is where they usually belong. *Source:* Buckingham (1979).

Then letters began arriving from the IBP project leaders, asking for more results from *their* model and requesting that we do similar exercises with their teams as participants. So we held a series of what were now called modeling workshops, and each case allowed us to test and refine various tactics to improve communication and programming. C.S. Holling suggested we try the process on resource management cases, and he arranged to do a series of workshops for the Canadian Department of Environment on problems ranging from spruce budworm dynamics in eastern forests to Arctic development to Pacific salmon management. These cases led to a growing demand, and the emerging AEA process was applied to literally dozens of cases. Hilborn's programming aids were developed into a widely used package called SIMCON (Hilborn, 1973), and the AEA process was turned on itself to assist in training new modeling teams. Four modeling teams were at work in North America and Europe by 1982.

Few of the models developed in AEA workshops have been used directly for policy analysis. Most have been put where they belong as mechanical instruments for prediction (Figure 3.1), having served the essential purpose of promoting clearer thinking by and communication among the workshop participants. Some have provided a starting point or broad framework of relationships for organizing sequences of more focused workshops and meetings, leading finally to serious policy recommendations.

Getting Started

Perhaps the most difficult step in AEA occurs during the first workshop day or in earlier "scoping sessions" involving the modeling team and key clients. This is the step of problem bounding, discussed in Chapter 2, where it is decided what basic components and space/time scales are to be considered. When the bounds are set in scoping sessions, it can be difficult to decide even who should be involved in the workshops. Politics and prejudices make it difficult to see a clear entry to the problem. Disciplinary specialists become defensive, worrying that their area of knowledge will not be represented in enough detail or will be seen as of questionable importance for continued research support. Managers insist on attention to practical questions and to the narrow objectives that they feel comfortable working toward. Policy people press for looking at the problem more broadly, since they must take account of social objectives and forces that extend far beyond the particular resource system being considered. The modeling team tries to make its work easier, promoting consideration of factors and processes that they already know how to represent in the computer.

Frustrations can build rapidly during the problem-bounding discussions, and final consensus is seldom reached except as a matter of sheer exhaustion or running out of time. Indeed, one of the original ideas behind building models during short workshops was to prevent the kind of "diseased introspection" or "paralysis through analysis" that has gripped some interdisciplinary teams, and to force movement through a series of steps that are quite likely in the end to reveal that many of the original or intuitive concerns were unwarranted.

Some of the early AEA workshops were conducted during a period when the Canadian bureaucracies were being encouraged to use systems approaches, and in particular the notion of "management by objectives." The idea was apparently that formal articulation and listing of objectives would make civil servants function more effectively. Our workshops were often attended by people fresh from meetings where such lists were produced, so we tried a few times to use their lists in problem bounding. The results were disastrous, basically because spelling out what you would like tells you very little about how to get there. In the end, we found it much more productive to ask for lists of (1) possible actions (policy instruments, regulatory measures, etc.), and (2) performance indicators (population size, revenues, employment, etc.) that would be used to measure attainment of objectives. These lists help directly to define a model structure, by pointing to input/output relationships that should be represented either directly or indirectly (as the consequence of interaction among other, so-called "state" variables).

We were often admonished to "stick to the facts" and try to model only those processes and variables for which a solid data base was available; that is, we were asked to represent some problems narrowly, but precisely. But by doing this we would simply be reinforcing concern with those well trodden paths that researchers and managers have found it convenient to follow, and the process would fail as a device to promote learning. Luckily, there were normally a few wiser heads among the participants, and they helped to convince others that we should worry first about understanding the problem system, rather than about what work happened to have been done to date on it.

Perhaps the most important lesson that we have learned about problem bounding is the value of deliberately looking at the system more broadly, and in somewhat more detail, than initially appears worthwhile. To encourage this, it is necessary to conduct the problem-bounding discussions as "brainstorming" sessions, with emphasis on getting people to throw in lots of ideas and factors while being careful not to make critical comments or judgments that may inhibit others from speaking. Scientists usually have a terrible time trying to behave this way, while policy people enjoy it. The following two examples show what effect it can have on later analysis and conclusions.

James Bay development

The James Bay development is an enormous hydroelectric project in northern Quebec, and we were asked to look at its environmental effects. Details of the AEA workshop and its results are given in Walters (1974); let me just trace here what happened because of one comment during the bounding exercise. Most of our attention had been focused on how to model the project elements (dams, diversions, etc.) and their obvious direct impacts on fisheries, wildlife, and the native Indian economy. Then one character asked "what about the camp cooks," and amidst groans of disgust from other participants went on to explain that the large construction camps would have cooks (and others) with some free time every day for fishing; they might take enough fish to make a significant dent in the very unproductive stocks, like lake trout, that are characteristic of northern lakes. To humor him, we agreed to include a few simple harvest calculations in the computer program. Later in the workshop, these calculations were discussed and some rough data on catches were presented. Suddenly the numbers did not look so small, and we began to worry about other "incidental" harvesting activity, for example by tourists stopping along the several hundred kilometers of access roads to the project from southern Quebec. It did not take long to realize that we were talking about fisheries and wildlife "impacts" that were likely (if uncontrolled) to be at least an order of magnitude larger than total direct impacts in the actual development area, where most of the monitoring and research had been concentrated!

This example alerted us to be very careful about spatial bounding for environmental assessment in general. As shown in Figure 3.2, assessment normally proceeds with the assumption, borrowed by analogy from physics and engineering, that effects of disturbances are greatest at the source and diffuse in space and time. It is easy to see how bad this assumption is, just by thinking about all the means we have for "transporting" some ingredients of a problem to other places, or "storing" them for someone else to worry about later.

Salmon fishing in British Columbia

Southern British Columbia supports a valuable sport and commercial fishery for chinook and coho salmon. In view of evidence that some stocks were declining, we were asked to examine various regulatory options for the complex gauntlet of fisheries involved (Argue et al., 1983). Shortly after the juvenile salmon go to sea, they become subject to "shaker mortality," when they are hooked and released by sport and commercial troll fishermen. Many are trapped and smashed in seine nets. As they migrate and grow, they become of legal size and are taken in various sport and trolling areas. Seines, gillnets, Indians, and river sportsmen take them as they mature and move into spawning rivers. The situation obviously calls at least for some careful quantitative book-keeping, since regulations at one point in the gauntlet may just make more fish available for harvest later. The fishing interest groups are not exactly in love with one another, and demanded assurances that conservation measures imposed on them would not benefit someone else.

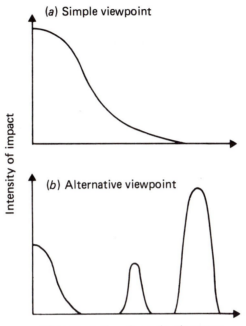

Figure 3.2. Alternative paradigms for the distribution of development impacts. *Source:* Holling (1978).

It is usual in modeling such situations to assume that fishing efforts are either constant or subject to full management control. Then the model is used to search for better, or at least mutually acceptable, effort combinations. But an economist in the workshop pointed out that all sorts of variables affect fishing effort, and that sport fishermen in particular may respond to changes in the abundance and size of fish available. He noted that the sport fishery is "open entry" with a trivial license fee, and involves a very large number of people (over 100 000) who could (and do) create political nightmares for anyone trying to regulate their effort directly. We were not convinced it would matter, but to pacify the economist we agreed to include a program option that made simulated sport effort increase with fish abundance.

That option came to dominate the later analysis, as we realized that the sport fishery is a significant mortality agent and response by it can completely cancel or even reverse the effects of most conservation and enhancement measures aimed at increasing juvenile and ocean survival. For example, suppose the shaker mortality can be reduced by making sportsmen and trollers use barbless hooks. Fish "saved" in this way may just attract more sportsmen later on. Sport effort responses may also exaggerate the "American plan" effects of enhancement on wild stocks (Chapter 2). In short, the sport fishery may induce an effectively unregulated "bionomic equilibrium" on the system, with all attempts to increase stock size just adding to the total sport harvest.

Similar concerns about sport effort response have become a policy issue in other North American fisheries, such as the billion dollar sport fishery developed in the Laurentian Great Lakes through salmon and lake trout stocking and efforts to control the parasitic sea lamprey (we will return to this case in Chapter 5). Effort responses should be a major worry in some wildlife situations as well. But effort responses are one of those processes mentioned in Chapter 1 as being very difficult to study experimentally.

Workshop Tactics

In the two examples above I skipped from the crucial first step of problem bounding to the end results, to show how apparently minor changes in the boundaries can make a drastic difference in what is finally considered most important. However, like many lessons that seem obvious with hind-sight, some careful and even tedious steps were required along the way. For each factor that emerges as important, usually another dozen fade from consideration as their possible effects are articulated. This section reviews the steps and tactics that follow problem bounding in AEA workshops.

Defining subsystem tasks

Problem-bounding discussions can lead to an amorphous set or listing of concerns. A first step in bringing some order to those concerns is the seemingly simple matter of classifying them into a few subsets or subsystems. Then later in the workshop each subsystem is analyzed more carefully by a working group consisting of one member from the modeling team and several participants. Generally the classification is most naturally based on areas of disciplinary concern and knowledge, so that each working group can concentrate on a "tightly interlinked" set of variables and relationships, which are "loosely connected" to other subsystems through a few variables of shared concern.

Unfortunately, the interactions that we call resource dynamics are not so easily decomposed into nice disciplinary clusters of roughly equal complexity (which would even the workload among the participant groups). Suppose, for example, that we are looking at a case where the obvious concerns include biological population dynamics, pollutants and their effects on the population, development of the harvesting industry, and the regional economy within which the industry is embedded. Here it would seem obvious to have two working groups consisting mostly of biophysical disciplinarians, and two with mostly social scientists. Now, which group should deal with (analyze, construct simulation rules for) harvest, that key link between population and industry? If the population group tries to model harvest, they may need all sorts of input information from the industry group, such as harvesting effort and measures of technological efficiency in capturing the organisms. This would violate the tactical objective of having working groups that share or exchange only a minimum amount of information. The situation might not be improved by making harvest a responsibility of the industry group, since they may need equally complicated biological information (on densities, seasonal distribution, spatial pattern of organisms available, etc.). As we shall discuss in Chapter 4, harvesting involves a tightly interdependent set of biological and economic factors, and some processes like the short-term "numerical response" of harvesters to resource density have not been studied carefully either by resource biologists or by economists. Such gray areas that fall between topics of traditional disciplinary emphasis arise very often in workshops, and are usually the source of key uncertainties. In practice, a workable approach has been to assign the difficult concerns, such as harvesting, to whichever subsystem working group appears initially to have the lightest workload. Then, as uncertainties and expert knowledge become clearer during working group discussions, responsibilities are reassigned by forming *ad hoc* new teams or committees to look at the most troublesome concerns. This adaptive approach to the workshop agenda is welcomed by most participants, though a few always find it confusing and even threatening. Indeed, to apply it in an ordinary setting (scientific meeting or management conference) would be to invite an unproductive diffusion and fragmentation of discussions, following lines of least resistance into old and comfortable topics. What makes the adaptive agenda work in AEA is the insistence on producing a working model during the time available; the developing model provides both a concrete framework for seeing where the discussions fit together, and an ever-present excuse to cut off irrelevant or repetitive conversations.

Looking outward

The initial problem breakdown into subsystems is followed by a specially structured discussion, called "looking outward," intended to define the working group responsibilities as precisely as possible. It proceeds by asking each working group to state what input information it will need from each other working group in order to represent whatever interactions or variables are assigned to it as internal concerns. At first this forces each group to do two things: (1) to think carefully about how detailed their submodel needs to be (to decide how detailed the information inputs from other groups need to be); and (2) to "bargain" with the other groups about what they can, in practice, provide. So a "fishing industry" group may initially ask a "population dynamics" group for information on total catches. Then the population group might reply with offers of detailed size composition of the catch. The ensuing exchange may reveal that there is some reason, not initially recognized, why the industry group should worry about such details. Or, the population group may realize that its initial assumption about the need for detail (as output from its submodel) is exaggerated.

After an initial round of defining and bargaining about input needs for all the subsystems, each working group is then asked to look systematically at all the outputs it has been asked to produce for the other groups. Then the question is asked "in view of these outputs requested from you, is there any other input information that you will need besides what was listed in the initial round? Or, even better, can you get by with fewer or simpler inputs than you originally thought?" These questions often result in quite surprising reappraisals by participants of what the problem is really about, and about the importance to other disciplines of measurements (variables) they are (or are not) prepared to supply.

In principle, the looking outward process can result in an escalating argument over details, with each group forced to request more inputs as requests for its outputs grow. To my knowledge, this difficulty has never occurred in practice; most often the discussions go to the opposite extreme, with participants too willing initially to oversimplify the interactions. Also, participants tend to "police" one another by referring back to the action/indicator lists developed during problem-bounding sessions, and insisting that each new request for inputs be justified as necessary for calculating indicators in relation to actions.

Occasionally the looking outward process is used in meetings where there is no time or commitment to produce an explicit model. This has been especially valuable for interdisciplinary research teams trying to plan joint studies. Too often, the members of such teams proceed into the field with a distorted or narrow notion of what the other members will need or can use, with a predictable result: a collection of reports that enhance each member's position in his discipline, and a paper or two about the whole system, but filled with arguments that bear little relationship to the data collected.

Submodel development

The action/indicator and looking outward discussions leave each working group with specific modeling responsibilities, as shown in Figure

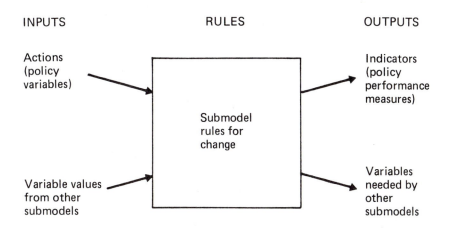

INPUTS RULES OUTPUTS

Actions
(policy
variables)

Variable values
from other
submodels

Submodel
rules for
change

Indicators
(policy
performance
measures)

Variables
needed by
other
submodels

Figure 3.3. Each submodeling group in an AEA workshop has a well defined job in terms of the information available to it and the outputs it must produce from its submodel's calculations.

3.3. The group must produce a collection of rules for predicting how a particular set of indicators and outputs to other groups will vary over time, in relation to actions and time-varying inputs from other submodels.

As a charge to each group before it begins to formulate its rules, we warn not to try anticipating which combinations of actions and time-varying inputs might be tried later in the workshop. In other words, each group must try to formulate its rules (submodel equations) so as to be valid for all possible values of its input variables. This has a way of shifting attention away from recent historical averages and trends, toward trying to make statements about functional relationships (how variables relate to one another, independent of particular times of observation). Almost immediately and inevitably, attempts to make such functional statements lead to the recognition that historical experience is limited, so that extrapolations (as in Figure 1.1) will be necessary.

Most AEA computer simulations are constructed with the basic logical structure shown in Figure 3.4. The state of the modeled system at any moment in time is represented in terms of the values of a collection of "state variables," such as population size and amount of harvesting equipment currently available. Then the simulation rules (model equations) try to predict new values for all the variables after some short time step (usually a year), in relation to (1) current variable values (e.g., current population affects population change), (2) policy actions specified for the step, and (3) values of external "driving variables" (random environmental effects, variables represented only as time trends) chosen for the step.

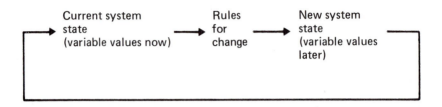

Figure 3.4. Logical structure of simulation models produced in AEA workshops. Long-term predictions are generated by repeatedly applying the rules for change to the new states resulting from the last application of the rules.

Long-term predictions are built up by applying the rules repeatedly, each time letting the last predicted state be the new starting state. It might seem that this iterative or "recursive" procedure just invites trouble, in the form of errors that accumulate over time. Unfortunately, there is no way to avoid such error accumulation in processes that the rules imply should exhibit positive feedback (geometric growth; larger variable values causing still larger changes). On the other hand, there is no other known way of deducing the dynamic consequences of whole collections of processes interacting to produce both positive and negative feedbacks.

The rules within each step may be quite complicated and involve all sorts of mathematical forms, such as differential equations, to represent variables that change continuously in time; but we find that most workshop participants can relate most easily to the idea of a basic clock or time step within which they are free to represent both continuous change and discontinuous phenomena like seasonal reproduction. Indeed, the rules for change within each time step are often organized to reflect the seasonal cycles of organisms and economic activity.

After this approach to simulation has been explained to workshop participants, it is usually easy for them to decide, at least generally, how their rules for change should be structured. With the help of its modeling team member, each group makes a list of the state variables for which they will need to construct rules of change. Then the change (per time step) in each variable is expressed as a tautology that decomposes it into more manageable components (for example, population change equals births minus natural deaths minus harvest plus immigrants minus emigrants). Then it is attempted to express each of the components as a function of whatever policy, state, and driving variables are thought to influence it. These functional relationships form the basic empirical assertions or scientific hypotheses of the model.

The business of developing functional relationships is difficult for participants who are used to thinking in terms of time series and trends, or who are not used to dealing with even simple equations. Here the modeling team

member lends his experience to the working group, suggesting possible relationships based on his experience and the quantitative literature. These
suggestions form "straw men" that the participants can criticize, and these
criticisms often point to better representations. Again we see an adaptive
process, remarkably similar in many ways to the model mentioned in
Chapter 2 about how children learn language. The effective modeler is a
child, and his baby talk is corrected through criticism toward the understanding that his parents, the participants, have. Some modelers I know consistently get this backward, and try to act like parents with people who, in
fact, are considerably more knowledgeable than they; the resulting confron-
tations are seldom an effective learning process.

The formulation of functional relationships is accompanied by specific
data analyses to get at least rough estimates of key parameters, and to pinpoint where extrapolations beyond historical experience will be necessary.
Also, there is usually an attempt to examine very long-term historical records
for clues to relationships that may have been overlooked initially. For example, very long records of relative fish densities can sometimes be obtained by
counting scales from cores of anaerobic bottom sediments (Soutar and
Isaacs, 1974); density cycles or inverse correlations between species may
point out possible ecological interactions (predation, competition) not evident from recent population statistics. Here modeling experience can again
be important: the modeler may see patterns predicted by other models; he
can then suggest to participants that they try (evaluate) similar causal arguments.

After developing functional relationships, the working groups face a
final set of sometimes difficult choices about how to represent policy actions.
The easiest approach is usually to specify what control theorists call "open-
loop" policies. An open-loop policy is simply a time series of actions,
imposed from outside on the model without reference to changing system
state. Though some actions may actually be imposed this way in practice,
the more common situation is to have some sort of "feedback policy" that
relates action to estimated system state. Then a key part of the modeling
exercise becomes the development of a set of alternative feedback policies
that relate simulated action to different state variables (or simulated sampling
estimates of state variables); these alternative policies can be compared to
one another as simulation "scenarios".

Programming and scenario development

When the logical structure of the submodels has been decided, the
modeling team closets itself to develop the simulation program. The participants work on parameter estimates, if needed, and then convene for scenario

development sessions. In these sessions, they try to define policies (combinations of actions) that will challenge the model's credibility and perhaps suggest directions for more careful policy analysis later. They are warned that the simulation program is not to be taken very seriously, especially at first, since there are likely to be programming errors and some ridiculous parameter estimates.

There is no fixed protocol for the actual programming; each team develops conventions and procedures that are comfortable for its members. Usually the programming starts with a short working session to (1) lay out (flow chart) the order of calculations for the overall program, and (2) decide on computer variable names and check consistency of measurement units for those variables that are shared among submodels (the subgroup inputs and outputs). Then each programmer follows his own procedures for making his working group's model. His only constraint is a time deadline, when he must be ready to have his subprogram interfaced with the others. Because of variable work loads, programming skills, and plain luck, the submodels are usually not completed in phase. This means they can be entered and "debugged" sequentially, and the programmers who finish first can help the others.

As the submodels are tested individually and interfaced, the programmers inevitably uncover logical problems with the initial formulation and parameter estimates. Populations grow geometrically without bound, or collapse even without harvest. Pollutants go to enormous concentrations, even when simulated sources are turned off. Harvesting effort oscillates violently, as the simulated harvesters alternately expect severe competition and none at all. Some of these logical problems can be traced back to simple programming errors, but there usually remains a subset that reflect errors in the original formulation and deeper systems issues (such as whether response time lags will actually cause harvesting effort to oscillate).

So even before any formal sessions are convened to explore policy options, the programmers must usually reconvene their subgroups. They show preliminary results that do not seem to make sense, explain why the equations imply these results, and solicit advice about whether to change or add new relationships. This is an exciting time for many participants, as they see the logical consequences of their own thinking and are forced to be even clearer and more precise. Some come to appreciate how fragmentary their thinking has been, and see new research questions and priorities. Management approaches are questioned as being dependent on one or another bad assumption, and imaginative new possibilities are suggested. Occasionally, implausible predictions are seen to be borne out by field data and to explain features in the data that had previously been ignored or explained in other ways. (This is dangerous; such lucky discoveries in no way imply that the model must be "correct;" see Chapter 6.)

With luck the model will go through several revisions in the last workshop day or two, with the results after each revision being harder to reject on the basis of intuition and available historical data. At this point many participants will develop a strong parental affection for the model, and insist on using it to develop at least a few policy scenarios. It becomes important that the model be "interactive:" it must easily permit a variety of policy changes, and must produce graphical results quickly (no more than a few minutes for the computer to produce each multistep scenario). With fast interaction, the model becomes a game, played for its own interest and to uncover flaws and policy possibilities.

In early workshops, we failed to recognize the learning value of game playing, and insisted instead that the participants develop just a few complicated scenarios and examine the results from each in detail. That is, we fell into the trap of treating models as instruments for precise prediction. Now most AEA workshops stress the use of tools, like microcomputers and interactive graphics packages, that make it easier to explore model results quickly and selectively. In a good workshop, we might end up running 20 or 30 scenarios, and then only examine the results from one or two in any detail.

Running down

Early AEA workshops ended with formal discussion sessions to itemize research and management priorities. The submodeling and scenario/gaming sessions usually make evident a number of gross uncertainties, and a few opportunities for policy improvement. We felt it was important to capture these pearls in some sort of workshop report, lest they be forgotten by participants after a few weeks "back at the office." It took a few years for us to learn that (1) forgetting times can be even faster, i.e., a few days; (2) reports are seen as a way to *complete* (and file away) analysis, not *keep it alive* in people's minds; (3) most of the top research priorities are about processes that no one knows how, or has the patience, to study (see Chapter 1), and so are ignored anyway; and (4) workshop models are too often deceptive, and priorities identified with them are usually changed after more careful analysis.

What these points imply is that the AEA process should be used as a starting point for analysis; there must be strong commitment to continue to bring key people together for model improvement and policy exploration. It seems, then, that the best way to end a first general AEA workshop is by deliberately *not* drawing and printing any formal conclusions. Those last hours are better spent in planning a timetable for further work. There should deliberately be no clear ending, just as in other human learning processes.

Compression for Understanding

After observing the behavior of an AEA model in gaming sessions, it is generally (perhaps always) possible to see ways to greatly simplify the representation of those variables (indicators) that are of key policy interest. Chapter 5 will discuss examples of the basis for and mechanics of doing such "compressions" (simplifications without loss of essential features). Here I will simply stress how model compression is promoted by and complementary to the AEA workshop process.

Reflect for a moment on the title of this section, which I lifted verbatim from Holling (1978). What do we mean when we say we "understand" something? Is it just that we can predict how the thing will behave? I think not; we might achieve good statistical prediction without ever asking how or why the thing works. Also, there are things everyone would agree are understood in some basic sense, but behave unpredictably (the weather, for example). "Understanding," then, has to do with our ability to describe the thing in simple terms, or place it in a "comfortable" category of similar things. Actually, I have never heard anyone give a satisfactory explanation of what understanding means, in spite of how important we all *feel* it is. But everyone would agree that simple description is somehow essential to understanding. Then, if compression was always successful in capturing the essential causality in and behavior of a system, we might want to title this section "Compression *results* in understanding." Instead, about the strongest justifiable assertion is that "compression is a *necessary, but not sufficient* step in understanding." Let us cut through this pseudophilosophical wandering with a more direct empirical assertion: if you want to be understood, be prepared to explain your system in simple terms; build this explanation from a careful compression of the original system representation.

Recall again that most resource policies are the outcomes of bargaining and debate in essentially political contexts. There is simply not enough time in such contexts to consider all the details of each disciplinary concern, even if it would help to do so. Recognizing this, resource scientists have too often tried to make themselves understood by posing only very simplistic causal arguments, such as "pollution will destroy the resource," or "current exploitation levels will decimate the stock," or "habitat improvement will provide more animals to harvest." These arguments may be quite understandable to other actors, but they are usually based on a kind of linear or single-factor causality that nature is very unlikely to follow. So when the arguments are convincing, they lead to unnecessary or extreme policies; when they are not, the resource scientists lose credibility, and even more simplistic recommendations may win favor.

Simple, but careful, reasoning based on compressing AEA workshop models has been used in various resource policy debates, with very mixed results. In several cases, the very fact that modeling had been used was a source of considerable debate (garbage in/garbage out, computers are taking over the world, etc.), in spite of the fact that the reasoning was strictly verbal and qualitative. In other cases there was considerable concern about why the resource scientists involved were not using the usual old arguments, and suspicions were voiced that the scientists must not agree among themselves on what to say (and therefore should not be listened to at all). From these cases, the cynical conclusion would be that rational arguments in general are neither welcome nor influential in resource policy debates. But we have seen enough cases where the arguments did matter, and significant policy changes were effected, to suggest that the cynical conclusion is unsound and should not be used as an excuse for inaction.

Building Modeling Teams

The key ingredient of AEA is an experienced modeling team. The team members must be broadly knowledgeable about how to model various processes, and must be adept at programming. They must have personality traits that are difficult to pinpoint, but include willingness to listen to participants, ability to articulate quantitative ideas clearly in verbal terms, and willingness to quickly admit and learn from mistakes. They must be able to take the stress and long working hours of the last few days of workshops. This combination of quantitative skill, personality, and drive is quite unusual, and recruitment of modeling team members has been a persistent problem in the development and spread of AEA.

It is often assumed that mathematicians make the best modelers, or at least that deep training in applied mathematics is necessary to become seriously involved. This misperception has often led to recruitment of inappropriate people by government agencies. As noted earlier in this chapter, modeling involves two phases: inductive (creative) and deductive (analytical). Too often, professional mathematicians are only trained and willing to engage in the analytical phase; they will happily take a completely silly model that someone else has proposed, and analyze it to death. But they may lack the experience and insight about natural systems that are necessary to be very critical or constructive. Some brilliant characters stand out as exceptions to these generalizations, mathematicians who learned other disciplines and have become deeply involved in both modeling phases. But the mathematical skill needed for model building (as opposed to analysis) is generally not that great, and includes only basic multivariable calculus, differential equations, linear algebra, and probability theory. Most universities offer courses in all these areas for students from other disciplines, so it is not at all difficult to find biologists and economists with the required skills.

Unfortunately, quantitative courses do not ensure that these people can proceed intelligently with model building. In recruiting new people for AEA teams, we look first for disciplinary knowledge (the particular area is not critical), then for evidence of systematic thinking and mathematical background. Finally, and most importantly, we look for that spark of imagination, confidence, and concern that marks individuals who will be willing to dare try and put together what they know in new ways.

Experience appears to be absolutely essential for AEA team members. Very few people can start afresh with a new problem, and immediately see how to give it quantitative form. Instead, we mostly build from past experience, seeking similarities and analogies between new and old cases that might permit mathematical formulations developed for the old cases to be applied afresh. Occasionally we may stumble upon or derive a quite new formulation, but most AEA models are built up from standard parts that have stood up to repeated tests. Some authors have correctly noted that model building mostly involves accumulating a "bag of tricks," and becoming adept at pulling out the right one for each new occasion. This is a tricky business, because it is all too easy for a modeler to try and force new problems into inappropriate mathematical frameworks that he happens to know well. A good example is the tendency for many operations researchers to force every problem into the optimization framework called "linear programming;" some do this even for problems where no optimization is called for in the first place. We try to avoid such difficulties, and the associated "railroading" of workshop participants, by supplying each AEA team member with the broadest possible bag of tricks, and accompanying the presentation of these with a range of examples of how they have been *misapplied* in the literature and in our own past workshops.

The AEA process has been turned upon itself several times to train new modeling teams. The agenda for training workshops involves a mix of lectures and demonstrations. plus a series of 2-4 mini-workshop case studies. Much of the lecture material is presented in this book and in Holling (1978). The miniworkshops take real cases of interest to the agency sponsoring the training, and go through all the workshop steps outlined above with a sympathetic participant group (usually personnel from the sponsoring agency). The case studies are selected partly to demonstrate particular dynamic patterns and interactions that turn up repeatedly in AEA workshops. Occasionally a training miniworkshop has stimulated so much interest among participants that a larger follow-up or workshop sequence has followed it.

A serious weakness in the AEA training process has been our inability to provide better guidance in tactics of handling workshop participants. We do not yet understand why the behavioral interplay between team members and participants sometimes results in remarkably creative efforts, but only in serious misunderstanding at others. Questions about how to deal more effectively with people should perhaps be the greatest emphasis in future development and experimentation with the AEA process.

Problems

3.1. Compare the "looking outward" approach as described in this chapter to the "working outward" approach to problem bounding described in Chapter 2. By looking at linkages among predefined subsystems, will looking outward promote broader problem definitions in the same way that working outward does? How can these approaches be used to complement each other?

3.2. Often in AEA workshops, a disciplinary subgroup will be bogged down in controversy about how to represent a particular process (detail required, form of relationship, etc.). Suggest alternative ways to avoid such deadlocks, and show how each of your suggestions might fail in practice.

3.3. Compare the learning process by participants in AEA workshops to the way scientists learn about a process by doing a sequence of experiments with it. Is it less "valuable" to see existing data in new ways (as often happens in workshops) than to spend the same time and effort gathering new data? Do you expect new concepts and perspectives gained during a single workshop to remain prominent in the minds of participants, without deliberate repetition and reinforcement later?

3.4. A useful tool for workshops is a simulation control program that sets up the time "loop" of Figure 3.4, provides for storage of key indicator variables as they are calculated at each time step, and allows plotting of variables over time and against one another after each simulation is completed. Develop such a program, so as to store the indicators in a two-dimensional array by code number and time and allow later plotting of them by code number.

3.5. Consider the common situation where there are two interdependent submodels, A and B, such that some new variable values (after one time step) in A depend on values of some B variables at the start of that step and some B values likewise depend on starting A values. What can go wrong in a computer program that updates variables in sequence (first submodel A, then B)? Suggest at least two ways to avoid mistakes in which variable values are used.

Chapter 4

Models of Renewable Resource Systems

Some people, notably poets and mathematicians,
use other tricks than words themselves can play
to convey meaning.

Beer (1978)

A deceptively simple way of describing adaptive management is to say that it involves trial and error learning. The difficulty with this description is that the phrase "trial and error" connotes for most people a process of blind probing where reward and punishment are easily recognized, i.e., a process of natural selection and evolution. It is certainly not the intent of this book to promote such a process; it should be possible to design resource management policies that are much less wasteful. In the first place, we can develop structural representations or models of how nature might respond to alternative actions, and then use these models to direct the learning trials more wisely. In other words, we should begin adaptive policy design by posing clear hypotheses based on previous experience and functional understanding. In the second place, the detection of and response to errors are not so simple in resource systems as they are in natural selection. We can always make foolish errors by adopting the simple model "nature will be the same next year as in the recent past," while restricting our learning process only to measuring that recent past more accurately. To step beyond this approach of "riding the trend," we must look more carefully and quantitatively at dynamic interactions—and when we do so, the errors become more subtle and difficult to detect. This chapter will examine some models of resource interactions, with emphasis on formulations that have been "successful" in the sense that they have predicted well for at least short periods, and so provide a good starting point for the more subtle learning that goes beyond simple trend analysis.

Later chapters will examine how model prediction errors can be formally used in adaptive learning processes, but there are some basic criteria for judging the credibility of dynamic models in relation to historical experience. These criteria should be applied in thinking about the models in formal policy design. So the first section below gives you some ammunition for thinking critically (adaptively!) about the models reviewed in later sections.

The review sections will look at progressively broader dynamics, beginning with single-species populations and ending with a look at how resource harvesting systems are embedded in larger regional economies. Each section begins with about the simplest plausible model for the processes in question, then proceeds to more detailed and realistic formulations.

Judging Model Credibility

There is great confusion in the renewable resources literature about how to recognize "good" or "bad" models when you see them. Previous chapters have noted that one common criterion, detail and precision of predictions, is of less value than it might seem intuitively. Another common criterion is whether a model represents "basic causal factors" correctly. But consider the following example. Suppose two analysts present you with models for a marine fish population, and you feel it is necessary to make some judgment about which to use for policy design. The first analyst has found that recruitment has been closely related to ocean temperatures when the recruits are spawned, and his model reproduces this pattern very precisely by making various calculations (based on experimental data) about how temperature affects juvenile growth and time exposed to various predators that take only small juveniles. The second analyst has not examined the mechanism of survival at all, and instead has simply derived empirical probability distributions for recruitment rates at different spawning stock sizes. If your background is in biology, your intuition will probably be in favor of the first model; it seems to represent the system better. But wait a moment! If you are concerned with management, perhaps you should worry more about *how* recruitment varies (in a statistical sense) than about why, especially if the variation is likely to be due to some environmental factor that you can neither control nor predict. If spawning stocks can be controlled to some degree through harvest regulation, you should worry much more about the average recruitment rates at different stock sizes (i.e., under different policies) than about the inevitable "random" variations. Indeed, if you want to use the first analyst's model for harvest assessment, you will need to assess (predict) a probability distribution for future marine temperatures—and when you use this distribution, you had better get the same results that the second analyst already gave you in a simpler form. (If not, the second analyst's distributions contain some source of variation that cannot be accounted for by marine temperature alone, and for which you should also account.) This example shows a very basic principle of modeling: the criteria and model you should use are very much dependent on what questions you are trying to answer.

The following subsections outline some model evaluation criteria that should be of particular concern to resource managers charged with evaluating the consequences of alternative harvesting policies. Some of these criteria may be of less interest to scientists charged with trying to understand why (in a functional sense) things vary. There is, of course, no clear distinction between these charges: without depth of understanding, the manager may be caught unawares by some perfectly predictable factor; likewise, the scientist may be directed to more interesting research questions by applying criteria that initially seem important only in management contexts.

Repeatability

As noted in Chapter 3, dynamic models consist of functional relationships that predict how various components of change will vary in response to changing system states. These functional relationships are assumed to represent the effects of processes that are somehow repeatable or stable over time. Consider, for example, a hypothesized relationship between fish spawning stock and subsequent recruitment, as in Figure 1.1. Suppose the spawning stock is maintained at one base level for some time and average recruitment from this base level is measured. Then suppose the stock is disturbed (reduced or increased) for some time, then returns to the base level again. The stock–recruitment relationship is said to be *strictly repeatable* if the same *mean* recruitment is obtained as before the disturbance, for all base stock/disturbance combinations. Probably no real stock–recruitment relationship would be strictly repeatable, since at least some extreme disturbances (stock reductions) would lead to irreversible loss of genetic and spatial substructures. However, it is reasonable to expect many functional relationships to be *weakly repeatable*, in the sense that the same response mean will be obtained after disturbances within a range of magnitudes that are most likely to occur in practice.

The importance of repeatability in model relationships is obvious. Without it, the model cannot be parameterized (quantified) by reference to historical experience, and we have noted in earlier chapters that there is no real substitute for that experience. In later chapters we will examine the possibility of estimating response parameters adequately, while assuming that they are in fact slowly changing. But the statistical methods involved in such parameter "tracking" do not perform well when several parameters are

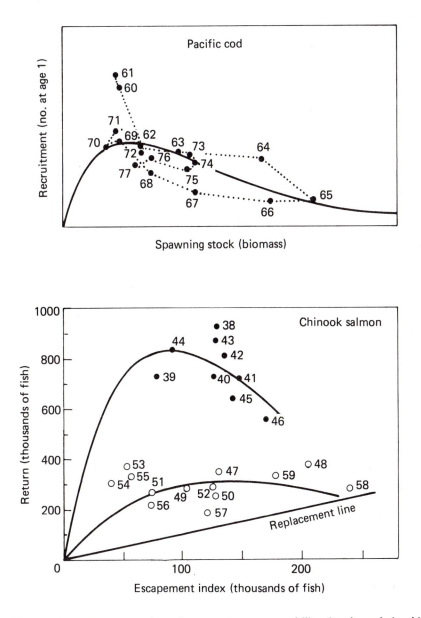

Figure 4.1. Two examples of apparent nonrepeatability in the relationship between spawning stock and subsequent recruitment. Pacific cod stock and recruitment estimates from cohort analysis (Walters et al., 1982). Columbia River chinook salmon data redrawn from Van Hyning (1973).

admitted to be simultaneously "drifting," so at least weak repeatability must be assumed for most relationships.

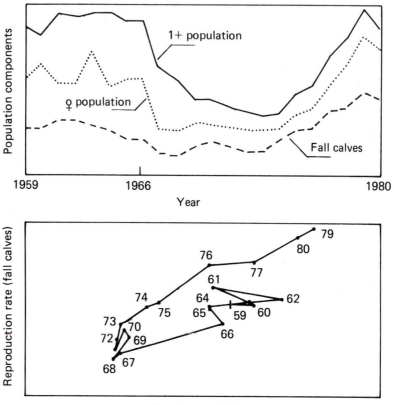

Figure 4.2. Trends in population size and the nonrepeatable relationship between calf production and female population, for a reindeer herd in northern Finland. Data provided by T. Helle (personal communication). The increased birth rates following decline are thought to be due to culling of older, less productive females from the population during the harsh winters (1967–68) when the decline took place.

Examples of nonrepeatable relationships are shown in Figures 4.1 and 4.2. In the Pacific cod example, juvenile survival rates are high during population increases, but low during decreases. The same is true of many ungulate populations, as in the reindeer example. In both of these cases, we would say that population density is not sufficient to predict reproductive rates. Disturbances leading to higher population densities result either in compositional changes within the population (for example, genetic selection for aggressive animals that do not care so well for their young), or in changes in the environment of the population (for example, deterioration of food supplies or increases in predator populations). In the third example, Columbia River salmon showed quite a repeatable stock–recruitment relationship until the early 1950s, when recruitment declined sharply, remaining at a lower level ever since. The decline may be due to habitat changes (hydroelectric dams on the Columbia), but a more likely cause is an increase in ocean mortality due to a growth in troll fishing that took place after 1950 and that probably harvests many Columbia River fish before they can be counted as returning recruits to the Columbia system itself.

Nonrepeatability in hypothesized functional relationships implies, in general, that the model is missing some basic state variables (such as food supply) or driving variables (such as high seas fishing mortality), whose effects are persistent over time. Thus the search, using historical data, for nonrepeatable relationships is a key step in trying to develop better models; it is an essential part of the learning process.

Stationarity

We expect some factors in the environment of any resource system to vary in an unpredictable or random fashion over time, without showing persistent changes in response to a changing system state. In simpler terms, we expect unpatterned variation around any functional relationship, as opposed to the persistent deviations of the examples in Figures 4.1 and 4.2.

When the probability distribution for unpatterned variation around a functional relationship is stable over time, we call the relationship a *stationary* stochastic process. In a stationary relationship, particular outcomes (for example, recruitment) may be quite unpredictable; we ask only that they be drawn from the same distribution whenever the same initial system state occurs. Stationarity may be viewed as a stronger assumption than repeatability, since for repeatability we require that only the mean response (but not the whole distribution of responses) remain constant.

Very few resource time series are long enough to provide even weak tests of stationarity hypotheses. An exception is the Skeena River sockeye salmon (Figure 4.3). In this case, intermediate spawning stock sizes have produced almost the same distribution of recruitments, even after a disturbance in the early 1950s resulted in very low spawning stocks for some years. Over the whole period of record, average recruitment at each spawning stock size has apparently been decreasing slowly (Ricker and Manzer, 1974).

It is reasonable initially to assume stationarity in the responses of most resource systems that exhibit extreme random variation naturally.

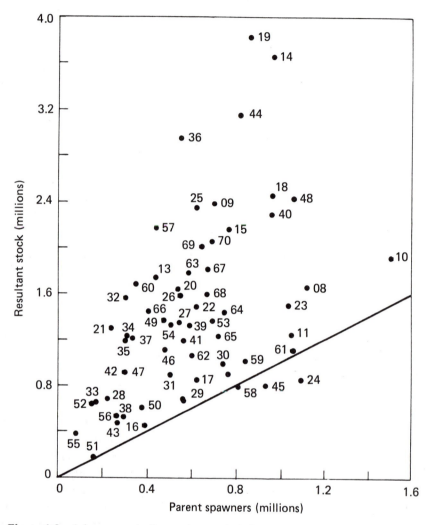

Figure 4.3. A long record of spawning stock and recruitment estimates for sockeye salmon in the Skeena River, BC. Such records are needed to determine patterns of "random" variation and the temporal stability (stationarity) of such patterns. *Source:* Shepard et al. (1964) and Walters (1975).

These systems would not be around today to worry about in the first place, unless they had quite strong mechanisms for repeatedly recovering from extreme states. The lie, of course, is given to this argument by many small clupeid fishes, such as herring and sardines, that have shown strong natural variation but have then collapsed, apparently irreversibly, after intense fishing.

Robustness

It is a common observation in model building that the same functional relationship or equation system can be derived from a variety of more detailed causal arguments. Functional relationships or models that have this property are said to be *robust* to uncertainties about underlying mechanisms. As can be shown even for very simple dynamic systems (Walters, 1980), robustness is a curse to the researcher who hopes that the ability of his model to fit data is a test of the assumptions used to derive it. By the same token, it can be a blessing to the resource manager who seeks a representation that is likely to have the right shape or pattern of response, no matter what the underlying causality.

Robustness will be a central concern in Chapter 5, where we will see that it is generally associated with very simple equations and graphical relationships. However, quite complex models can also be robust in the sense that all of their many functional relationships can be derived from a variety of even more detailed arguments. But they are generally not robust to changes in the structure or pattern of connection among relationships, or to the addition of new details to the model structure. My favorite example of this is the British Columbia salmon model discussed in Chapter 3. That detailed model had a myriad of relationships involving recruitment, mortality, growth, fishing, and so forth. Then we added the "little detail" of making the sport fishing effort respond to fish abundance, and the whole character of our predictions about fishing regulations was changed dramatically. With hindsight we were able to strip the big model of most of its complexities, leaving behind a simple and robust representation of the basic dynamics of stock production in the face of unregulated effort. But we could not see how to design that robust model in the first place. Thus robustness is not guaranteed by starting out simple; it can only be assured by much penetrating and critical analysis.

Consistency and completeness

No model can completely represent the world, but it is reasonable to ask at least that it be dimensionally consistent (in the sense of introductory physics or chemistry texts), and reflect basic accounting or conservation principles. Thus a model that seeks to represent dynamic changes in a population of N animals should not have its reproductive rate expressed in biomass units; if it seeks to represent how the N animals move about in space, it should not implicitly create or destroy any when moving numbers from area to area. It is surprising how often such very basic logical requirements are violated by models reported in the resources literature; usually the justification is statistical convenience or unwillingness to extrapolate (until conclusions are drawn!) from units of direct measurement. It is common, for example, to see recruitment relationships in fisheries represented in terms of recruitment and spawning stock "indices" (like catch per hour of test fishing gear) whose relationship to actual historical system states is tenuous at best.

Identifiability

It is easy to make a model appear very realistic just by decomposing its parameters into simpler, explicit components. For example, any survival rate over a finite time period can be written as the product of several survival rates over shorter time periods or "stanzas" of life. This can make sense, when the component rates have been measured directly and are functionally related in the model to state variables or policy actions. But, too often, models just string together a collection of constants. In the statistician's terms, these constants are not separately *identifiable*, in the sense that their individual effects cannot be distinguished from data on the overall process. It is fine as a scientist to say that $y = b_1 b_2 x$, where b_1 and b_2 are constants; but if only y and x are observed, then only $b' = b_1 b_2$ in the overall relationship $y = b'x$ is identifiable.

There is even some danger in stringing together nonidentifiable constants in models. Errors in the initial or baseline estimates of such constants do not always "average out" across the whole collection. So, an apparently realistic representation is quite capable of giving quantitatively much worse predictions than an aggregated model whose parameters are more easily estimated from historical data.

Systems properties

Analysts and theorists about dynamic systems in general have learned to watch for a number of basic model properties that are difficult to see by examination of individual functional relationships. The most important of these properties are:

- feedback loop pattern
- time delay structure

- thresholds and limits
- inflected functional relationships

The feedback loop pattern is discerned by examining the collection of positive and negative effects that state variables have on one another. Positive feedback (exponential growth) interactions are examined with the suspicion that some key ingredient might be missing from the model. Negative feedback interactions are examined as regulators or limiters of system behavior.

The time delay structure can be particularly important in resource system models, in terms of both biological and economic variables. Students of population dynamics are acutely aware of how time delays can introduce various overshoots and oscillations to population behavior. Similar economic oscillations are a daily fact of life for everyone.

Thresholds, limits, and inflections (sigmoid shape) in functional responses can also have profound effects on stability properties of the model system. The classic example is the sigmoid feeding functional response of vertebrate predators to prey density. At low prey densities, acceleration in predator attack rates as prey density increases can result in regulation of the prey population at low levels. On the other hand, satiation (limiting, saturation) of predators at higher prey densities can result in "release" of the prey to increase explosively until some other regulatory agent becomes important.

Stocks as Units of Analysis

The basic biological unit of management is most often assumed to be the single-species "unit stock," a collection of individuals who interbreed and exhibit an aggregate production process dominated by reproduction, growth, and mortality rather than by migratory exchanges with other collections. It was once thought that such unit stocks would be relatively easy to identify, in terms of morphological similarity among individuals and similarity in basic behavioral patterns, such as selection of reproductive locations and migratory paths. However, recent studies of genetic composition (using electrophoresis) and dispersal patterns (using large-scale marking and tagging) have revealed considerable "fine structure" within some collections that were thought to be homogeneous units (STOCS, 1981). In other cases, greater dispersal among collections has been discovered than was originally expected. Also, the spatial boundaries of management are often dictated by political or administrative practicalities. In particular, it is often not practical or economical to establish separate monitoring and regulatory activities for every genetic "substock" revealed by electrophoretic or other methods.

So a "managed population" is usually an arbitrarily defined collection of organisms with at least some internal heterogeneity and exchange with other populations. In the models discussed below, we will assume that the unit is large enough so that changes in abundance are dominated by processes other than dispersal to and from other units; internal heterogeneity will be seen as a source of nonrepeatability and slow parameter change over time in the basic production processes of recruitment, growth, and natural mortality.

Population dynamics models can be classified into four groups, based on the complexity of representation of production processes. *Surplus production* and *stock-recruitment models* look only at total population changes as a function of stock size, without regard to how particular processes (reproduction, growth, etc.) are varying. *Dynamic pool models* represent the population state in terms of age structure, with "submodels" for recruitment, growth, and mortality. The *Deriso model* combines these approaches, using only total stock size as a state indicator, but with component processes explicitly represented and some age structure effects implicit in the model equations. *Life history stanza models* usually consider age and sometimes spatial structure, and decompose recruitment and survival processes into variously detailed steps or life history stanzas.

A common feature of population dynamics models of all complexities is the attempt to represent a variety of ecological interactions as implicit functions of stock size. That is, stock size is treated as a surrogate or index for the intensity of factors such as intraspecific competition for food, cannibalism, mortality due to diseases, aggressive interactions leading to dispersal, and so forth. Thus the danger of nonrepeatability is always present, since various external variables (such as food supply and predator abundance) may not respond immediately to changes in stock size.

There have been two basic schools of thought or paradigms about the importance of various factors in the response of populations to harvesting. Proponents of the "annual surplus" paradigm, which arose mainly from studies of birds and small mammals, argue that many populations produce an annual "reproductive excess" that will later be lost through competition for food or breeding space if it is not harvested. In other words, the habitat has a basic carrying capacity that is exceeded each year due to high reproductive rates; the excess can (and should) be harvested, and this harvest is unlikely to outstrip the reproductive potential. Therefore, management should focus mainly on maintenance and improvement of the habitat carrying capacity, and not be too concerned about harvest regulation.

Proponents of the "logistic production" paradigm, which arose mainly from studies on fisheries and large mammals, argue that population size is the most important determinant of the annual surplus that can be taken on a sustained basis. Birth and/or death rates change smoothly as population density increases, so that net production is maximal at an intermediate density well below the carrying capacity. Since harvesting may drastically affect the population size (and hence future surpluses), harvest regulation should be the central concern of management. As we shall see in Chapter 5, these paradigms are really not that different; general models may produce either extreme of qualitative behavior (or something intermediate), depending on reproduction and survival parameter values. In this chapter, we will mainly review models arising from the logistic production paradigm.

Surplus production models

Surplus production models usually look at the harvestable biomass B_t present at any time t, and describe rate of biomass change as the difference between two functions:

$$\frac{\mathrm{d}B_t}{\mathrm{d}t} = P(B_t) - C(B_t, E_t) \tag{4.1}$$

where $P(B_t)$ is the net rate of biomass production as a function of current biomass, and $C(B_t, E_t)$ is the rate of biomass harvest (yield) as a function of current biomass and some measure of the amount of effort E_t directed at harvesting. Usually it is assumed that $P(B)$ is a dome-shaped function with a single maximum at an intermediate stock size B^*.

The most common assumption about production is that $P(B)$ is logistic:

$$P(B) = rB \left(1 - \frac{B}{k} \right) \tag{4.2}$$

where the parameter r is an intrinsic or low density rate of production and k is the natural equilibrium stock size in the absence of harvesting. Under the logistic assumption, the maximum net production rate is achieved at $B^* = k/2$, and the rate at this stock size is $rk/4$. Another common formulation is the beta function

$$P(B) = rB^\alpha \left(1 - \frac{B}{k} \right)^\beta \tag{4.3}$$

where r and k are as in the logistic and the parameters α and β can be adjusted to give $P(B)$ various shapes over the biomass interval $[0, k]$.

The harvest function $C(B, E)$ has been modeled in various ways, or treated simply as a "policy variable" subject to direct control. The most common assumption is that effort E_t can be measured as a rate, in terms of the area (or volume) swept by harvesters per unit time, and that this area is distributed at random with respect to the spatial distribution of biomass.

Then the catch rate is modeled simply as

$$C(B, E) = qB_t E_t \tag{4.4}$$

where q, the "catchability coefficient," represents the proportion of B_t taken by one unit of effort. A generalization of this "random search" model is to use a Cobb–Douglas type production function (as is common in economics), of the form

$$C(B, E) = qB_t^\alpha E_t^\beta \tag{4.5}$$

where α and β again represent shape parameters to account for effects of nonrandom searching, competition among units of effort, saturation of the harvesting gear, and so forth.

　　In practice, biomass time series are almost never observed directly. Thus, to evaluate and calibrate surplus production models against historical data, it is necessary to hypothesize a second *observation model* about how observable abundance indices are related to the actual biomass. The most widely used assumption comes from equation (4.4), namely, that the catch per unit effort y_t is linearly related to stock size:

$$y_t = \frac{C_t}{E_t} = qB_t \tag{4.6}$$

where the catch and effort rates are integrated or averaged over a short time period around t. This observation model has been very widely and rightly criticized, because harvesting efforts are seldom applied at random with respect to the stock distribution. The effect of nonrandom searching is to make y_t remain high as B_t decreases; that is, the catch per effort fails to reflect stock declines until they are well under way. We shall return to this point later.

　　On a global basis, probably the most widely applied assessment procedure in fisheries is the so-called "Gulland method" (Gulland, 1961), which is based on equations (4.2), (4.4), and (4.6). The method begins with the assumption that effort changes are gradual enough so the stock always remains near equilibrium. Then, if the production function is logistic and catch is proportional to stock and effort, it follows that

$$rB\left(1 - \frac{B}{k}\right) \approx qBE \tag{4.7}$$

and the equality is exact at equilibrium ($P = C$). This equation can be solved for equilibrium stock B_e as a function of effort:

$$B_e = k - \frac{qk}{r} E \tag{4.8}$$

Then, assuming [equation (4.6)] that catch per effort is proportional to stock, substituting $y_e = qB_e$ into (4.8) gives the simple prediction:

$$y_e = qk - \frac{q^2 k}{r} E \tag{4.9}$$

that is, catch per effort should be linearly related to effort ($y = b_1 + b_2 x$), with intercept $b_1 = qk$ and slope $b_2 = -q^2 k/r$. An alternative of way of expressing this prediction is by saying that catch should be quadratically related to effort; if $C_e = qB_e E$, then substituting (4.8) for B_e gives

$$C_e = qkE - \frac{q^2 k}{r} E^2 \tag{4.10}$$

which is of the form $C = ax + bx^2$ with $a = qk$ and $b = -q^2 k/r$. To allow for disequilibrium effects and response delays in processes such as recruitment, Gulland recommends plotting y and C against a running average of past efforts rather than just E_t. Hilborn (1979) has shown that this approach is deceptive in most cases, resulting in overestimates of the equilibrium catch except when fishing efforts have changed very slowly over very long time periods (15–20 years).

Equation (4.10) has sometimes been confused with the so-called "catch equation" that results from looking at the cumulative catch Y_t achieved each year when an initial fixed and nongrowing biomass is depleted rapidly each year. In that case, it is reasonable to approximate the biomass depletion and catch accumulation by

$$\frac{dB}{dt} = -qEB \qquad \frac{dY}{dt} = -\frac{dB}{dt} \tag{4.11}$$

The solution to these equations over a short fishing season of length T, given an initial biomass B_0, is

$$Y = B_0(1 - e^{-qET}) \tag{4.12}$$

The exponential form represents "exploitation competition," in which later units of effort partly sweep areas from which biomass has already been removed by earlier effort units. Unfortunately, equations (4.10) and (4.12) both predict a curved relationship between catch and effort for lower effort levels; equation (4.10) predicts that catch will fall off at higher efforts, while (4.12) predicts that it will reach an upper limit (B_0). If a data set containing many years' catches fits (4.12) better than (4.10) at high effort levels, the implication is that B_0 has remained high from year to year, i.e., biomass production between harvest seasons somehow results in B_0 each year independently of how much was left behind after last year's harvesting.

Schaefer (1957) and later Schnute (1977) have tried to avoid the equilibrium assumption of Gulland's method. Schaefer looked at the logistic equation in terms of a discrete time approximation:

$$B_{t+1} = B_t + rB_t \left(1 - \frac{B_t}{k} \right) - qE_tB_t \tag{4.13a}$$

Solving the observation equation (4.6) for B_t gives the "state reconstruction" $B_t = y_t/q$, then substituting this into (4.13a) and rearranging gives

$$y_{t+1} = (1 + r)y_t - \frac{r}{qk} y_t^2 - qE_ty_t \tag{4.13b}$$

This equation is in the form of a multiple linear regression ($y = b_1x_1 + b_2x_2 + b_3x_3$) to predict next year's relative abundance y_{t+1}, with $b_1 = 1 + r$, $b_2 = -r/qk$, $b_3 = -q$, $x_1 = y_t$, $x_2 = y_t^2$, $x_3 = E_ty_t$. Note the similarity of this regression to that of the Gulland method [equation (4.9)]. Schnute (1977) began with the differential equation

$$\frac{dB}{dt} = rB \left(1 - \frac{B}{k} \right) - qEB \tag{4.14}$$

and then noted that this can be written as

$$\frac{dB}{B} = \left(r - \frac{r}{k} B - qE \right) dt \tag{4.15}$$

Integrating equation (4.15) over a one-year time step gives

$$\log B_{t+1} - \log B_t = r - \frac{r}{k} \bar{B} - q\bar{E} \tag{4.16}$$

where $\bar{B}$ and $\bar{E}$ are the mean biomass and effort over the period t to $t + 1$:

$$\bar{B} = \int_0^{t+1} B \, dt$$

Then again, using the observation model (4.6) and state reconstruction $B = y/q$, but approximating $\bar{y}$ by $(y_{t+1} + y_t)/2$ and $\bar{E}$ by $(E_{t+1} + E_t)/2$, Schnute also arrives at a multiple linear regression to represent the dynamics of relative biomass change and to estimate the parameters r, k, and q. Unfortunately, it is easy to show that both Schaefer's and Schnute's methods fail badly when there is significant observation error (so y_t is not exactly equal to qB_t) or nonlinearity in the observation relationship (Uhler, 1979). The direction of bias in parameter estimates is often such as to encourage the manager to think that the stock is now near its most productive level B_t, no matter where it really is relative to that level.

The greatest difficulties in applying surplus production models have come from their stringent assumptions about the relationship between stock size and observed abundance indices, such as catch per effort. Nonlinear relationships are probably the rule, and q tends to change over time as harvesting technologies improve. Other problems have arisen because of time delays in component production responses (particularly recruitment), and persistent or cyclic disequilibrium in internal stock structure. For example, a single, large year class or cohort can enter a stock, then have its growth and natural mortality rates dominate the stock's overall production over a period of several years; net production is dominated first by recruitment and growth, then later by natural mortality. At no point will the stock then exhibit its long-term average productivity for each of the stock sizes it passes through.

Stock–recruitment models

Reproduction in most organisms is highly seasonal, and the adult stock often suffers high or total mortality after reproducing. Such stocks are usually harvested shortly before the reproductive season, as "recruits" that will not likely undergo much more growth or natural mortality before reproducing. It has been an irresistible temptation to model such stocks in terms of the relationship between stock after harvest in one generation, and subsequent recruitment to the harvestable stock of the next generation. Letting S_t represent the reproducing stock of one generation, and R_{t+1} be resulting recruitment, most models have been of the form

$$R_{t+1} = \alpha S_t h(S_t) \tag{4.17}$$

where α is the maximum recruitment per S_t, and $h(S_t)$ is a decreasing function of S_t such that $h(0) = 1$. $h(S)$ is intended to represent density-dependent effects of adults on juvenile survival or intraspecific competition among the juveniles. The two most common examples are the "Ricker model" (Ricker, 1954), where $h(S) = e^{-\beta S}$, and the Beverton–Holt model (Beverton and Holt, 1957), where $h(S) = 1/(1 + \alpha S/\beta')$. Both β and β' are parameters for the equilibrium, unharvested stock size. The Ricker model always generates a dome-shaped relationship, while the Beverton–Holt model predicts that recruitment will increase to an asymptote. Various other formulations, such as $R = aS^b$ and $R = aS/(k + S)^c$, have given marginally better fits to particular data sets, or seem to be based on sounder biological derivations (for discussions, see Paulik, 1973; Ware, 1980; Shepherd, 1982). But, for reasons to be discussed below, it is seldom worthwhile using complicated formulations with many parameters that are each assigned much biological significance.

Recruitment rates usually appear to vary rather drastically due to factors other than spawning stock, and this variation has led to three reactions from management analysts. First, some analysts have argued that effects of spawning stock on recruitment can be effectively ignored, since recruitment can be better *predicted* by other factors. This argument reflects serious confusion about the objectives of management: it assumes that year-to-year prediction is somehow more important than understanding average responses to exploitation (as measured by stock remaining after harvest), and that average recruitment will remain healthy no matter how high the exploitation rate.

Second, many biologists have tried to decompose the survival function $h(S_t)$ into a sequence of survivals through various life history stanzas,

$$h(S_t) = s_1 s_2 s_3 \ldots s_n \qquad (4.18)$$

and then have tried to explain or predict how these component survivals vary in relation to stock size and key environmental factors. So in marine fishes it is often found that egg and larval survival rates are correlated weakly with larval densities, and more strongly with physical variables, such as water temperature and the spatial pattern of water transport (Sharp, 1980). Survival rates in freshwater and anadromous fishes are usually related to water flows (or rainfall or runoff) during one or another life history stage. Occasionally, survival rates have been shown to vary in relation to biotic factors that are less random from year to year than most environmental factors, such as densities of predators. There have been three basic difficulties with the disaggregation approach: (1) lovely correlations have a nasty way of suddenly breaking down, never to reappear; (2) there is usually at least one survival stage that is quite variable, but for which no clear explanatory factors can be found; and (3) the environmental and ecological explanatory variables are not themselves predictable, so even knowledge that they are causes of variation does not help much in predicting average future responses to policy options.

Third, it has been suggested that recruitment should be modeled as a fundamentally stochastic and unpredictable process, with the stock–recruitment function viewed as a collection of probability distributions (one for each level of spawning stock). In this view, the stock–recruitment curve is a description of how average recruitment varies with reproducing stock (and is therefore useful for analysis of long-term average responses to various harvest policies), without pretense that the curve is at all useful for short-term predictions. The curve should be described simply, with relatively few parameters, and reflect little more than basic density dependence and the reasonable assumption of *continuity*—similar (nearby) reproducing stocks should result in similar recruitment distributions. Also, in this view it is important to have realistic models for the distributions of recruitments around the curve of averages. In the 1970s, some workers (Allen, 1973; Walters, 1975) noted that these distributions often appear to be log-normal —bell shaped around the average but with a long upward tail so that occasional very large recruitments are produced. Rinaldi and Gatto (1976) noted that there is a good theoretical reason for the distributions to be log-normal, and Peterman (1981) has provided further empirical justification. The Rinaldi–Gatto argument is very simple, and begins by noting, as in the previous paragraph, that

$$h(S) = s_1 s_2 s_3 \ldots s_n$$

where n is potentially very large. Now, each of the s_i is likely to be a random variable, and many of the s_i's will be essentially independent of one another (involve factors operating at different life stanzas, etc.). If s_i is a random variable, so is ln (s_i), and the sum

$$\ln h(S) = \ln s_1 + \ln s_2 + \ln s_3 + \cdots + \ln s_n \tag{4.19}$$

is of course a sum of random variables. When such sums are not dominated by a few elements that take only a few extreme values, they tend to be normally distributed by the Central Limit Theorem of fundamental statistics. Thus ln $h(s)$ for any fixed s is likely to be normally distributed, hence the product

$$h(s)e^{v'} \tag{4.20}$$

is log-normal where v' is a normally distributed random variable representing the combined effects of all random survival factors. The overall stochastic recruitment model is then

$$R_{t+1} = \alpha' f(S_t) e^{v_t} \tag{4.21}$$

where α' is estimated so as to include the mean of v', $f(S_t)$ is a deterministic "kernel" of density-dependent effects, and v_t is normally distributed with mean zero. This is particularly convenient when $f(S) = e^{-bS}$ and we define $\alpha' \equiv e^a$, to give the stochastic Ricker model

$$R_{t+1} = S_t e^{a - bS_t + v_t} \tag{4.22}$$

The above theoretical arguments imply that the linear regression formula

$$\ln \frac{R_{t+1}}{S_t} = a - bS_t + v_t \tag{4.23}$$

will have the nice statistical property of additive, normally distributed errors, so a and b can be estimated (and uncertainty about them interpreted) using simple classical formulas. The basic predictions of equations (4.21) and (4.22) are that variation around the recruitment curve will increase with increasing mean recruitment, while variation in ln (R/S) will be similar for all values of s; these predictions accord well with most historical data sets.

Unfortunately, there is a serious empirical difficulty in applying any of the approaches outlined above, which statisticians call the "errors-in-variables" problem. We shall examine this problem further in Chapter 6, but it is worth highlighting the main difficulty at this point. The problem is

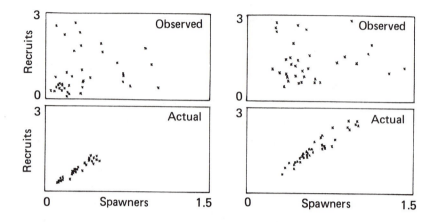

Figure 4.4. Two examples of the distortion in apparent stock–recruitment relationships when the spawning stock is measured with random error. The "actual" relationships are straight lines from the origin, with small scatter. The "observed" relationships are plotted after adding random counting errors to the spawning stocks. *Source:* Walters and Ludwig (1981).

that reproducing stocks S_t are usually measured with considerable random error. Walters and Ludwig (1981) and Ludwig and Walters (1981) have shown that these measurement errors smear out or bias the apparent $S-R$ relationship in the worst possible way from a management viewpoint: they make recruitment appear to be independent of reproducing stock. Figure 4.4 shows two examples where the stocks are actually severely overexploited so that $R \approx \alpha S$; the apparent relationship is $R = $ constant $+$ random errors. In general, errors in S assessments cause the estimate of α to be too high, and of the equilibrium stock parameter (b, β, β', etc.) to be too low. A manager looking at $S-R$ data from an overexploited stock will think that there is no recruitment problem; a researcher will seek other variables besides S to explain variation in R. There is no good statistical procedure to correct for these biases, since the measurement errors essentially destroy information about the functional relationship. The problem is not alleviated in any way, and in fact is not even addressed, by methods for studying correlated random variables, such as Ricker's (1973a) GM regression.

The errors-in-variables problem has made it difficult to study key issues in stock–recruitment modeling, particularly the usual assumption that $f(S)$ decreases monotonically as S increases, so R/S is highest at very low stock sizes. There is crude evidence of depressed productivity at low stock sizes in some populations, due either to difficulty in finding mates (the Allee effect), breakdown in social structure and migration patterns, or increased

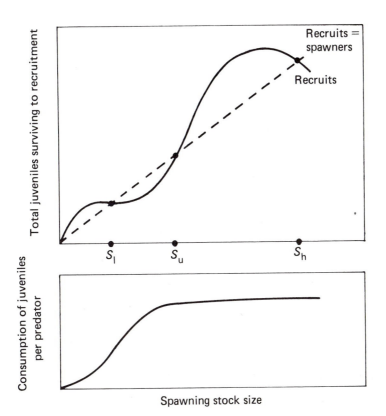

Figure 4.5. A relationship between spawning stock and subsequent recruitment that can arise when predators follow a sigmoid functional response in their consumption rates of juveniles. If the predator population remains fixed over time, the prey stock may display stable equilibria at stock sizes S_l and S_h; the stock size S_u is an unstable level from which the stock will either collapse to S_l or grow toward S_h.

(depensatory) mortality rates due to predation. A particularly bad situation can be created by predators that show behavioral "switching" to other prey types as the stock of concern decreases. Such predators are said to have a type III or sigmoid functional response to prey density (Holling, 1965; Peterman, 1977). Mortality rates due to such predators are highest at prey densities just above the switching point, and are less significant at both very low and very high prey densities. This results in recruitment curves for the prey of the shape shown in Figure 4.5, with a "predator pit" and the possibility of two stable equilibria in stock sizes. Once depressed by harvesting, such stocks may be "caught in the pit" and maintained at low levels by

Table 4.1. Average age composition of the catches and average body weights for four British Columbia fishes.

Age a	Pacific cod 1960–80 (21 yrs)		Rock sole 1956–80 (25 yrs)		Herring 1951–80 (30 yrs)		Pacific ocean perch 1966–80 (15 yrs)	
	P_a	W_a	P_a	W_a	P_a	W_a	P_a	W_a
1	0.0156	0.175	–	–	0.0212	0.013	–	–
2	0.2732	0.914	0.0303	0.33	0.1249	0.054	–	–
3	0.3752	2.00	0.1008	0.44	0.3473	0.085	–	–
4	0.2202	3.00	0.2137	0.55	0.2459	0.115	–	–
5	0.0789	4.14	0.2574	0.66	0.1501	0.145	–	–
6	0.0241	5.21	0.213	0.77	0.0722	0.165	0.0105	0.485
7	0.00824	6.19	0.1028	0.87	0.0289	0.173	0.0143	0.529
8	0.00324	7.20	0.0482	0.97	0.00781	0.187	0.0249	0.572
9	0.00105	8.31	0.0204	1.07	0.00156	0.189	0.0523	0.614
10	0.0003	9.2	0.0096	1.15	0.00014	0.231	0.0790	0.654
11			0.0037	1.22			0.1077	0.694
12							0.1228	0.732
13							0.1302	0.770
14							0.1050	0.806
15							0.0754	0.841
16							0.0545	0.876
17							0.0445	0.909
18							0.0409	0.942
19							0.0382	0.973
20							0.0335	1.00
21							0.0227	1.03
22							0.0149	1.06
23							0.0113	1.09
24							0.00873	1.12
25							0.00367	1.15
26							0.00260	1.17
27							0.0012	1.20
28							0.00087	1.22
29							0.000133	1.30

predation in spite of reduced harvest rates. Indeed, as noted in Chapter 2, unregulated harvesting may itself be related in a sigmoid pattern to stock size, thus preventing stock recovery after accidents. See the anchoveta history in Figure 2.1, and theoretical arguments in Jones and Walters (1976). Even an apparently small, but actively switching, fishery may maintain a stock at depressed levels.

Dynamic pool models

Surplus production and stock-recruitment models do not explicitly represent how net production results from the balance of recruitment, growth, and natural mortality rates. Dynamic pool models make specific predictions about these rates, usually with respect to the population age structure as a basic state description (rather than total biomass or numbers). Early workers usually dealt only with equilibrium rates, assuming constant recruitment and simple models for growth and survivorship. These so-called "yield per recruit" models were a great source of confusion for scientists and managers with weak mathematics backgrounds, because they were presented as elaborate equations that seemed very complicated and realistic and had all sorts of magical symbols such as ω, $\varkappa$, t_λ, and Ω. In fact, the basic biological assumptions underlying these calculations were trivial or even deceptive, and could safely be used to predict little more than the best "age at entry" or minimum size of animals that should be allowed in the harvest.

Equilibrium analysis

The basic conclusions of traditional equilibrium yield analysis can be obtained very simply with a hand calculator or microcomputer, without wading through any elaborate equations. Let us illustrate the steps with four typical data sets (Table 4.1), on the average proportions of the catches and body weights at age a for four fish that are commercially important on the Canadian Pacific coast. Before analyzing these data, we must make some independent assessment of the average natural mortality rates, as a proportion of the fish that die naturally each year between fishing seasons, for each of the stocks. It is usual to assume that this rate is independent of the age of the fish, and we shall stick with this assumption (but note how the data in Table 4.1 hint strongly at decreasing survivorship, or perhaps reduced vulnerability to fishing, in the older ages!).

Chapter 6 will deal with some of the difficulties in obtaining natural mortality estimates; for the stocks in Table 4.1, various procedures have given rates in the following ranges:

Stock	Natural mortality rate (proportion dying/year)
Pacific cod	0.4-0.8
Rock sole	0.2-0.5
Herring	0.3-0.6
Pacific ocean perch	0.05-0.15

Note that these rates are highly uncertain compared with the other data available, and one might be tempted to just pick some conservative base figures using intuition or major summaries of published estimates, such as Pauly (1979), which indicate that survival rates can be predicted nicely from basic data on growth and longevity.

To estimate equilibrium yield, let us first note that this yield is a sum over ages of average number (N_a) of fish reaching each age times average body weight (W_a) at the age times exploitation rate at the age. If recruitment is at equilibrium near the historical average, say $\bar{R}$, then the number of fish reaching each age can be written as $N_a = \bar{R} \, l_a$, where l_a is the total survivorship up to age a:

$$l_a = s^{a-1} \prod_{k=1}^{a-1} (1 - h_k)$$

where s is the average natural survival rate (i.e., $s \approx 0.85 - 0.95$ for Pacific ocean perch), and h_k is the exploitation rate at age k. Using this shorthand, the equilibrium yield $\bar{Y}$ can be expressed as

$$\bar{Y} = \bar{R} \sum_{a=1}^{\infty} l_a W_a (1 - h_a) \tag{4.24}$$

Now, the key policy variables in this yield model are the h_a, the age-specific harvest rates. Usually these cannot be controlled separately, but may be modeled as $h_a = \bar{h} v_a$, where $\bar{h}$ is the average exploitation rate, and the v_a are average age-specific vulnerabilities that depend on the harvesting technology (net mesh sizes, etc.) and space/time pattern ("nursery areas", etc.) of fishing allowed. The yield per recruit, from equation (4.24), is

$$\frac{\bar{Y}}{\bar{R}} = \sum_{a=1}^{\infty} l_a W_a (1 - h_a)$$

and this sum can be computed easily once we fix s and a harvest policy $h_a = \bar{h} v_a$ for all ages a.

So what to do about the unknown $\bar{R}$, average recruitment, in equation (4.24)? Let us estimate $\bar{R}$ using a simplification of the method most widely used in fisheries, which is called "cohort analysis" (Pope, 1972; Jones, 1981). This method uses the idea that we can solve the basic survival equation

$$N_{a+1} = (N_a - C_a) s \tag{4.25}$$

(where C_a is the catch at age a) backward in time if we know N_{a+1} at some terminal age. Solving (4.25) for N_a, we get

$$N_a = \frac{N_{a+1}}{s} + C_a$$

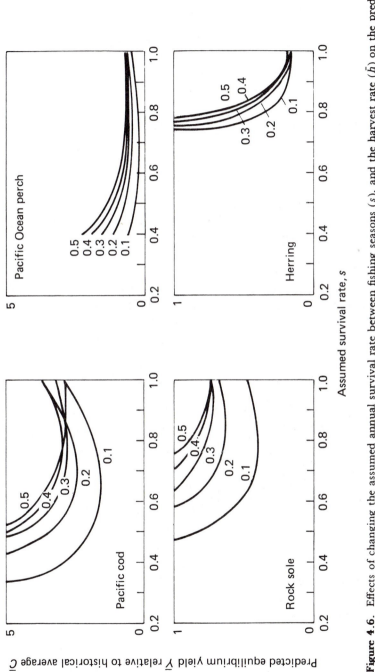

Figure 4.6. Effects of changing the assumed annual survival rate between fishing seasons (s), and the harvest rate ($\bar{h}$) on the predicted equilibrium yield ($\bar{Y}$) relative to the historical average ($\bar{C}$), for four British Columbia fish stocks. Each curve shows how the prediction $\bar{Y}/\bar{C}$ varies with assumed survival rate, while holding the harvest rate h fixed.

Applying this equation repeatedly beginning with some maximum age $a = m$ for which $C_{a+1} = 0$, we get

$$N_m = C_m$$

$$N_{m-1} = \frac{C_m}{s} + C_{m-1}$$

$$N_{m-2} = \frac{C_m}{s^2} + \frac{C_{m-1}}{s} + C_{m-2} \qquad (4.26)$$

$$N_1 = R = \frac{C_m}{s^{m-1}} + \frac{C_{m-1}}{s^{m-2}} + \cdots + \frac{C_2}{s} + C_1$$

If catches at age a have been reasonably steady over time, then equation (4.26) implies that average recruitment $\bar{R}$ can be expressed as a weighted sum of the average catches at age $\tilde{C}_a$, where the weighting factor associated with each a is $1/s^{a-1}$. These factors get bigger (fast!) as age increases, so the catches of older animals have a strong influence on $\bar{R}$; this makes sense, considering that for each animal who survived to be caught at an old age, many more must have been born and died naturally prior to that age (especially if s is low). As a further step, let us next write $\bar{R}$ in terms of the historical average total catch $\tilde{C}$ and average proportions of catch at age $\bar{P}_a$ (Table 4.1), by substituting $C_a = \tilde{C}\bar{P}_a$ into equation (4.26) and rearranging:

$$\bar{R} = \tilde{C}\left[\bar{P}_1 + \frac{\bar{P}_2}{s} + \frac{\bar{P}_3}{s^2} + \cdots + \frac{\bar{P}_m}{s^{m-1}}\right] \qquad (4.27)$$

The basic point of this equation is that average recruitment is estimated by looking at catch $\tilde{C}$, the catch age distribution $\bar{P}_a$, and natural mortality $(1 - s)$ which takes some of the recruits. In the following, we shall take $\tilde{C} = 1$ as the unit of catch measurement, in order to simplify the discussion.

Let us return now to equation (4.24), which says that average yield is $\bar{Y} = \bar{R}(\bar{Y}/\bar{R})$. We see that the estimation of $\bar{R}$ depends on s [equation (4.27)], such that $\bar{R}$ increases if we decrease the estimate of s. $\bar{Y}/\bar{R}$ depends on s also, in exactly the opposite way: as s decreases, we predict that fewer animals will reach the ages where the W_a are larger, and so we predict that $\bar{Y}/\bar{R}$ will decrease. So an immediate and important question arises: just how sensitive is the overall assessment of $\bar{Y}$ to estimates of s, which are notoriously difficult to obtain with any confidence? For the four examples in Table 4.1, the disturbing answer to this question is shown in Figure 4.6, which plots $\bar{Y}/\tilde{C}$ computed from equations (4.24) and (4.27) as a function of the assumed survival rate s, for increasing $\bar{h}(0.1, \ldots, 0.5)$ and reasonable v_a values for each species considering current harvesting technologies. In each case we see that there is a range of survival rate assumptions for which the assessment of average yield obtainable (at intermediate harvest rates h) increases very sharply as s is further decreased. Thus there is a basic asymmetry, which in one sense simplifies the assessment problem: either the yield assessment is quite good (actual s almost surely greater than critical value for increase), or is in grave doubt with the potential $\bar{Y}$ perhaps being very high relative to the historical average $\bar{C}$.

We could, of course, elaborate the equilibrium yield calculations outlined above in much more detail by using age-varying survival rates, more realistic vulnerability schedules, seasonal or instantaneous growth data, and so forth. I will leave it as an exercise for the reader to show that such details add very little to the basic predictions. In most practical cases, it is silly to worry about 5 or 10 or even 20% changes in the assessment of $\bar{Y}$, considering that changes in fishing effort or recruitment rates (Shepherd, 1982) may have much larger effects in the long term.

Dynamic predictions

While equilibrium analysis can give insights about the importance of some rate estimates, it is usually important to study the temporal behavior of age-structured stocks, especially in relation to various assumptions about the stock–recruitment relationship, effects of random variations in recruitment, and effects of the time delays implied by having harvest and reproduction concentrated in older age classes. Such studies can be done efficiently with simulation models that keep track of the age structure over time using two basic equations:

$$N_{a+1,t+1} = N_{a,t}s(1 - h_{at}) \qquad a = 1, \ldots, m$$

$$N_{1,t+1} = E_t f(S_t)e^v$$

(4.28)

Here the second equation is a stock–recruitment relationship, where the egg production or total relative reproductive output in year t is

$$E_t = \sum_{a=1}^{m} F_a N_{at}$$

(4.29)

and F_a is the relative or absolute fecundity of an animal of age a. Density-dependent effects on juvenile survival are represented by $f(S_t)$, where

$$S_t = \sum_{a=1}^{m} G_a N_{at}$$

(4.30)

and G_a is the relative effect of an animal of age a on juvenile survival. The most usual case is to assume $G_a = kF_a$, implying either that reproducing adults impact juvenile survival in proportion to fecundity, or that density dependence in survival is due to interactions among the juveniles. e^v is a log-normal random survival effect, as in the previous section on stock and recruitment. The age-specific harvest rates h_{at} can be modeled as $h_{at} = \bar{h}_t v_a$, where $\bar{h}$ is the simulated overall exploitation rate, and the v_a are age-specific vulnerabilities. Annual yield is calculated as either

$$\sum_a W_a N_{at}(1 - h_{at}) \quad \text{or} \quad \sum_a W_a s N_{at}(1 - h_{at})$$

depending upon when the harvest takes place in the annual cycle [equations (4.29) and (4.30) implicitly assume that N_{at} is the stock measured at the time of spawning in year t].

Equations (4.28) can display all sorts of fancy and interesting behavior, especially if $f(S)$ is of the form e^{a-bS} or some other equation that results in a dome-shaped recruitment relationship (for an analysis of the basic stability theory involved, see Botsford, 1979). The simulated population may cycle, overshoot then damp to the equilibrium, or even show almost chaotic time behavior. The key parameters that determine stability are those that set the maximum recruitment rate per adult [a in $f(S) = e^{a-bS}$], and the time delay from birth to recruitment and first harvest.

A potentially very important factor not usually included in age structure simulations is density dependence in growth rates, and associated changes in reproduction and vulnerability schedules (Botsford, 1981). Growth rates are higher in most organisms when densities are lower, and fecundities tend to be more closely related to body weight than to age. In fact, it seems that most stocks have a fixed weight at maturity, with fecundity increasing almost linearly with weight afterward. This weight usually corresponds roughly with the growth curve inflection point, where growth rate begins to decelerate (food energy is put into reproduction instead). Yield per recruit analyses usually indicate that the best minimum size for harvesting is also at about the point where growth rates start to decelerate. It is quite possible for the entire mechanism of population regulation to operate through density-dependent effects first on growth, which then affects fecundity and survival, without there existing any sort of density dependence in juvenile survivals [$f(S) = $ constant]. For an example of the modeling details involved, see Walters et al. (1980). The few existing models that represent this mechanism of population regulation have predicted pronounced temporal cycles in abundance, due to the long time delays for feedback between growth rate changes and the subsequent reduction in fecundity. A variation on the theme is to assume that impacts of older animals on juvenile survival [the G_a parameter of equation (4.30)] are related to adult size, which can happen, for example, when the larger animals become cannibalistic. This additional feedback can result in multiple population equilibria (Botsford, 1981).

Deriso's model

A dramatic step in connecting the simple surplus production and stock-recruitment theories with the more detailed predictions of age structure analysis was published by Deriso (1980). He showed that the biomass dynamics of age-structured populations can often be represented by a simple delay-difference equation that looks much like a surplus production model, but contains the same parameters usually used in dynamic pool models. His derivation is based on noting that the harvestable population biomass B_t at any time can be written as

$$B_t = \sum_{a=k}^{\infty} N_{at} W_{at} \tag{4.31}$$

where N_{at} and W_{at} are numbers and body weight at age a, and k is the "age at entry" or first harvest. If we then assume that all animals in B_t are equally vulnerable to harvest at rate h_t (so-called "knife-edge selection" at age k) and have the same natural survival rate s, so that the total survival rate is $l_t = (1 - h_t) s$, then equation (4.31) can be written in terms of the previous year's numbers and new recruits (N_{kt}) as

$$B_t = l_{t-1} \sum_{a=k+1}^{\infty} N_{a-1,t-1} W_{at} + N_{kt} W_{kt} \tag{4.32}$$

Next, Deriso assumes that the age at entry k is sufficiently large so that the body growth rate is decelerating, and can be approximated by the Brody equation, $W_{at} = W_k + \varrho W_{a-1,t-1}$ (for $a > k$), where ϱ and W_k are empirical growth parameters. Substituting this growth model into equation (4.32) and noting that

$$\sum_{a=k+1}^{\infty} N_{a-1,t-1} W_{a-1,t-1} = B_{t-1}$$

we get

$$B_t = l_{t-1} \varrho B_{t-1} + l_{t-1} W_k \sum_{a=k+1}^{\infty} N_{a-1,t-1} + N_{kt} W_{kt} \tag{4.33}$$

Note that the summation in this equation is just the total number of vulnerable fish N_{t-1} at time $t-1$. This total can be written as $N_{t-1} = l_{t-2} N_{t-2} + N_{k,t-1}$, and multiplying it by $l_{t-1} W_k$ gives $l_{t-1} W_k N_{t-1} = l_{t-1} l_{t-2} W_k N_{t-2} + l_{t-1} W_k N_{k,t-1}$. Note that the first element on the right-hand side of this equality is l_{t-1} times the second term in equation (4.33), but shifted back one year, to $t - 2$. Equation (4.33) can be solved for this second term, while shifting all indices back one year:

$$l_{t-2} W_k N_{t-2} = B_{t-1} - l_{t-2} \varrho B_{t-2} - N_{k,t-1} W_k$$

Thus $l_{t-1} W_k N_{t-1} = l_{t-1}(B_{t-1} - l_{t-2} \varrho B_{t-2} - N_{k,t-1} W_k)$. Substituting this into equation (4.33), rearranging, and shifting the time indices forward one year, gives the basic Deriso model:

$$B_{t+1} = (1 + \varrho) l_t B_t - \varrho \, l_t l_{t-1} B_{t-1} + W_k N_{k,t+1} \tag{4.34}$$

where the first term represents the growth and survival of B_t, the second term "corrects" for age structure/growth changes, and the third term represents the biomass of new recruits at time $t + 1$.

The key point about Deriso's model is that it represents age structure effects on biomass dynamics exactly (the algebraic result is not an approximation) when the three basic assumptions of knife-edge recruitment, age-independent survival, and Brody growth curve are met. The assumption of knife-edge recruitment can even be relaxed, by interposing a "prerecruit pool" between recruitment and B_t, with recruitment to B_t coming from this pool (see Deriso, 1980, for details). This is extremely convenient for modeling, since equation (4.34) is much easier to deal with than the age structure accounting of the previous section. Also, any model that you would like can be substituted for the recruitment term $N_{k,t+1}$ in equation (4.34); for example, $N_{k,t+1} = S_{t-k+1} f(S_{t-k+1})$, where S_{t-k+1} is the spawning biomass in the year when $N_{k,t+1}$ was produced. If we define B_t as the biomass just before harvest and $S_t = B_t(1 - h_t)$ as the biomass after harvest (Deriso calls this "spawning stock"), equation (4.34) can be written as

$$B_{t+1} = (1 + \varrho) s S_t - \varrho s^2 \frac{S_t}{B_t} S_{t-1} + W_k N_{k,t+1} \tag{4.35}$$

Another way of writing this equation is

$$B_{t+1} = s S_t + \varrho s S_t \left[1 - s \frac{S_{t-1}}{B_t} \right] + R_{t+1} \tag{4.36}$$

where R_{t+1} is the biomass of new recruits. In this form it is more obvious that the biomass dynamics will have components due to survival ($s S_t$), growth corrected for age structure changes $[\varrho s S_t(1 - s S_{t-1}/B_t)]$, and recruitment ($R_{t+1}$). The connection to surplus production and to stock–recruitment models is also clarified by equation (4.36): it predicts essentially trivial effects due to growth and survival, but very important effects due to R_{t+1} and its responses to past exploitation (i.e., to S_{t-k+1}). Indeed, if we hold R constant and examine the equilibrium catch $C = hB$ [solve (4.36) with the time subscript dropped], we obtain

$$C = \frac{hR}{1 - \beta_1(1 - h) + \beta_2(1 - h)^2} \tag{4.37}$$

where $\beta_1 = (1 + \varrho) s$ and $\beta_2 = \varrho s^2$. This equation either increases asymptotically (toward R) as h increases toward 1.0 (if $\beta_1 < 1$; i.e., low growth relative to survival), or can have a weak peak at an intermediate harvest rate (if $\beta_1 > 1$; i.e., fast growth in early ages). This is the typical pattern of equilibrium predictions from dynamic pool models that assume constant recruitment. To get a nicely quadratic relationship between surplus production and stock size (as in logistic surplus production models), the Deriso and other dynamic pool models agree in predicting that recruitment changes (decreases at high h) must be involved; growth and survival changes alone will not produce a strongly curved production function.

One serious problem with the Deriso model as presented above is that body growth of many animal species cannot be well approximated by the two-parameter Brody growth equation; the Brody equation does not admit that many organisms start out small, then grow slowly for many years. In such cases, a three-parameter growth equation is needed, and Schnute (1985) has shown how the Deriso model can be generalized to incorporate this equation. Schnute also shows precisely how various simpler production models are special cases of his generalized delay-difference equations.

There have been various attempts to estimate Deriso model parameters using only total catch and relative abundance data, as are used in surplus production analysis. However, usually at least five parameters are involved: ϱ, s, two recruitment parameters, such as a and b of the Ricker equation, and a catchability or observation parameter q in the observation model $y_t = qB_t$, where y_t are the observed relative abundances (usually catch per effort). Unfortunately, most single-input, single-output (effort → catch, etc.) time series can be approximated well by three- or four-parameter models. This means that there will not be enough information in the time series to estimate more than three or four of the original Deriso parameters, so that at least one must be fixed through independent assessments. The obvious candidate is ϱ, which can be estimated from the body growth data that is available for most organisms. Indeed, one of the strongest features of the Deriso model is that it invites the use of independent assessments of various parameters, especially ϱ, s, and q. The value of having such assessments (which may be very costly to obtain in the field) can be evaluated in terms of how much they improve (reduce variance in predictions from) the surplus production analysis obtained by fitting the model (instead of simpler surplus production models) to catch–abundance time series, while holding the independently assessable parameters constant.

Because of its simplicity and flexibility in the use of various data sources, the Deriso model may gradually replace both more simplistic and more detailed models as a basic standard or starting point for analysis. The main deterrent to its general use is that it requires nonlinear parameter estimation procedures. Such procedures are always tricky to work with (Chapter 5), and are extremely tedious to do without a computer. In a way this is good, since it forces the analyst to look very carefully at his or her data and parameter estimates (and not rely on the nice, but often deceptive, cookbook recipes that are available for surplus production models).

Effects of population structure

The modeling approaches outlined above, from logistic to Deriso, all assume some homogeneity or similarity of response by the organisms making up the "unit stock" under consideration. In practice, harvesting activities are usually nonrandom with respect to the population's spatial distribution, highly selective with respect to behavioral and growth phenotypes (the big, dumb ones are easier to catch), and aggregated with respect to spatial substructure generated by homing behavior and natural selection for specialized local characteristics. The basic implication of these heterogeneities is that parameters measured as averages across the overall stock are likely to change over time, as less productive and/or more accessible individuals (or substocks) are depleted during periods of heavy harvest, or recover slowly when protected. There is no simple way to predict in advance how fast these changes will occur. Consider, for example, a situation where the recruitment $f(S)$ function is decreasing in S because organisms are forced to breed in suboptimal habitats when S is large (imagine a salmon stream where some spawners may be forced, through territorial behavior by others, to lay their eggs too near the river bank or in tributary streams subject to flooding). If the breeding habitat can produce a total of K juveniles, then for large S, $f(S) \approx K/S$. Now include some homing or limited dispersal in the picture, such that even those animals produced in the marginal areas will tend to return or stay there. They may well form persistent subpopulations in the absence of harvesting. Next, imagine a harvesting process that hits both the marginal and more productive areas nonselectively. Subpopulations in the marginal areas will, of course, be reduced most rapidly, and $f(S)$ will increase as S is reduced in the ratio K/S; but even if S is held fixed, K will then appear to fall as individuals fail to return to some areas at all. How fast this will occur will depend on the details of the time patterns of local depletion. Then imagine that the harvesting is suddenly stopped, so S increases sharply. Will the marginal areas be reinvaded (recolonized) immediately? If not, total juvenile production may build in fits and starts as accidental reinvasions take place, genes for dispersal behavior are favored, and so forth. Each reinvasion accident may be followed by a relatively rapid local recovery. In any case, it is quite unreasonable to hope that the details of either the depletion or the recovery process can be predicted, or indeed that they are even repeatable from one depletion event (experiment!) to another.

There have been two approaches to the representation of population substructure. The most common has been to try and develop more detailed monitoring, modeling, and regulatory systems for the component subpopulations. The alternative has been deliberately to try to track overall parameter changes treated as statistical phenomena, by methods such as discarding older data or weighting the older data exponentially less back in time. The second approach will be discussed more fully in Chapter 7, as a problem in adaptive parameter estimation. The first approach deserves a bit more discussion, since it is an area of intense research interest by biologists and involves considerable expenses for management agencies that adopt it.

Attempts to develop detailed models of stock structure have run into three main difficulties. First, there is the tactical problem of where to begin. We know, for example, that all the salmon of British Columbia should not be treated as a single stock, and, on the other hand, that it would be hopelessly complex to model all the several thousand discrete groups that have been identified on the basis of spawning locations, run timing, electrophoretic patterns, etc. There is no ideal stopping place between these extremes, and even the smallest groupings noticed so far may have substructures not detected by existing survey methods.

Second, it is seldom economical to gather long-term data on the relatively fine space/time scale needed to detect composition changes and construct accurate substock models. In particular, harvest data are usually recorded on units of political or economic convenience, and historical changes in substock structure cannot be inferred from the aggregate data without assuming the very patterns that one seeks to measure in the first place. Where very massive data sets have been gathered, as, for example, for Pacific herring in British Columbia (Hourston and Nash, 1972; Hourston, 1981), changes in sampling methods over time and space have made detailed analysis almost impossible.

Finally, there is the functional problem of how to model the dispersal/colonization processes associated with maintenance and reestablishment of structure. Functional questions that must be addressed include: how does stock size effect dispersal tendency or rate? How do dispersers select sites at which to settle? What is the survival rate through the dispersal process? How does selection act on colonizers to increase their fitness in a new environment, and how fast does their performance improve? These questions are about events on spatial and temporal scales that have proved exceedingly difficult and costly for biologists to study. They are certainly not answered (or even addressed) by the routine data usually gathered in resource monitoring programs.

A few theoretical studies have simply assumed an arbitrary substock structure, then asked what effect this structure would have on the overall relationships (i.e., stock–recruitment) that are usually analyzed. Ricker (1973b) has shown, for example, that as a fishery develops on a collection of stocks such as salmon, the "a" parameter of his stock–recruitment model [equation (4.22)] for the whole collection will appear to increase, and the equilibrium stock will decrease. Hilborn (1983) has examined similar systems in greater detail, and has shown that an effect like the errors-in-variables regression bias may develop: recruitment may appear to be independent of spawning stock, so it will appear best either to fish harder or to keep the system at its current level, even if most of the substocks are actually being overexploited. Studies such as these are valuable, and should be conducted as a guide to possible problems with aggregated analysis, even when it is not possible to specify more than a few reasonable hypotheses about the unknown substructure.

Detailed models of spatial structure have proven quite valuable in the study of migratory organisms, where movements from area to area can be established through methods such as tagging. Particularly in cases where the organisms move from one political jurisdiction to another, or are subject to different (and effectively competing) harvesting gear as they move about, even some simple bookkeeping or allocation models can be of considerable value in policy analysis. Good examples are the "gauntlet" models used to study harvesting of Pacific salmon as they move along their ocean migration pathways (Paulik and Greenough, 1966; Argue et al., 1983). The typical result of such bookkeeping models is to show that conservation measures imposed at one stage in the migratory pathway are likely to be canceled by increased harvest of the organisms later on, so that some relatively complicated and balanced policy must be implemented for the migration system as a whole.

Stocks as Ecosystem Components

A bewildering variety of models have been constructed for exploited ecosystems, ranging from simple predator–prey models with harvesting terms, through to massive simulations of interactions among hundreds of species. There are models for gross energy flux between trophic levels of the ocean, and detailed day-by-day predictions of how plankton blooms develop and are fed upon by local fish. Relationships between ungulate populations and their predators and food supplies have been represented by models ranging from a few equations for biomass transfer between trophic "compartments," to detailed calculations of foraging patterns and nutritional balances of individual animals. A first step in making sense of the various approaches is to distinguish between models that concentrate on "fast variables" (such as plankton populations) and interactions whose outcome may be essentially random from year to year, and models that concentrate on long-term, more persistent changes in production processes. These longer-term models are the most interesting from an applied viewpoint, since they focus on interactions that might cause single-stock analyses to fail. As with effects of spatial structure, one approach to the analysis of interactions would be to view them as unpredictable generators of change in stock parameters, which we might hope to track by means of adaptive estimation procedures rather than explicit modeling of how the changes arise. This section will concentrate on the alternative approach of trying to model the interactions explicitly.

From the viewpoint of any species stock, the ecosystem generates four general inputs or categories of interaction:

(1)   a basis for production in terms of food supply;
(2)   sources of mortality in terms of parasites and predators;
(3)   a physical structure that moderates trophic interactions by providing safe places to breed, refuge from predation, more or less favorable microclimates for physiological processes, and so forth; and
(4)   a chemical environment that provides vital compounds (such as O_2) and may contain various threatening pollutants of human and biogenic origin.

Ecosystem modeling has concentrated on the obvious trophic inputs (1)–(2), with only passing or implicit concern for how these interactions are shaped by physical structures of biological origin (3). Modeling of chemical interactions, particularly pollution effects, has been discouraged by lack of data on the chronic (as opposed to directly lethal) effects of various compounds on processes such as growth, migratory behavior, and avoidance of predators. Even major pollution sources, such as acid rain, have proved extremely difficult to model (except when grossly lethal levels are involved), since they often generate both positive (fertilizers) and negative effects that may accumulate quite slowly over time.

In the absence of long historical time series, there has been a tendency in ecosystem modeling to assume that the correct model structure should produce a stable equilibrium or "balance of nature." Forms of functions and parameter values have then been chosen to guarantee this result. But a variety of recent studies have cast serious doubt on the very idea that there exists any balance of nature, except as a purely statistical consequence of averaging dynamic changes over large areas. To illustrate this point, I like to use the example of a wildlife biologist concerned with the savanna ecosystems of Africa. Suppose the biologist chooses to look at ecosystem interactions on a one hectare plot of ground at the edge of the famous Serengeti plains. On this plot he will observe highly unstable dynamics, with seasonal rainfall stimulating grass growth, the grass being depleted by passing herds of ungulate grazers or fires that burn away shrubs and accumulated dry grasses, trees being knocked down and browsed by passing elephants, and so forth.

If he enlarges his perspective to the area covered by seasonal migrations of the major ungulate species, a more stable pattern will emerge; fires, grazing, and rainfall generate a vegetation mosaic that the animals move across and "sample" in sequences that allow them to maintain relatively steady feeding rates and exposure to predators. But on a longer time scale the biologist will find temporal instability even over this larger area due to droughts, disease outbreaks, changing human activities and shifts in woody vegetation cover due to larger fires and changes in the elephant population. In frustration he may turn to the whole of Africa. On this scale he is likely to see considerable temporal stability; effects of droughts in one area will be averaged against effects of high rainfall in others, and so forth. Dispersal of ungulates between major regions or migration pathways will ensure that such areas are repopulated even after major disasters. Only a few very large-scale ecological events, such as locust outbreaks, will alter such overall statistics as total grass production.

The point of this example is that there is no "natural" perspective for the biologist to adopt; he can see any dynamic pattern that he wishes, from chaos to robust "balance," just by changing his window of observation and aggregation. It is therefore not surprising that ecologists have argued, sometimes bitterly, about whether ecological interactions are likely to result in such unstable phenomena as predator–prey cycles or instead in "prudent" behavior leading to equilibrium. The modern synthesis is that both views are correct: at smaller spatial scales (or over large homogeneous areas), interactions involving competition and predation are likely to produce unstable population sizes in the species involved (and models of homogeneous areas should behave accordingly); at larger scales, sources of spatial heterogeneity (barriers and refugia) and local renewal (dispersal and colonization) lead to statistical stability (and models of large areas should represent these mechanisms).

Models of trophic interactions

Three basic approaches have been taken to the modeling of trophic interactions (food supplies and production). None of these approaches has dealt adequately with the questions of spatial scale raised in the previous paragraph, nor with questions about how unmodeled ecosystem components may set the stage for (and modify over time) the obvious trophic interactions that are represented. As an example, keep in mind that the aquatic macrophytes growing near the shores of a lake may contribute little to the trophic base for a predator–prey interaction involving two fish species; but by providing a refuge for the juveniles of one or both species, these macro-phytes may be a key determinant of predation rates.

The simplest approach has been to extend surplus production models to include species interaction terms in the so-called Lotka–Volterra frame-work. The two-species prototype would then be

$$\frac{dB_1}{dt} = r_1 B_1 \left(1 - \frac{B_1}{k_1} \right) - a_{12} B_1 B_2 - q_1 B_1 E_1$$

$$\frac{dB_2}{dt} = r_2 B_2 \left(1 - \frac{B_2}{k_2} \right) + a_{21} B_1 B_2 - q_2 B_2 E_2$$

(4.38)

Here the logistic terms $rB(1 - B/k)$ represent the effects of intraspecific competition for unmodeled resources, the $a_{ij} B_i B_j$ terms represent competition or predation effects across species, and the qBE terms represent human exploitation ($E_1 = E_2$ if the same harvesting process takes both species). A key assumption in such formulations is that species interactions are like "mass action" random contacts in chemical reactions, so increases in either species will increase the interaction rates $a_{ij} B_i B_j$.

The most important consequence of this assumption is most easily seen by comparing the two isocline patterns in Figure 4.7, which shows combinations of B_1 and B_2 where $dB/dt = 0$ and gives a general picture of how the system will change from any starting combination. Case A shows the behavior of equation (4.38), and the basic prediction is of a single stable equilibrium (though one or the other species may vanish if the efforts are too high). Case B shows an apparently similar model, except that the predator feeding rate is assumed to satiate. Instead of the predation rate per predator being $a_{12} B_1$, this rate is modeled as $a_{12} B_1/(\alpha + B_1)$, where α is the density of B_1 at which the predators get half of their maximum ration and a_{12} becomes the maximum ration per B_2. Parameter values for case B were chosen to give the worst possible situation, which involves two potentially stable equi-libria (Bazykin, 1976; Bazykin et al., 1981): at point S^1, both stocks are low and the predator–prey interaction is the dominant stabilizing mechanism. At point S both stocks are high, the predators are satiated, and predation is relatively unimportant. Harvesting mortality rates affect the locations of the isoclines, and changes in them can make one or other equilibrium disappear suddenly, with the result that stock size will move catastrophically toward the other equilibrium. Thus an apparently minor change in functional assump-tion leads to qualitatively different predictions. Even more complex patterns are possible if the predators are considered to have type III (sigmoid) func-tional responses to prey density. However, the essential point is that model-ing species interactions is not like approximating more complex population models by quadratic (logistic) production functions; qualitatively new

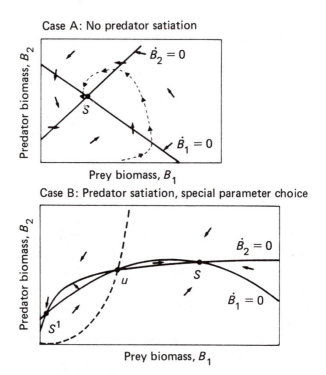

Figure 4.7. Qualitative behavior of simple trophic models can often be deduced by looking at "isoclines," which are combinations of biomasses where one species is expected to have zero rate of change ($\dot{B}_i = 0$). In case A, the predator B_2 captures B_1 at a rate $a_{12}B_1B_2$ (i.e., does not satiate) and turns captures into B_2 biomass at a rate $a_{21}B_1B_2$; for the parameter combination chosen, there is a stable equilibrium at the combination S. In case B, the predation rate by B_2 on B_1 is $a_{12}B_1B_2/(\alpha + B_1)$ (i.e., the predator satiates); for a small domain of parameter combinations, there can be two stable equilibria (S, S^1) and one unstable one (u), with S^1 having a "domain of attraction" shown by the dotted line. The isocline equations are given by

Case A: $(a_{12} < 0$, $a_{21} > 0$, $r_2 > 0)$

$$\dot{B}_1 = 0 : B_2 = \frac{1}{a_{12}}\left[r_1 - q_1 E_1 - \frac{r_1}{k_1} B_1\right]$$

$$\dot{B}_2 = 0 : B_2 = \frac{k_2}{r_2}\left[r_2 - q_2 E_2 + a_{21}\right]$$

Case B: $(a_{12} < 0$, $a_{21} > 0$, $r_2 > 0$, $\alpha > 0)$

$$\dot{B}_1 = 0 : B_2 = \frac{1}{a_{12}}\left[r_1 - q_1 E_1 - \frac{r_1}{k_1} B_1\right](\alpha + B_1)$$

$$\dot{B}_2 = 0 : B_2 = \frac{k_2}{r_2}\left[r_2 - q_2 E_2 + \frac{a_{21} B_1}{\alpha + B_1}\right]$$

concerns are introduced depending on precisely how the interactions are represented. For further analyses of harvesting in the simpler interaction models, see Beddington and Cooke (1982), Brauer and Soudack (1979), and May et al. (1979).

Attempts to fit multispecies surplus production models like equation (4.38) to time series data have not met with much success. The interaction terms generally turn out not to be statistically significant. [For an exception see Rinaldi and Gatto's (1978) work on shrimp–fish interactions.] This has been interpreted as evidence that single-stock models are acceptable for management purposes. But it is quite possible for the harvestable individuals measured in B_1 and B_2 to interact weakly, yet have strong impacts on one another's juvenile survival. If the juveniles are recruited to B_1 and B_2 only after considerable time delays, the biomass changes may even appear to be positively correlated (imagine high B_1 resulting in low survival of species 1 juveniles; then if B_2 decreases over the next few years, while the impacted juveniles are recruiting to B_1, it will appear that the fall in B_2 has led to decreasing production rates in B_1). Thus, it would seem that potential species interactions should be examined in more detail than models like (4.38) will permit.

The second approach to trophic interactions has been to connect collections of dynamic pool models together through interaction terms in stock–recruitment relationships, growth rates, and natural mortality rates. This approach at least avoids the problem, just noted, of time delays when older animals of one species impact juveniles of another. However, these models have usually made dangerously simplistic assumptions about the interaction terms. Anderson and Ursin (1977), for example, assumed that instantaneous mortality rates in prey age–species categories are proportional to the abundance of predator categories; this is essentially the same as the mass action assumption of equation (4.38) and leads to similar conclusions about stability (Figure 4.7, case A). At another extreme, Levastu and Favorite (1977) assumed that predator rations are constant, and their model can produce highly unstable dynamics; some species are assumed essentially to take quota harvests from others. It is easy to confirm with simple simulation experiments that one must be just as careful about how predator functional responses (to prey density) are represented in complex models, as in the simpler surplus production models like equation (4.38).

A third approach to trophic interactions has been to start with a careful "experimental components" (Holling, 1965) analysis of the functional responses of predators to changes in their prey density, then to surround this analysis with whatever bookkeeping is needed to keep track of age structures, etc. The basic starting point for such analysis is Holling's disk equation, which is derived by noting that a feeding animal must split his time between searching for and handling prey. If he can devote a total time T_t

per day to feeding, this total will consist of searching time T_s and handling (pursuit, capture, manipulation, rest when gut is full) time T_h:

$$T_t = T_s + T_h$$

If the average time required to handle each prey is h, and if he captures N_a prey per day, the predator will spend $T_h = hN_a$. Now suppose that he sweeps an area a' per time spent actually searching, and has a probability p_c of recognizing and successfully attacking any prey that is in the swept area. If the average prey density in the area swept is N (prey per area), he should on average attack $a'p_cN$ prey per time spent searching—in other words, $N_a = a'p_cNT_s$. Thus we can reexpress T_s as $T_s = N_a/(a'p_cN)$. Substituting this and the expression $T_h = hN_a$ into the basic time budget gives

$$T_s = \frac{N_a}{a'p_cN} + hN_a$$

If we then solve this for N_a, the number of attacks per time, the result is Holling's disk equation:

$$N_a = \frac{aT_tN}{1 + ahN} \tag{4.39}$$

where $a = a'p_c$, the "rate of effective search." The derivation can be extended in several ways, for example, by assuming that a depends on prey density through search image formation (to give a type III response). When there are many prey types (i.e., age classes of a prey species, where each age class is encountered at different rates), the multispecies disk equation for the attack rate on prey type i is just

$$N_{ai} = \frac{a_iT_tN_i}{1 + \sum_j a_jh_jN_j} \tag{4.39a}$$

where the sum is across all prey types taken.

Given the so-called "instantaneous" attack rates per predator produced by equations (4.39, 4.39a), it is then necessary to solve the exploitation equations $dN_i/dt = -N_{ai}P$, where P is the predator density (or $-\sum_k N_{aik}P_k$ if there are many predators). A good approximation for short time intervals over which the N_i do not change too much ($<50\%$) is to take

$$N_{i,t+1} = N_{it} \, e^{-N_{ai}P/N_{it}} \tag{4.40}$$

where N_{ai} is evaluated from equation (4.39a) with $N_i = N_{it}$.

An alternative "exploitation equation," which implicitly assumes some spatial effects in the sense that encounter patterns may not be random, is Mace's (1983) model for a single average prey type:

$$N_{t+1} = N_t \left(1 + \frac{a[P - h(N_t - N_{t+1})] N_0^{1/k}}{k} \right)^{-k} \tag{4.41}$$

where a, h, and P are as defined for equations (4.39)–(4.40), and the "search pattern" parameter k is defined as $k = 1/(n - 1)$, where $0 < n < 1$ implies that the search is concentrated where prey are scarce, $n = 1$ implies a random search, and $n > 1$ implies that the search is concentrated in places were prey are clumped. Mace suggests ways to estimate n and to extend equation (4.41) to cases where several prey types are present. Since N_{t+1} appears on both sides of (4.41), the equation must be solved iteratively in each time step; this extra effort is quite worthwhile when the search is thought to be highly nonrandom.

There is a simple reason for describing the disk equation derivation in some detail: it is usually necessary to estimate the feeding rate parameters from behavioral information rather than from field monitoring data, such as stomach contents. The handling time h can usually be obtained rather easily, from experimental data on maximum feeding or growth rates (maximum rate $= 1/h$). Rates of successful search (a) are more difficult to estimate. One approach is to note that $a = p_c \times$ (distance moved) $\times$ (width of reactive field). Thus, it can be estimated roughly from data on the combined movement speeds of prey and predator, along with laboratory or field assessments of the predator's reactive distance to prey (note that these speeds and distances will depend on prey and predator sizes). The capture probability p_c is usually quite low, in the range 0.1–0.3, and should be estimated whenever possible from field observations. Obviously, big errors can arise from chaining together assumptions on distances, speeds, and so forth. But the estimates obtained in this way will at least give reasonable bounds for interaction rates, and, more importantly, they force the analyst to think more clearly about the spatial and temporal structure of the process. It usually becomes obvious, for example, that searching patterns are rarely random in time and space, and at least some of the prey will be distributed in partial refuges that are searched rarely or not at all. Examples of experimental component calculations with comparisons to other field data on feeding rates can be found in Haber et al. (1976) and Clark et al. (1979). Experimental component analyses of predation interactions have indicated that the strongest effects should usually be concentrated in small or juvenile prey categories. This agrees with the general field observation in fisheries and wildlife that mortality rates in older age groups tend to be quite stable over time, in spite of major changes in predator populations. An exception to this rule has been observed in some cases involving introduced parasites, such as sea lamprey feeding on trout in the North American Great Lakes (see references in SLIS, 1980).

Given estimates of mortality patterns associated with trophic interactions, it remains for the modeler to represent how food intakes are translated into rates of growth, reproduction, and mortality in the predator categories. Here a useful observation, noted earlier, is that reproduction rates tend to be related to body size; thus, it is relatively simple to model changes in reproduction output once the growth effects have been determined. Changes in mortality rates are more difficult; if predation rates are represented as size-dependent, then some mortality effects will be implicit in the other calculations. But there may also be mortalities due to starvation and increased exposure to various risks (including predation) associated with increased movement (foraging and dispersal); this syndrome of mortality effects when food is short has never been clearly understood by ecologists.

The above discussion has barely touched on the possibilities and problems associated with taking that one obvious step beyond single-stock modeling, to at least represent trophic interactions among key stocks. Yet we see that this step is not a small one; it involves a geometric increase in the number of assumptions and parameters that must be considered. This observation has led many scientists to reject the possibility of ever constructing credible "mechanistic" models of interaction, and to concentrate instead on either improving single-stock models or looking at ecosystems in terms of other variables like energy flows and diversity.

Compressed representation of ecosystem effects

It would be nice if we could talk about ecosystems as productive units, without ever becoming involved in the biology of individual species. Then we could speak in such terms as Regier and Henderson's (1973) "stress–response" model, about how various stresses, such as harvesting and eutrophication, tend to

(1)   reduce spatial heterogeneity and species diversity;
(2)   increase temporal variability;
(3)   drive the body size distribution toward smaller forms (species) that turn over more rapidly and erratically;
(4)   shift energy and nutrient flows from maintenance pathways (decomposer food chains, etc.) into net output pathways, sometimes with deleterious effects in the long term; and
(5)   selectively favor species and forms that are of less value to humans, but which are competitively replaced by more valuable forms when the stresses are reduced.

But the key difficulty with this approach is in the last point (5): we value productive resources not in terms of overall measures like total biomass production, but rather in terms of particular species that stand out in quality as food, fiber, or sport. So discussions that begin with general ecosystem concerns have a way of deteriorating quickly into arguments about how valued species fit into whatever general response patterns are being described; the specific interactions that were to be ignored then reassert themselves as central topics of debate and inference.

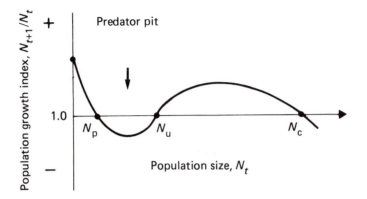

Figure 4.8. A simple model for classifying populations and predicting responses to some disturbances. N_p is a lower equilibrium maintained by predation, N_c is a higher competitive equilibrium, and N_u is an unstable stock level. The "predator pit" survival depression may be absent in some populations.

Although discussions of individual stocks seem unavoidable, we might at least seek prototypical or synoptic frameworks within which to discuss all cases as they arise. Southwood and Comins (1976) and others have suggested beginning with the simple gain relationship $N_{t+1}/N_t = g(N_t)$, where g has the form shown in Figure 4.8. The key features of g are:

(1)   the intrinsic rate of growth when N_t is small;
(2)   the presence or absence and depth of the "predator pit" representing effects of depensatory predation processes at intermediate N_t; and
(3)   the decline in g due to intraspecific competition when N_t is large.

We may use this compressed model in two ways. First, species may be classified according to the three basic dimensions of maximum growth, vulnerability to predation, and competitive ability. Second, some general inferences about ecosystem interactions can be developed by asking how the qualitative form of g should change over time (increased harvesting will lower the whole curve, fertilization will move it upward, increases in predator populations will deepen the pit, and so forth).

Classification is aided by the fact that no species can be good at everything, so there are trade-offs among maximum growth rates, predation

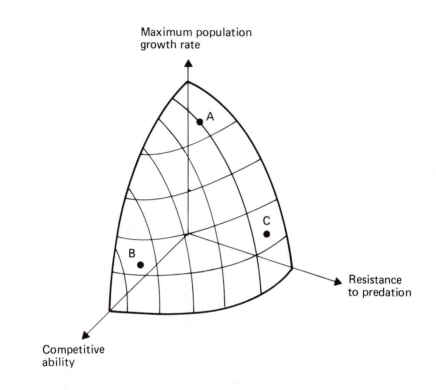

Figure 4.9. Populations may be classified according to maximum growth rate, tactics for predation avoidance, and competitive ability. Populations with parameter combinations below the constraint surface should disappear over time. A, B, and C represent three specialized strategies. After Cody (1976).

avoidance, and competitive abilities. But we expect natural selection to push these abilities toward some constraint set or surface in what Cody (1976) has termed the "strategy space" (Figure 4.9). Thus, small species tend to have high growth rates and vulnerability to predation, and mixed responses to competitors. Large species tend to have low growth rates, low vulnerability to predation, and to outcompete smaller species in the long term. Ecosystems can be viewed as collections of such "strategists," occupying various points on the surface in Figure 4.9. Strategists A are favored after major disturbances or under continued hard exploitation; their *g* function shows high maximum growth and a deep predator pit (when predators are present). Strategists B and C are favored in the absence of disturbance: B types sacrifice growth rate for competitive ability, and C types sacrifice growth for the ability to avoid predation.

As an example of how the synoptic model might be used for dynamic prediction, consider how a type B species would respond to increasing then decreasing harvest rates. Suppose the stock is initially at its upper (competitive) equilibrium, point N_c in Figure 4.8. Increasing harvest rate will lower the g function, moving the equilibrium N_c down and the unstable point N_u upward to meet it. If harvesting becomes intense enough, the N_c and N_u points will vanish, and the stock will collapse to the predation-maintained equilibrium N_p. Then, if harvesting is reduced, the stock will remain down around N_p (!), even if the other equilibria reappear as potentials. Suppose, then, that exploitation rates on predator stocks increase, or that habitat structure alters so that juveniles have more refuges from the predators. In either case, the predator pit will become shallower, so N_p will increase and N_u will decrease; if they coalesce (vanish), the stock will begin increasing toward N_c (perhaps spectacularly). So we see from this example that a variety of interactions can be "simulated," at least qualitatively, as changes in the synoptic g function. The main danger in such simulations is in failing to keep track of coupled changes in variables that have similar speeds of response; it was assumed tacitly (and perhaps incorrectly) in the above example that predators would not respond quickly and strongly to the initial collapse.

Compressed or synoptic models of how stocks respond to various factors are likely to become more popular in applied ecology, because they are robust to uncertainties about the details of production relationships. As we shall see in Chapter 5, they can be constructed initially from fairly rigorous and quantitative analysis of key interactions, such as predation. Then qualitative analysis is performed by asking how the initial quantitative parameters should change in relation to factors, such as spatial structure, that are difficult to model explicitly.

The Harvesting Process

Let us turn now to the interface between harvested populations and the human harvesting system. This interface, where stock size and its biological attributes combine with harvesting effort and its technological attributes to produce the harvests we observe, represents a process that is crucial to two aspects of adaptive management. First, the process is never completely regulated, and it is important to understand how practical controls (such as closed seasons, area closures, and effort limits) may produce varying results over time. Second, it is generally through the harvesting process that we attempt to measure the resource itself. Since independent sampling and monitoring systems are very costly for regulatory agencies to maintain, the harvesters are treated as samplers of attributes, such as relative stock size and age composition. We must at least understand the dangers in treating such samples as representative of system state.

The mass action assumption

As noted earlier, a starting point for the analysis of harvest rate (C_t) in relation to stock size (N_t) and harvesting effort (E_t) is the simple mass action model

$$C_t = qN_tE_t \tag{4.42}$$

where the catchability coefficient q is interpreted as the proportion of N_t taken by one unit of harvesting effort. If the stock N_t is distributed over some area A and each unit of harvesting effort sweeps an area a' with probability p_c of capturing each organism in the area swept, and if the sweeps are taken at random with respect to the spatial distribution of N_t (which need not be random), then q has the interpretation $q = p_c a'/A$. That is, it is exactly the same as the rate of effective search parameter a discussed above for natural predators. Note that there is a basic symmetry which makes the model more general; we can just as well view the harvesters as sitting still, while the organisms sweep out search areas for the gear. We expect p_c to vary with the size of organisms (and a' as well, if it is the organism that moves rather than the gear), to give a "mesh selection" pattern. In looking at actual catch and effort data, we must take care that effort is measured in units that reflect the area swept; such measures as the total number of fishing boats or hunters will obviously not do, since these boats or hunters may spend variable amounts of time actually searching and may change technologies so as to sweep larger areas (Mangel, 1982). Clearly it is a tricky business to translate time series of harvesting statistics into effort measures such that q will remain constant with respect to those measures.

In the following subsections we will look more deeply into the harvesting process by questioning some of the assumptions underlying equation (4.42). As you read through this section, you may well wish to join some prominent analysts in arguing that catch and effort statistics are basically useless as practical measures of changing system state, so it is imperative to establish independent sampling systems. I will argue in the last subsection that there is another option, namely to seek incentives or regulations that will induce the harvesters to spend some of their time acting and reporting as more representative samplers.

Exploitation effects

The first and most obvious thing wrong with equation (4.42) is that successive units of harvesting effort may overlap one another during short periods of intensive effort or along migration routes. Ignoring other factors like natural mortality during such periods, the simplest model for the cumulative catch Y_t in relation to the cumulative effort is the catch equation mentioned earlier [e.g., (4.12)], which in the notation of (4.42) becomes

$$Y_t = N_t(1 - e^{-qE_t}) \qquad (4.43)$$

where N_t is the stock size at the start of the period of harvest. This model is just the solution of $dC = (qN)\, dE$, with $dN = -dC$. For small E, this relationship approaches equation (4.41); as E increases, it predicts that Y_t will approach the asymptote N_t.

When several agents of mortality or types of effort E_i are involved in the exploitation episode, the catch by each type is given by

$$Y_i = N_t \frac{q_i E_i}{z} (1 - e^{-z}) \qquad (4.44)$$

where $z = \Sigma_j q_j E_j$ is the total "force of mortality" or area swept by all the agents. Here the total harvest is $N_t(1 - e^{-z})$, and each agent gets a fraction $q_i E_i/z$ of this total.

Equations (4.43) and (4.44) are troublesome as "observation models" for N_t, since they imply that the state N_t cannot be reconstructed given estimates of q just by taking $\hat{N}_t = C_t/qE_t$. However, they imply that the common abundance index C/E approaches N/E when effort (or q) is large, so the state N_t can be estimated directly as $\hat{N}_t \approx Y_t$. This, of course, does not help much in management unless it can be safely assumed that N_{t+1} is independent of N_t, in which case there is little need for management in the first place.

Nonrandom searching patterns

Harvesting effort is seldom distributed at random with respect to N_t. In the worst cases, harvesting involves a two-stage process in which searching gear (airplanes, etc.) map out the stock distribution (find large schools, etc.), and effort is concentrated entirely on the largest aggregations. In this case the harvesters may all achieve capacity loads or bag limits except when the total stock is very low, and their catch rate is approximately $C_t = cE_t$, where c is the capacity or bag limit. Obviously, then, the catch per effort is completely independent of stock size ($C/E = c$), and no signal is received about N_t by monitoring C/E until the stock is so low that some harvesters are unable to reach capacity due to local exploitation effects.

When all large aggregations can be easily found and harvested completely, then $C_t = p^*N_t$, where p^* is the proportion of the stock that is included in the aggregations that are large enough to be economical to exploit. If p^* tends to be a constant proportion of N_t (no change in aggregation behavior as stock size is reduced), then $C/E = p^*N/E_i$ is certainly not a stock size index and it would be better to use C instead. But who would believe that p^* is stable over time?

Actual searching processes involve all cases between purely random encounters and the extremes just mentioned. Harvesters may sample in partially nonrandom ways by using physical features that attract aggregations, by working in areas that have historically been more productive, by following spiral searches to retarget on aggregations after a first "strike," and by watching one another so there is effectively some cooperation in finding aggregations. To make matters still worse, all of these behaviors are likely to develop and evolve over time, so the process becomes steadily less random and more efficient. One need only glance through a fisheries trade magazine to see how inventive fishermen have been about making the search for aggregations more efficient; the development of nonrandom search and the competitive edge it gives to clever harvesters has been one of the major thrusts of technological development in fisheries.

A particularly nasty pattern of nonrandom search occurs when spatial depletion moves outward from major bases of operation, such as fishing ports. Often, heavy harvesting near to home (inshore, etc.) quickly results in the depletion of local substocks, so the harvesters are forced to move progressively further away. Catch per effort measures may remain high until the depletion "wave" hits the spatial limits of the stock, or else decline steadily if the effort measure is crude and includes the nonsearching time associated with movement to areas where depletion has not yet occurred. Competition among harvesters can be fierce during such developments, and a premium is placed on investment in larger and more mobile gear; some workers have claimed that this technological development is economically "healthy," confusing it with the economies of scale that can lead to more efficient harvesting. In fact, the depletion process may in no way represent efficient use of the resource. Similar situations also occur in forest wildlife management, when development of road networks into more remote areas (for mining and forest harvesting) keeps making more big game populations accessible to depletion by hunters.

No general and simple theoretical models have yet been developed to represent the spectrum of possible nonrandomness in search processes. An equation with some potential is Mace's exploitation model, used to model nonrandom search by natural predators [see equation (4.41)], but this model cannot adequately represent some of the extreme (and economically most important) cases. At this point it remains necessary to monitor in some detail the spatial pattern of harvesting relative to stock distributions, on a case-by-case basis. Thus, detailed logbooks and other recording systems for effort and harvest distribution are a key component of resource monitoring programs, as are the rather elaborate data-processing facilities necessary to make use of this information.

Gear competition and saturation

Especially with stocks that are highly aggregated or schooled, there can be direct interference (gear) competition among the harvesters; the presence of one or more units already in an area may prevent others from searching or casting their gear. Further, the units already in place may not move on until their "gear is saturated" (nets or holds filled, bag limits achieved). An extreme example of this competition occurs in some fisheries for Pacific salmon, where it is not uncommon to see a dozen large seine vessels "lined up" waiting to set at particularly good locations, such as headlands and closed area boundaries.

Unless effort is monitored very precisely in terms of actual areas swept, nets set, etc., the basic effect of gear competition is to make catchability coefficients decline as effort increases. This may be misinterpreted as a decrease in stock size, if catch per effort is used as an abundance index.

Gear saturation has a potentially more dangerous effect, since it implies that catch per effort will be a nonlinear, saturating function of stock size. When gear saturation is prevalent, actual stock size may decline considerably before there is a noticeable drop in catch per effort. In multispecies harvesting, an associated problem is "discarding:" when more valued species are abundant, harvesters may discard others to make room in their holds or bags. Since the sorting is a function of the densities of preferred species, landed catches per effort of the less preferred species decline —this creates the appearance of biological interaction among the species, especially when the less preferred ones are small enough to be potential prey of the others.

Let us make a small theoretical digression. Suppose you can observe the equilibrium catch (C_e) from a stock, and this catch exactly follows a logistic production function, so that $C_e = rB_t(1 - B_t/k)$. But suppose that instead of observing B_t or qB_t, your monitoring system is actually giving an abundance index Y_t that saturates as B_t increases: $Y_t = aB_t/(1 + ahB_t)$. This disk equation model might represent gear saturation. If you plot C_e versus Y_t, you will get the correct quadratic curve only if $h = 0$ (i.e., $Y = qB$). Otherwise, the "production function" will appear to peak at higher stock sizes than $k/2$. As h becomes larger, the peak moves closer and closer to the value of Y associated with $B = k$. This is not a management problem if you have "sampled" the C_e function thoroughly, since you will know what index Y to shoot for in order to maximize C_e (or whatever). But if you have only observed the right-hand limb of the function (production at high apparent stock sizes), then act as though the peak were going to occur at Y equal to one half of its maximum, you will get a very nasty surprise! Your prediction of very high production at that point will be grossly in error.

Effort versus quota regulation

If it were possible to measure q, and if one could be sure that $C = qEN$, then a simple feedback policy would be to fix the effort E at some desirable and safe level. Then, if the stock size were small, the per capita net production rate should exceed the mortality rate qE, and the stock should increase. Likewise, it should move down toward equilibrium from unproductively high levels. Under this system, it would be enough to have a reasonable estimate of the optimum exploitation rate, and periodic fixes on q to make sure that it was not changing rapidly. There would be no need to measure or estimate stock size directly, except when reassessing q.

But this system will obviously not work if stock and effort are nonrandomly distributed or if there is gear saturation. Under fixed effort, decreases in stock size will not necessarily result in decreases in catch. If catch remains nearly constant as N decreases, exploitation rates will increase in a depensatory (predator pit) pattern. Recognizing this, many management agencies have invested in expensive monitoring systems to make annual estimates of stock size as accurately as possible, and to keep a running track of catch as it accumulates so that harvesting can be shut down when a desired quota (relative to the estimated stock size) has been reached. In such situations there is a tendency to ignore harvesting effort entirely, or to treat it only as a very short-term predictor of the catch accumulation. Then, if the harvesting is profitable and efforts continue to grow over time, the quota is reached earlier each year, and it becomes progressively more costly and difficult to monitor and regulate the annual catch accumulation. As the harvesting season is shortened, intense gear competition develops (as discussed above); to stay in the race, harvesters must invest in fundamentally uneconomical technologies for getting there faster, holding the catches on board (so as not to waste time unloading), and so forth.

So at one extreme we have the inefficiencies, but potential stability, of random searching systems, and, on the other, a pathological quota system that is expensive, requires good stock monitoring, and generates another kind of inefficiency. It is a major challenge in resource management to design short-term (in season) policies that avoid these extremes, or perhaps to take another approach altogether.

Fishing for information

Harvesters of common property resources may be tightly regulated for conservation or economic reasons, but by tradition they have been free to make the most of their own decisions about where and when to exert harvesting effort. Regulatory agents view their task as the negative one of preventing depletion or waste, and they see the harvesters as antagonists who should not be involved in management or be imposed with any more regulations than necessary. Harvesters are rarely viewed as active cooperators or collaborators in gaining better understanding of the resources, and, indeed, there are usually strong economic incentives for them to sample nonrandomly and to withhold information that might reach competitors and tax collectors.

In terms of monitoring abundance changes, it would be valuable to have harvesting effort more evenly or randomly distributed in space and time. This is especially true in the early development of resources, when harvesting effort would otherwise be concentrated on the densest and most accessible stock aggregations. Why not make it a condition of participation in the harvesting (part of the license fee, if you will) that each harvester devote some fraction of his effort to searching, netting, etc., on a sampling grid predetermined by the regulatory authority? With modern computer systems, it would certainly be simple enough to assign each harvester a sampling path that would not inconvenience him too much or place him at a severe disadvantage relative to others. The main objection to this idea of obligatory "fishing for information" is that too many of the harvesters would either cheat (fake the observations) or record their sample catches too inaccurately. This is a fair objection, but we should compare the costs of monitoring and policing the harvesters to the costs that would otherwise be incurred by the regulatory authority having to run its own sampling system; government survey and research vessels, helicopter charters, etc., are notoriously expensive.

There has not yet been a large-scale test of any fishing for information system by common property harvesters. But there are scattered examples of voluntary cooperation which suggest that the rewards to both harvesters and managers (i.e., the public) of such systems can be substantial. The key problem now is to break down the traditional adversarial attitudes on both sides, and to design incentive systems that make the sampling effort at least equitable.

Dynamics of Harvesting Effort

It has been noted repeatedly above that the harvesting process is never completely regulated. Even when strong limits are placed on the development of harvesting capacity and short-term deployment of that capacity as searching effort, the harvesters are at least free to choose not to go out if they are unlikely to meet operating costs or see better opportunities elsewhere. This means that there will be some "natural" feedback between stock size and harvesting effort. This feedback may make some management actions unnecessary or redundant, but more important for the adaptive manager it may place severe constraints on how much informative variation can be deliberately introduced into the harvest rate patterns over time.

It is tempting to look initially at the dynamics of harvesters as though they were just another natural predator, and to use existing ecological models to describe their behavior over time. The previous section did essentially this, in viewing their functional responses to prey density as involving experimental components of searching, handling, and learning. We could go further, and talk about investment as a birth process, depreciation as mortality, and so forth. However, I am uncomfortable with this approach for two reasons:

(1)   the short-term (within one year) "numerical response" of harvesters to prey density is likely to be much stronger than most natural predators would exhibit, since harvesters have many more options for surviving through short periods of prey scarcity by engaging in activities other than harvesting (they can do a lot more than switch prey species or move to other habitats in search of alternative prey); and

(2)   their response "parameters" may change quite rapidly in time due to technological developments and changes in the economic environment (prices, wages, leisure time, etc.) that they face. For example, the "reproductive process" for a particular fishing fleet depends not only on how many boats there are and how well they are already doing, but also on the development status of other fishing fleets that might be rapidly converted (through minor technological investments) to join the fleet of interest.

In more vivid terms, fishing fleets as stocks may be much more tightly coupled to the economic system surrounding them than are natural predator stocks coupled to their surrounding ecosystem. This difference is, of course, a matter of degree, but it is simply not worth ignoring in favor of sloppy arguments by analogy.

Short-term effort responses

Let us consider how to model the response of harvesting effort over time periods sufficiently short so as to preclude significant change in harvesting capacity. That is, let us assume that the total fleet size, the population of sportsmen who know about and are equipped to pursue the stock of concern, or any other capacity measure is constant, and examine how much of the short-term potential represented by this capacity will actually be exhibited. Here we must begin by being very careful to distinguish between arguments about how harvesters are likely to behave as individuals, versus arguments about how the statistical aggregate of their responses will look to us if we plot or model this aggregate (total effort) as a function of key variables like stock size. Individuals have basically two decisions to make: whether to go out at all, and how long to keep searching. Both of these will be a function of perceived abundance, which may depend in a complicated way on total stock size, the behavior of other harvesters, and various regulations.

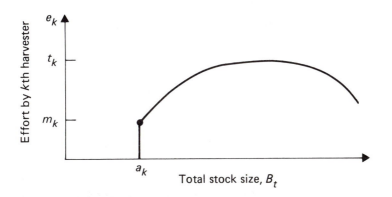

Figure 4.10. Expected pattern of effort exerted by a single harvester as a function of the resource stock size. Each harvester $k(K = 1, \ldots, N)$ may have different parameters a_k, m_k, and q_k.

In general, we expect the effort expended by the kth harvester, e_k, to vary with stock size, as shown in Figure 4.10. There will be an apparent stock size a_k below which he will perceive that it is not worth going out, and a minimum effort m_k that he will need to expend just to determine whether it is worth staying out for longer. Then his effort may increase with stock size to some maximum t_k, beyond which it will fall due to gear saturation, competition with other harvesters, and so forth. Obviously, the parameters a_k, m_k, and t_k will vary greatly from harvester to harvester, and will not be stable over time. a_k may be zero for some harvesters, especially where there are alternative stocks that attract them to the area of concern (m_k is then the "incidental" effort on the stock of interest).

But the key management interest is in the total effort measure

$$E_t = \sum_{k=1}^{N} e_k$$

where the harvesting capacity measure N is fairly large for most resources. Let me repeat: E_t is a statistical aggregate, and it will therefore be a more stable function of stock size than are the individual e_k responses. Nor need it bear any simple relationship in its shape to the individual response pattern of Figure 4.10, since it is the sum of many such shapes. A key determinant of how E_t will look as a function of stock size is the probability distribution $p(a_k)$ of stock sizes at which different harvesters will begin going out. Four examples of this distribution are shown in Figure 4.11, along with the resulting total effort response curves generated by assuming that t_k is not much larger than m_k. In case A, most of the harvesters will go out even if stock size B_t is zero (catch from B_t is incidental to their interests), and only a few more are attracted as B_t increases. In case B, the harvesters vary widely in their perceptions of abundance, and effort increases smoothly over a wide range of stock sizes. This is often the case, for example, in sport fisheries. Case C shows a dangerous management situation, where the harvesters have similar technologies, expect to cover costs even if the stock is very low, and perhaps share information that gives them all nearly the same perception of stock size. In this case, of course, effort rises rapidly over the narrow range of a_k values held by most harvesters, then decreases at high B_t due to gear saturation. bag limits, or even plain indifference to further catch. Case D shows a mixed situation, where there is a small but efficient "local" harvesting community with mostly low a_k values, and a larger "distant" community that is attracted to harvest only when B_t is large. This situation is probably common in wildlife management (local hunters versus hunters from the cities), and has become a serious marine fisheries problem since 1950 with the development of high-capacity, distant fishing fleets by a few nations.

It is extremely difficult to study effort response patterns, as in Figure 4.10, directly from historical data. The probability distribution of a_k values will change from year to year, even if technologies are stable, due to changing perceptions of abundance. The number of potential harvesters is likely to change on about the same time scale as stock size. Rapid changes in stock size (the x axis of the effort response function) are usually accompanied by regulatory actions that distort the response in various ways. Finally, there is the familiar errors-in-variables statistical problem: even if the harvesters have responded smoothly to changes in stock size, errors in its measurement will make efforts appear to be independent of it. For examples of empirical assessments of effort responses in fisheries, see Buckingham and Walters (1975) and Argue et al. (1983).

In spite of measurement problems, it is important to think of the short-term effort response as a basic component in analysis of stability of the resource system, and as a "limiting factor" in attempts to deliberately vary harvest rates more rapidly than would happen naturally (adaptive probing policies). Let us examine the stability implications briefly here. There are two of these: fast effort responses imply a greater stabilization of stock sizes than simple predator–prey models would predict, and they introduce the

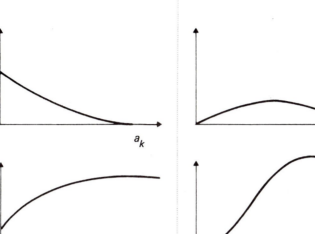

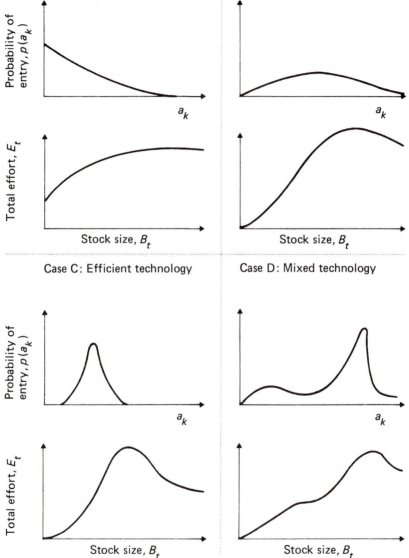

Figure 4.11. Total short-term harvesting effort E_t is a sum of responses like those in Figure 4.10, and this sum can have various shapes when plotted as a functional relationship (E_t versus stock size B_t). Cases A–D are explained in the text.

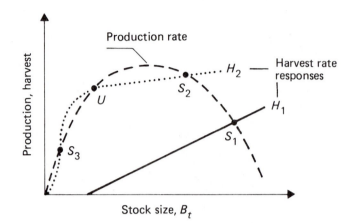

Figure 4.12. The short-term balance between production and harvest rate, as a function of stock size. The effort response H_1 (like case B, Figure 4.11) leads to temporary equilibrium at the point S_1. The response H_2 (like case C, Figure 4.11) leads to two possible stable equilibria (S_2, S_3).

possibility of multiple equilibria in stock sizes. The first of these implications is obvious and hardly needs discussion. The second can be seen most easily by thinking about the simple logistic production model, as shown in Figure 4.12. Equilibrium occurs when the production rate (dome-shaped curve) is balanced by harvest rate (curve that increases with stock size). Inefficient harvesting or uneven technology will produce the short-term harvest response marked H_1, which has a stable equilibrium at the point S_1. If the harvesters become more efficient, but the catch capacity is constrained by quotas or physical availability of the organisms, the short-term response may shift from H_1 to H_2. If the response H_2 is stable over time, the stock may be held at either of the stable equilibria S_3 (low) or S_2 (high). The stock will move up toward S_2 if it is initially (or accidentally) above the unstable point U; it will move down toward S_3 if some disturbance pushes it below U. Of course, all of the equilibria S_1, S_2, and S_3 may move about or be unstable on longer time scales due to changes in harvesting capacity and efficiency (see Jones and Walters, 1976).

In models aimed at understanding qualitative patterns of effort and stock size change, the key issue about the short-term effort response is whether it has a sigmoid shape (positively accelerated) at low stock sizes. When you consider the probability distributions $p(a_k)$ shown in Figure 4.11, and the individual responses shown in Figure 4.10 for stock sizes above a_k, it is clear that we should expect sigmoid E_t responses whenever the harvesters are seeking a single stock. The inflection point may occur at very low stock size, especially if stock and effort are not randomly distributed in space, but it should exist nonetheless. The situation is less clear for incidental harvesting in multispecies systems; efforts may follow any pattern from no response at all to the classic sigmoid pattern of switching, as exhibited by natural predators.

Dynamics of capacity and efficiency

On longer time scales, we expect the "population" of potential harvesters to develop through capital investment and diffusion of information about harvesting opportunities. Here it is worth making an initial distinction between sport and commercial harvesters.

Let us first take a passing look at sport harvesters. On large spatial scales, populations of sport harvesters develop through demographic changes and a complex cultural process of information exchange between generations [most hunters learn how from their fathers; new entrants are likely to have high a_k values (see Figures 4.10 and 4.11) which decrease over time as the individual gains experience]. In recent years, a rash of "how to" books on sport fishing and hunting may have speeded up this learning process considerably. Also, there have obviously been increases in the amount of leisure time available to most people, and in the financial resources available for sportsmen to make "high-technology" investments, such as larger boats and four-wheel-drive vehicles. These technological developments have resulted in nasty surprises for some managers, who are used to thinking in terms of quite stable sport harvesting methods based on a tradition of tight regulation. With respect to smaller spatial scales, such as single lakes and local hunting areas, the development of sport effort again involves a complex and obscure process of information exchange. Good spots (high stocks) may go almost untouched for many years, then be depleted rapidly as "word gets out" and diffuses rapidly through larger communities. In many places, public information services have made the diffusion process much faster, but the effect of this has not been documented quantitatively.

Long-term changes in commercial harvesting capacity and efficiency have been a subject of intense investigation by resource economists and biologists concerned with using catch and effort statistics in abundance estimation. Classical economic theory has it that we should expect capital capacities for harvesting to grow when returns on capital investment (profits) are above the average for investments in general in the economy, and to decline through attrition of less efficient harvesters when average profits are negative. It is easy enough to construct a simple dynamic model of this argument (Figure 4.13), if we assume that short-term profitability is related to stock size only (no direct gear competition). Figure 4.13 identifies three critical parameters:

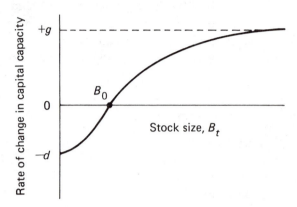

Figure 4.13. Expected pattern of change in the growth rate of capital capacity for harvesting, as a function of stock size. At stock size B_0, the addition of new capital stock is just balanced by attrition. Capacity grows if the stock is above B_0 and declines if the stock is below B_0.

(1) B_0, the stock size at which the average harvester already working will be barely taking enough harvest to cover his costs (including labor and equipment replacement);

(2) d, the rate of depreciation or permanent conversion of capital to other uses when the average catch per harvester is very low; and

(3) g, the maximum rate of growth as determined by the availability of investment capital and new investors, and by the magnitude of each new unit of investment (fleets based on very large vessels, and hence larger risks, are expected to grow more slowly).

Unfortunately, simple investment ap(de)preciation models cannot explain some critical features of the development process. As noted in Chapter 2, they do not explain how development gets started in the first place, or the initial penetration of new technologies into an established industry. Such initial developments seem to involve either great risk taking by individuals, or sheer desperation by people confronted with deterioration of other opportunities. These arguments imply that the initiation of development is going to be practically impossible to predict in most cases. Another difficulty is that models like Figure 4.13 are like the effort response curves of Figure 4.11: they represent the aggregate behaviors of a collection of potential investors. This has two implications. First, there is no reason to expect that investors even care about the average rate of profit among harvesters already at work; many new investors will feel that they have a special edge in technology, intelligence, or whatever over the "old crowd," while other new investors will be equally concerned about the disadvantage that they may face due to lack of experience and established contacts with related industries (processors, etc.).

Second, new investors must make some forecast of expected harvests, based on imperfect knowledge. They are likely to use some weighted average of past catches or profitabilities as a prediction. A simple way to model this is to assume that investment is related not to B_t, but to a running average

$$\sum_{r=1}^{L} \alpha_r B_{t-r}$$

where the α_r are weights placed on different past abundances (or catches, or profitabilities) and chosen such that $\Sigma_r\, \alpha_r = 1$ (the simplest weighting would be $\alpha_r = 1/L$).

In present times still another difficulty with the simple investment models has become increasingly important. This problem relates to prediction of the maximum growth rate g, considering that there has been an accumulation of cheaply convertible capital equipment in various harvesting fleets around the world. There have been a lot of capital "overdevelopments," followed by retirement and "mothballing" of many harvesting units. While these units do deteriorate over time and are scrapped, there nevertheless exists at most times a substantial pool of equipment that can be purchased cheaply and reequipped to fuel new resource developments. Thus, recent developments, such as coastal upwelling fisheries for clupeid fish off Africa and South America (Glantz, 1983), have proceeded much more rapidly than was predicted on the basis of earlier fisheries. In one dramatic example involving the Pacific herring, most of the vessels entering the fishery were local salmon boats that required only minor conversions, such as smaller-mesh nets (Pearse, 1982).

To deal with some of these complexities, capacity and efficiency modeling can initially be approached by brute force using a state representation analogous to age structure modeling. This approach begins by assuming that the population of harvesting units (vessels, individuals, etc.) can be partitioned into a collection of subpopulations or classes, each relatively homogeneous with respect to three parameters:

(1) a_i, the stock size at which units of class i can catch just enough to meet their operating costs (see Figure 4.10);

(2) q_i, the catchability coefficient (fraction of stock taken by one unit of effort) per i unit; and

(3) c_i, the unit capacity (maximum catch per vessel when stock size is large) or bag limit for class i units.

The state of the industry at any time is then represented by $N_i (i = 1, \ldots, n)$, the number of units in each class. The number of classes need not be large, especially considering that a, q, and c are usually correlated: low a_i values are associated with units having high q_i and, in fisheries at least, high c_i. Note that if we can simulate effort E_i by categories, we will automatically account for changes in overall efficiency, as measured by the average catchability coefficient

$$q = \sum_{i=1}^{n} q_i \frac{E_i}{E_t} \quad \text{where} \quad E_t = \sum_{i=1}^{n} E_i$$

The number of harvesting units N_i in any efficiency-capacity class can change over time due to four economic decision processes:

(1)   investment in construction of new units;
(2)   investment in movement and conversion of units from other harvesting systems;
(3)   investment in conversion of units to (and from) other classes within the system as initially defined; and
(4)   depreciation (irreversible scrapping or emigration).

In trying to model these processes, note that an investor usually cannot prescribe *a priori* the efficiency q_i that will result from his decision, although he may influence this considerably through his choice of technology (horsepower, vessel size, etc.). Thus, if we scan through the (a, q, c) classes seeking that class with the highest current "attractiveness" (catch times price less operating costs less fixed costs including interest payments on capital invested), it should not be assumed that all new investments and conversions will be to that class. Investments and conversions "aimed" at the most attractive class should be smeared into surrounding classes using a reasonable probability distribution. The process of incremental decision making, where single, large, and risky investments are avoided, can be modeled as a progressive conversion of units toward more attractive classes; such moves reflect the possibility of transferring both physical assets (equipment) and "know-how" among classes.

While it is easy enough to construct realistic models to represent the incremental processes of conversion and depreciation, there remains the difficult question of how to predict rates of new investment. Particularly when the new investments are likely to involve large capital outlays, the decision process may involve many actors (corporations, governments, banks, etc.) who can engage in various contractual arrangements to spread their investment risks. Also, the large investments can involve considerable time lags. Thus, the best approach from a management viewpoint may be to view investment as an unpredictable "driving force" to be monitored adaptively (and perhaps regulated) as it proceeds.

Options for economic regulation

We have noted in passing that various economic processes can be influenced by regulations that enhance or detract from natural feedbacks between stock size and harvesting effort. Let us classify these regulations a little more precisely in terms of the dynamic parameters that they influence. There are four basic categories of economic actions:

(1) Entry taxes (license fees, etc.) and subsidies that influence investment decisions and the stock sizes a_k at which various harvesters will initiate effort within each regulatory period. One-shot (first entry) taxes and subsidies tend to promote overinvestment in large and efficient harvesting units.

(2) Operating taxes and subsidies that influence short-term incomes (i.e., landing taxes reduce the effective resource price) and costs. These actions affect short-term effort responses, and tend to discourage major new investments and conversions among classes.

(3) Unit quotas and rights that set the capacities c_k of individual harvesting units. As long-term rights, such regulations can promote harvesters to take a longer view of the resource (and therefore become active in helping to manage it), and to invest accordingly.

(4) Restraints and subsidies on new technologies, which prevent or encourage changes in a_k, q_k, and c_k. Such actions may be introduced either to protect existing harvesters from new competitors, to maintain smooth feedbacks between stock size and effort, or to encourage more efficient harvesting in spite of impacts on existing actors.

Interest among economists has recently centered on category (3), since unit quotas have other effects besides breaking up the short-sighted decision making that results in the "tragedy of the commons." Unit quotas, if they could be assured in the face of changing stock sizes, would eliminate the incentive to invest in technologies that do not directly promote cheaper harvesting, but are instead directed at improving competitive positions (getting there first, getting a bigger share of the overall quota, etc.). The incentive would, of course, remain for harvesters to take their quotas as cheaply as possible, which might even involve sharing equipment rather than placing it in direct competition.

Rather bizarre and pathological situations have developed in a few resources, due to failure to coordinate various economic and biological management programs. For example, the Canadian Pacific salmon fishery during the 1970s was managed under biological regulations that resulted in progressively shorter fishing seasons in fewer areas, while the efficiency of harvesting vessels was increasing rapidly. This growth in efficiency was fueled by a license limitation and vessel "buy-back" program that reduced fleet size and increased profits available for reinvestment, by profits from the lucrative roe herring fishery that many salmon fishermen entered, and (amazingly) by continued government subsidies and tax incentives for construction and technological upgrading of vessels. As improvements in efficiency occurred, further restrictions were made in areas and seasons until the industry "suddenly" felt itself strangled by regulation. Such situations might be avoided by involving the various government and industry actors in adaptive modeling exercises that allow them to demonstrate to one another at least the qualitative consequences of the interactions promoted by their actions.

Harvesting Industries as Components of Economies

I noted in Chapters 2 and 3 that there are no natural boundaries to the definition of natural resource systems as units of analysis and management. This becomes particularly apparent and bothersome when we attempt to model the development of resource industries. Consider for a moment just the problem of estimating the profitability of harvesting as a determinant of short-term effort responses, technology conversions, and entry of new investors. It is easy enough to say that profitability is expected to be catch times price minus operating costs as a function of effort expended and fixed costs as a function of previous investment, interest rates, and so forth. But will increases in catch drive the price to harvesters down, and what can we say about how basic "factor prices" (labor costs, capital costs as measured by interest rates, etc.) will change over time? Just as there is no natural limit for the analysis of ecosystem interactions surrounding a stock of interest, it is clear that we could extend the analysis of economic interactions far beyond the harvesting industry. Rather than presuming to sketch out alternative programs for large-scale economic modeling in this text, let me restrict the discussion to two topics that are of particular concern in the design of adaptive policies for renewable resource management. The first topic concerns variations in basic factor prices over the long time scales (20–50 years) that are of interest in resource dynamics, due to forces that are largely independent of how any single renewable resource is developed. The second topic is the temporal pattern of economic infrastructure and interdependence that develops as a result of resource harvesting, and that greatly extends the community of actors who influence and constrain the decision-making process.

The backdrop of factor prices

Scenarios for resource development are almost invariably constructed with the presumption that economic parameters (prices, costs, interest rates) will either remain stable or grow (in the case of demand for products) smoothly into the future. This has been encouraged by some economists, whose thinking about the dynamics of economic growth as a smooth process was encouraged by the period of rather smooth development from World War II until the early 1970s. But the economic "shocks" of the 1970s and early 1980s have led to concern that economic policies and "tinkering" have not really been all that effective, and to renewed interest in "disequilibrium" theories that emphasize the existence of various waves or cycles in economic systems. With periods of around 50 years, there are the so-called "Kondratieff" cycles that mark major depressions, and are thought to be driven by human demographic waves ("baby booms," etc.), sequential depletion of basic resources, and replacement of capital equipment in major industries. On shorter time scales, various production and recession cycles have been recognized (and debated!) in economic time series. Much of the debate has been about whether there are true, rhythmic cycles or just the appearance of regularity that the human eye will associate with any autoregressive or random walk process. This debate need not concern us here: the important point is that factor prices (and hence the profitability of harvesting) can be expected to vary considerably over time, even if resources themselves are managed so as to produce stable yields.

Variations in factor prices can be profoundly important in resource management, since their effects are greatest in industries with low profit margins, which is the normal state of affairs at "bionomic equilibrium." So price increases, or cost decreases, are expected to encourage investment in harvesting (or at least upgrading of technologies), and the windfalls are quickly dissipated. Then, when prices fall again, or costs rise, the industry enters a "crisis mode" that is discouragingly familiar to practicing resource managers. There is strong pressure on government to allow short-term overharvesting to prevent immediate "collapse of the industry," and harvesters are encouraged to cheat the system by ignoring regulations and failing to report catches. But then as economic conditions improve, there is a sense of well-being and a window of opportunity to introduce innovative management schemes without strong opposition from an industry preoccupied with survival. Thus, economic variation should be viewed not as a discouraging source of unpredictability, but rather as a generator of opportunities for adaptive management.

Development of infrastructure and dependence

There is an interesting contrast from the theoretical point of view between how ecosystems and economies respond to the development of resource harvesting. We expect increasing exploitation, and measures like fertilizers and pest control introduced to help sustain it, to result in simplification of ecosystems (and with this an assortment of risks). Yet we expect the opposite response in economic systems; as a resource is developed, a progressively more complex set of related industries and activities may develop around it. These range from community services for harvesters to processing and marketing operations to businesses that build, repair, and even invent the equipment for harvesting and processing. Thus, the basic harvesting industry induces a sometimes complex infrastructure of other economic activities, and the businesses involved can become progressively more specialized and therefore directly dependent on the continuation of stable harvests. A simple economic measure of infrastructure is the number of jobs induced in the economy per job in the primary harvesting industry; generally this statistic is on the order of 1-3.

From the resource manager's viewpoint, infrastructure development means that there will be at least one additional person to bring pressure for or against management reforms, for every harvester that he deals with as a direct client. As the community of specialized actors and interest groups grows, and their very real conflicts of interest become more evident, the pressure for stable harvests and protection from further economic change (i.e., competing new technologies) is likely to make policy change progressively more difficult, except during the brief windows mentioned at the end of the previous subsection.

The overdevelopment (from a manager's viewpoint) of infrastructure is often promoted by other government agencies, through various subsidies and capital investments like road and harbor development. Continued variation in, and uncertainty about, harvests has a way of holding back these developments. But natural and inevitable patterns of variation are often forgotten during even brief periods of stability and prosperity. This short "adaptive memory" means that the management system will never reach a blissful equilibrium with its clients.

Problems

4.1.   Discuss the validity of modeling resource-harvester interactions as a simple prey-predator system. We can obviously identify things that correspond to the ecologist's functional and numerical responses of predators, prey selection, prey avoidance behavior, and so forth. But do natural predators always have alternative prey, i.e., ways to make a living? Do natural predator populations accumulate equipment for prey capture and processing (capital) that does not just die if left idle, and may reenter the system later on?

4.2.   Suppose you are handed a 30-year data set on survival rates in a population, and the data suggest a 10-year cycle that is not related to obvious factors, such as population density. Does this pattern indicate nonrepeatability or nonstationarity as defined in this chapter? If you find a factor (for example, predator abundance) that seems to explain the cycle, and include this factor in an "extended model" of the population, what happens to your conclusions about repeatability and/or stationarity?

4.3.   When we plot equilibrium yield as a function of harvest rate for an age-structured population model containing a stock–recruitment (density dependent birth–survival) relationship, the result is usually a dome-shaped curve. This suggests that the equilibrium pattern can be approximated by a simpler surplus production model, such as the logistic. Identify key reasons why this suggestion is often unwise. What if there is a need for nonequilibrium (transient, stochastic) predictions? What policy variables will the surplus production analysis ignore? Which of these difficulties can be avoided by using the Deriso model?

4.4.   Consider a simple logistic production model written as

$$B_{t+1} = (1 + r)\, S_t - \frac{r}{k}\, S_t^2$$

where B_t = biomass available for harvest, and $S_t = B_t - H_t$ is the "spawning biomass" after harvest H_t. Compare this model to the Deriso model, which can be written as

$$B_{t+1} = \left[(1 + \varrho)\, s - \varrho s^2\, \frac{S_{t-1}}{B_t} \right] S_t + R(S_{t-k})$$

where $R(S_{t-k})$ is a function relating the biomass of new recruits to spawning biomass k years earlier. What happens to the Deriso growth–survival term $[(1 + \varrho)\, s - \varrho s^2 S_{t-1}/B_t]\, S_t$ if the stock approaches equilibrium, and what does this mean about its relationship to the logistic $(1 + r)$? What terms represent density-dependent effects in both models? What form must the Deriso recruitment term $R(S_{t-k})$ have in order that the Deriso dynamics will "look" logistic?

4.5.   Use a microcomputer to simulate the logistic and Deriso models in problem 4.4, with the following parameters and relationships

$$S_t = (1 - h_t)B_t \qquad h_t = \text{harvest rate; see below}$$

$$r = 0.2$$

$$k = 1.55$$

$$\varrho = 0.5$$

$$s = 0.8$$

$$R(S_{t-k}) = \frac{0.5 S_{t-k}}{1 + S_{t-k}} \qquad k = 2$$

$$S_0 = S_{-1} = S_{-2} = B_0 = 0.05$$

How do the population growth predictions (over, say, 50 years) differ when $h = 0$? What happens if you set $h_t = 0.5$ for $t = 30, 31$ only? Using $h_t = 0$ for $t = 1$–30, then $h_t = 0.2$ for $t = 31$–50, plot the resulting net production measure $B_{t+1} - S_t$ versus $S_t (t = 1$–$50)$ for the two models; why does the Deriso model show nonrepeatability in this relationship?

4.6.   Prepare a computer program to demonstrate the distortion in apparent stock–recruitment relationships when the spawning stock is measured with large random errors. Generate the population dynamics with a Ricker model $N_{t+1} = S_t e^{1 - S_t + W_t}$, where W_t is normally distributed with mean zero and standard deviation 0.1; set $S_0 = 0.5$ and take $S_t = 0.6 N_t$ (i.e., 40% harvest) every year. For each year, generate measured spawning stock $\hat{S}_t = S_t e^{v_t}$, where v_t is normal with mean 0 and standard deviation σ_v; take the measured recruitment to be $\hat{R}_t = \hat{S}_t + 0.4 N_t$ (i.e., catch measured exactly). Plot $\hat{R}_{t+1}$ versus $\hat{S}_t$ for various sample sequences of W_t and v_t (each sequence is a "Monte Carlo trial"), $t = 1, \ldots, 20$, while varying σ_v from 0.0 to 1.0. Then repeat the trials for a variable harvesting regime, where S_t ranges from 0.1 R_t to 1.0 R_t. Does variability in harvesting rates help to compensate for the biases due to bad measurement?

Chapter 5

Simple Balance Models in Applied Population Dynamics

*Single-species models need not be as naive
as they appear.*

Caughley (1981)

This chapter is intended primarily for readers who are dissatisfied with the rather vague and general discussion about simple versus complex models in the previous chapter. I shall now focus more precisely on the ecological side of resource dynamics, and develop some arguments that can be applied to the economic side as well. As a preamble, let me reiterate that managers are facing increasing demands to produce quantitative predictions of population responses to disturbances, such as harvesting. Prediction requires some sort of model, whose development is annoying but challenging. A tempting approach has been to construct the most detailed possible calculations based on life table information, in the form of a computer simulation (Walters, 1969; Gross et al., 1973; Lett and Benjaminson, 1977). A few workers have questioned the wisdom and need for such complicated models (Burgoyne, 1981; Goodman, 1981; Ludwig et al., 1978; Deriso, 1980), pointing out that the complex calculations can often be "compressed" into simple models without great loss of accuracy. Simplified models have obvious advantages: their parameters can be more easily estimated from lumped (across age classes, etc.) field data, they are easier for everyone involved (scientists and policymakers) to understand and evaluate critically, and they are more readily incorporated into broader frameworks, such as optimization and "ecosystem" models. Indeed, the search for sound but simple models is a basic objective of science, as important in its own right as the search for detailed understanding.

We can go a step beyond just saying that simple models are easier to understand and evaluate: it appears that simplicity is essential to adaptive learning. All of us think about the world in terms of images (metaphors, analogies) that embody experience with causality; we do not really think very logically. Even textbook exercises in logic must be presented in terms of vivid imagery (Venn diagrams, etc.) before most of us can initially grasp them. With experience, we seem to become better logicians, but that is precisely the point: this experience represents adaptive learning through a set of quite modest steps, and we must become psychologically comfortable with each step before proceeding. Even when we construct complex models through apparently systematic procedures, we maintain overview and judgment about what to do along the way by visualizing the emerging model in terms of simpler images (submodels, hierarchies, etc.). So what we will try to do in this chapter is to strike more directly at mathematical representation of those images that are understandable (and are therefore a basis for further learning), using population dynamics as a prototypical problem.

The following sections present just one approach to the construction and analysis of simple "balance" models for predicting year to year changes in animal population sizes and for estimating average or equilibrium stock sizes and harvests. The approach is illustrated using examples ranging from deer to lake trout, cases where more complex models already exist for comparison of predictions. Simple algebraic arguments are used to show why some common complications, such as population age structure, usually have little effect on model predictions. Then I discuss some deceptive aspects of assuming simple and repeatable relationships involving population density; such relationships are usually assumed even in detailed simulations, and are central to concepts such as maximum sustained yield.

Population Balance Models

This section describes development, empirical testing, and equilibrium analysis of simple balance models that involve only a single variable to describe population state from year to year. It is argued that single-variable models can realistically represent some biological complexities. When fitted to population time series data, they can provide insights about what additional factors need to be considered. Equilibrium analysis helps to define optimum harvest strategies and average responses to other disturbances.

Developing the balance equation

Several approaches can be used in developing simple population models. The easiest and most obvious is to begin with some existing formula, such as the logistic equation, then try to interpret its parameters in terms of the particular population of concern. This approach fails on two counts: precise interpretation of parameters is impossible and the formula always contains "hidden assumptions" that were used in its derivation, but are not apparent in the final recipe. A more thoughful approach is to begin with a tautology that specifies additive components of population change (next year's population is this year's plus births minus deaths plus immigrants minus emigrants), then elaborate how these components are related to population size and other factors. The problem with this approach is that population rate components are not simply additive (i.e., total deaths depend on total births since some juveniles will die; winter deaths depend on how many animals remain after harvest; and so forth), so it is tricky and cumbersome to define each component correctly.

The approach recommended here is a variation on the theme of beginning with components of population change. I find that the simplest algebraic formulation usually results from the following initial statement: next year's population is the survivors from this year's population, plus survivors of the recruits added this year. Algebraically, the basic balance equation is then

$$N_{t+1} = s_{at}N_t + s_{jt}R_t \qquad (5.1)$$

where N_t is a well defined measure of population size taken at a particular point in the annual cycle (and N_{t+1} is measured at the same point next year), s_{at} is the total survival rate to $t + 1$ of animals present at time t, R_t is new recruits during the cycle between t and $t + 1$, and s_{jt} is the survival rate of the new recruits from the time when they enter the population until they are measured as part of N at time $t + 1$. Models involving additive components of population change can always be converted into the framework of equation (5.1), and vice versa, by replacing death rates with survival rates. The main advantage of equation (5.1) is that it allows simple representation of the effects of sequential, independent mortality agents. For example, if N_t is defined as the spring population, the annual survival rate s_{at} can be elaborated as

$$s_{at} = s_s(1 - h_t)s_w \qquad (5.2)$$

where s_s and s_w are summer and winter survival rates, and h_t is a fall harvest rate. Note that the resulting prediction of $s_{at}N_t$ does not depend on the temporal ordering of mortality agents; it applies whether the harvest is taken in spring, fall, or after winter mortality. To express correctly the survivorship pattern of equation (5.2) in terms of additive mortalities, it would be necessary to keep track of exactly when the mortalities occur; summer deaths would be $(1 - s_s)N_t$, the fall harvest would be $h_s s_s N_t$ since it comes from the summer survivors (not N_t), and the winter deaths would be $(1 - s_w)s_s(1 - h_t)N_t$. The resulting balance equation would be unnecessarily complex, and would reduce algebraically to equation (5.2). Worse, it would be less general, since different additive terms would be needed depending on the time of harvest.

Equation (5.1) simply expresses a tautology, since the survival rates and recruitment can be defined so as to include all possible effects, such as immigration–emigration and temporal changes in age structure. The procedure for building a predictive model from it involves two basic steps. First, and most critical, it is necessary to settle on a clear and precise definition for N_t in terms of *which* animals are to be included, at *what* time of year. Examples would be "all deer present just before young are born in the spring," or "all yearling and older harp seal females alive at the time of calving," or "all 5-year-old lake trout alive in early summer." Without such a clear definition, it is impossible to correctly define appropriate survival rates and dependences such as the effect of N_t on recruitment.

The second step is to define how the survival and recruitment rates are to be calculated each year. At this point, historical data and hypothesized density-dependent mechanisms become important, and it is often instructive to proceed through a sequence of increasingly complex model versions. The simplest hypothesis would be to assume density independence for all specific rates, so that each s is constant and $R_t = bN_t$, where b is a constant birth rate per animal. In this case, the balance equation for an unharvested population measured just before the young are born is

$$N_{t+1} = s_a N_t + s_j b N_t$$
$$= (s_a + s_j b) N_t \qquad (5.3)$$
$$= R N_t$$

where $R = s_a + s_j b$ is the annual rate of population increase ($R = 1.1$ implies 10% annual growth, etc.). Notice that this model has no balanced population level N_t such that $N_{t+1} = N_t$; N_t stays where it is in the unlikely event that R is exactly 1.0, but this is true for any starting N_t. Such behavior would normally not be considered credible in a population model, and the model would be complicated by having at least one rate component vary with population size. About the simplest hypothesis at this next level of complexity would be that the birth rate decreases linearly with population density ($b = b_0 - b_1 N_t$), as, for example, in McCullough (1979). This results in the predictive model analogous to equation (5.3):

$$N_{t+1} = s_a N_t + s_j(b_0 - b_1 N_t) N_t$$
$$= (s_a + s_j b_0) N_t - s_j b_1 N_t^2 \qquad (5.4)$$
$$= R_0 N_t - R_1 N_t^2$$

where R_0 is an "intrinsic" rate of increase and R_1 is a density-dependence factor. This is a classical logistic growth model for discrete time prediction, but with the logistic parameters defined in terms of more meaningful component rates ($R_0 = s_a + s_j b_0$ and $R_1 = s_j b_1$).

Balance equations and subsequent simplifications exactly like equation (5.4) result when survival rates are assumed to depend linearly on population density. A variation on this theme is to assume that the birth rate (b) and adult survival are density independent, while juvenile survival follows the relationship

$$s_{j_t} = \frac{s_0}{1 + s_1 N_t} \tag{5.5}$$

where s_0 and s_1 are empirical parameters. This results in the balance model

$$N_{t+1} = s_a N_t + \frac{s_0 b N_t}{1 + s_1 N_t} \tag{5.6}$$

The recruitment component of this equation increases to a limit $s_0 b / s_1$ as N_t increases, thus representing a "bottleneck" for juvenile survival. If both juveniles and adults are indiscriminately subject to the bottleneck, so

$$s_a = s_j = \frac{s_0}{1 + s_1 N^*} \tag{5.7}$$

where N^* is the number of animals entering the bottleneck, the balance model becomes

$$\begin{aligned} N_{t+1} &= \frac{s_p(N_t + bN_t)\, s_0}{1 + s_1 s_p(N_t + bN_t)} \\ &= \frac{R_0 N_t}{1 + R_1 N_t} \end{aligned} \tag{5.8}$$

where s_p is the survival rate from spring until the bottleneck period begins. Again, R_0 is an intrinsic rate of increase [$s_p s_0(1 + b)$], and R_1 is a density dependence factor [$s_p s_1(1 + b)$]. The maximum bottleneck population size N_{t+1} in this case is $R_0/R_1 = s_0/s_1$.

The above examples only hint at the complexity that can be incorporated into balance models, then hidden through redefinition of "lumped parameters" such as R_0 and R_1. The survival rates can be decomposed into arbitrarily detailed seasonal components, with density dependences that are nonselective or operate only against juveniles. Especially interesting cases arise where N_t is defined as the number of animals m years old and older, where m represents some age of interest, such as first maturity or first vulnerability to a mortality agent like fishing. In these cases, recruitment R_t represents the number of juveniles produced m years ago by N_{t-m}, multiplied by the total survival rate (or l_m survivorship) over the m years of life prior to recruitment. If the average annual survival rate over the pre-recruit period is s_I, and all rates are density independent, the balance equation becomes

$$N_{t+1} = s_a N_t + s_j s_I^m b N_{t-m} \qquad (5.9)$$

If the population growth rate R has been relatively constant for the past m years, so $N_t = N_{t-m} R^m$, then equation (5.9) can be simplified to

$$N_{t+1} = s_a N_t + \frac{b s_j s_I^m b N_t}{R^m}$$

$$= \left(s_a + \frac{b s_j s_I^m}{R^m} \right) N_t = R N_t \qquad (5.10)$$

Notice here that $R = s_a + b s_j s_I^m / R^m$, and this transcendental equation cannot be solved directly for R as a function of s_a, s_j, and b; it is analogous to Lotka's equation for the intrinsic rate of population increase, r.

It is difficult to define a purely mechanical recipe for constructing balance models, such as equations (5.3)-(5.10). In most cases, the appropriate terms are obvious when the population definition and desired survival–birth rate assumptions are spelled out very clearly. In several years of requiring undergraduate biology students to develop various balance models, the only common error that I have noted is the substitution of incorrect variables into density-dependence relationships. For example, in equations like (5.7), students often use incorrect terms for N^* [N_t instead of $s_p(N_t + bN_t)$], thus involving the wrong number of animals in survival and reproductive processes; again, this is just a matter of not thinking precisely about definitions.

Parameter Estimation and Model Testing

The construction of simple balance models involves the deliberate neglect of many factors that may influence population change. This in no way implies that such models will not give sufficiently accurate predictions for management purposes. Even serious failures can be useful, since by carefully examining the pattern of prediction errors it is often possible to identify critical directions for model improvement. But in seeking more accurate predictions and informative failures, one should at least try to make each model give its best possible performance, i.e., fit the data as well as it can. This section deals with the problem of choosing balance model parameters that will give best fits (in a simple least-squares sense) to historical population time series. Chapter 6 will take a more critical look at the whole idea of parameter estimation in decision-making contexts. Here I treat estimation only as an aid to the scientific development of better models.

The simplest estimation case arises when the observed population growth (or decline) has been purely geometric, with apparently constant R. In this case only one balance model parameter (R) can be estimated, by obvious statistical procedures. However, the balance model may still be useful in gaining further insights about some parameters when others have been measured, independently. For example, Cooper and Smith (unpublished) used equation (5.10) to develop better estimates of annual survival rates in Northern elephant seals (*Mirounga angustirostris*). They had data on R ($\approx$ 1.15), m (3), and b (0.95 per 3-year-old and older female). They were willing to assume $s_l = s_a = s$ except for the first year of life, and they wished to know how different estimates of first year survival s_j would affect their estimates of s. Their balance model reduced from equation (5.10) to

$$R = s + \frac{0.5 s^{m-1} s_j b}{R^m}$$

By solving this equation for s with the other factors given, it was possible to show that s must be greater than 0.9 when reasonable estimates of s_j are assumed. Similar calculations are becoming common in the literature (e.g., Smith and Polachek, 1981; Eberhardt, 1981).

More interesting estimation problems arise when the observed population series shows some evidence of density-dependent rates. Here the usual approach has been to estimate density-dependence parameters from plots of rates versus population size, without regard to whether the rate data might be biased or unrepresentative for N_t as a whole. Inserting such estimates indiscriminately into a balance model is hardly a fair test of the model's predictive ability. If one elects, instead, to estimate model parameters directly from the abundance (N_t) time series, the first necessary step is to reduce the model to the simplest possible algebraic form so that the fewest possible lumped parameters (like R_0 and R_1 in the above example) are included. More detailed parameters are likely to be statistically confounded (e.g., if $s = s_1 s_2$ and only the total effect s is observed, then one can choose any value for s_2 provided the constraint $s_2 = s/s_1$ is maintained).

Given a lumped model, two extreme estimation procedures are possible. In the first, or "process error" procedure, one may assume that erratic, short-term fluctuations in the abundance data are due solely to natural processes (bad winters, etc.) and not to errors in measuring N_t. In this case, the balance equation can be treated as a linear or nonlinear regression, predicting $\hat{N}_{t+1}$ from observed N_t and unknown parameters R_0, R_1, etc. For example, R_0 and R_1 in the logistic model

$$\hat{N}_{t+1} = R_0 N_t - R_1 N_t^2$$

are the linear and quadratic coefficients of a polynomial regression with zero intercept.

In the second, or "observation error" procedure, one assumes that erratic fluctuations are due solely to errors in measuring N_t, and that the actual dynamics followed some deterministic path. Standard regression procedures should not be applied in this case, due to the "errors-in-variables" problem (Walters and Ludwig, 1981); further, if one assumes the dynamics are deterministic, one should be willing to predict $\hat{N}_{t+1}$ from the previous predicted value, $\hat{N}_t$, and to estimate $\hat{N}_1$ or $\hat{N}_0$ as an additional unknown parameter. The observation estimation procedure always involves nonlinear, iterative searches for best parameter estimates.

Appendix 5A describes a simple estimation algorithm that can deal with both the process error and observation error procedures. It can be easily implemented on a programmable calculator or personal microcomputer that has enough memory to store the population time series twice, plus a few small tables of intermediate calculations.

Figure 5.1 shows results of applying the estimation algorithm to McCullough's (1979) data on the George Reserve deer herd (Table 1.1). He gives data on the prehunt population N_t and harvest H_t. Following his "recruitment model" arguments, and ignoring weak evidence for density-dependent mortality, results in the balance equation

$$N_{t+1} = s_a(N_t - H_t) + s_j[a - b(N_t - H_t)](N_t - H_t)$$

$$= R_0 P_t - R_1 P_t^2$$

(5.11)

where $P_t = N_t - H_t$. He estimated the recruitment rate parameters $a \approx 1.0$, $b \approx 0.005$. Applying the estimation procedures of Appendix 5A results in the estimates:

	Process error	Observation error
$R_0 = s_a + as_j$	1.93	2.04
$R_1 = bs_j$	0.00494	0.00608

Both procedures give surprisingly good fits to the data, and suggest that s_a and s_j are similar and are at least 0.95; this agrees with McCullough's observation that natural mortality rates have generally been very low. Obviously, we cannot reject the simple balance model on the grounds that its prediction errors are too large on average.

Stronger tests for model failure can be devised by examining the temporal pattern of prediction errors, as in Figure 5.1. For the George Reserve, these errors clearly depict a pattern that has been widely recognized, but could be easily confused with the effects of harvesting: the population performed better than predicted when it first peaked, then more

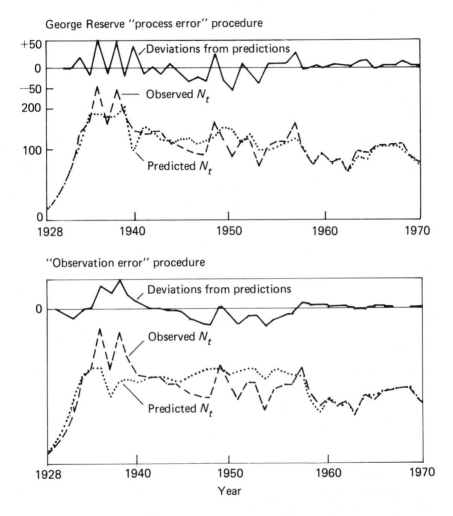

Figure 5.1. Least squares fit of a simple balance model to the historical time series of population numbers in the George Reserve deer herd. Data from McCullogh (1979). For an explanation of alternative fitting procedures, see text and Appendix 5A.

poorly for a longer period afterward. The obvious explanation in this case is that early high productivity was sustained by an excess of accumulated forage, while poor forage production later resulted from "overbrowsing" during the peak years. However, other mechanisms (such as genetically based changes in behavior) could produce similar effects. Note in Figure 5.1 that the observation error procedure, with its deterministic prediction of population changes (except for effects of time-varying harvest), results in a smoother pattern of prediction errors with stronger autocorrelations (next year's deviation is similar to this year's). Such autocorrelation is expected if there actually is considerable process error, since each natural variation will have a somewhat persistent effect on actual population size (but not on predictions based on earlier predictions rather than on earlier actual numbers). Thus, patterned deviations from deterministic model fits do not necessarily reflect any fundamental weakness in the model, since they can be generated by ephemeral, unpatterned, and generally unpredictable natural events.

While examination of patterned prediction errors may help to identify additional factors worth considering, the lack of pattern must not be taken as evidence that the correct density-dependent factors have been identified. The problem is that the same lumped balance model may result from a variety of alternative assumptions, which in turn may ultimately have different management implications. For example, an alternative model for the George Reserve deer herd would be that adult survival is linearly density dependent ($s_a = s_0 - s_1 P_t$), while birth rate is density independent:

$$N_{t+1} = (s_0 - s_1 P_t) P_t + s_j b P_t$$

$$= R_0 P_t - R_1 P_t^2 \tag{5.12}$$

where $R_0 = s_0 + b s_j$, and $R_1 = s_1$. Without independent data concerning density dependence in birth rates, one might erroneously conclude that survival is density dependent [since equation (5.12) gives the good fits of Figure 5.1], and that increased harvest rates would be compensated by improved survival. There is no substitute for good independent data on population rate processes.

More sophisticated estimation procedures and tests for model failure can be devised in special cases. When the observation and/or process error effects have a known probability distribution, the least squares fit may be replaced by some stronger criterion (Ludwig and Walters, 1981). When population size has not been measured directly, so only a time series of abundance indices (catches per effort, crowing cock counts, etc.) is available, models can be fitted to this series provided one is willing to assume a fixed functional relationship between the index and actual population (for example, index $= k N_t$, where k is unknown). Unknown parameters of this functional relationship, or "observation model," can sometimes be included as part of the set to be estimated, though usually they are confounded with those balance model parameters that determine maximum population size or carrying capacity. Fisheries workers routinely estimate the "catchability coefficient" q in the relationship (catch per effort) $= q$ (stock size), while otherwise assuming logistic stock growth (see Chapter 4). However, it has been my experience that various elaborate procedures seldom give more insight than simple process and observation error estimation based on independently derived estimates of total population.

Equilibrium Analysis

Equilibrium analysis involves calculation of how population size, harvest, and perhaps other performance measures vary under the special condition that $N_{t+1} = N_t$. The analysis may, at least, indicate directions of population change, and likely average population levels, even in cases where year-to-year predictions are unreliable due to environmental effects and other complications.

The analysis proceeds in three steps. First, explicit terms are inserted into the balance equation to represent "control factors" of interest, such as exploitation rates. Generally, it is best to express harvesting in terms of exploitation rate rather than absolute harvest quota (even when there is no practical way to hold the exploitation rate constant), in order to simplify the subsequent algebraic manipulations. For example, one might represent harvesting of yearling and older animals in a deer population having linearly density-dependent reproduction as

$$N_{t+1} = s_a(1 - h) N_t + s_j(a - bN_t) N_t \tag{5.13}$$

where N_t is the spring population and h is the exploitation rate. In this case the predicted fall harvest H_t would be $H_t = hs_sN_t$, where s_s is the survival rate from spring to fall of the harvestable animals. In order to include fawns in the harvest, one would simply multiply the recruitment term by $1 - h$ as well, and the predicted harvest would then be $H_t = h[s_sN_t + s_s'(a - bN_t) N_t]$, where s_s' is the fawn survival rate from spring to fall.

The second step is to solve for the equilibrium population by setting $N_{t+1} = N_t = N_e$, then manipulating the balance equation algebraically to get N_e on one side of the equals sign, and everything else on the other. For the above example [equation (5.13)], and for many other cases, the algebra is trivially easy:

(1)   Substitute N_e for N_t, N_{t+1}, etc.:

$$N_e = s_a(1 - h) N_e - s_j(a - bN_e) N_e$$

(2)   Divide both sides by N_e:

$$1 = s_a(1 - h) + s_j(a - bN_e) \tag{5.14}$$

(3)   Rearrange to get N_e alone:

$$N_e = \frac{1}{s_j b} [s_a(1 - h) + s_j a - 1]$$ (5.15)

The result is an equation that predicts how equilibrium population should vary in relation to all the rate factors considered. An obvious extension is to predict equilibrium harvest H_e, for example, if $H_t = hs_s N_t$, then substituting N_e for N_t we get

$$H_e = \frac{hs_s}{s_j b} [s_a(1 - h) + s_j a - 1]$$ (5.16)

Notice that this apparently complicated yield equation follows from a series of simple steps; there is nothing mystical or incomprehensible about it if these steps are kept in mind.

The final step in equilibrium analysis is to develop tables and graphs to display how the equilibrium population and associated harvest vary with parameters of interest, such as exploitation rate. This just means substituting an appropriate range of numerical values into equations like (5.15) and (5.16). Elementary calculus can be employed to find exact solutions for key quantities (find maximum H_e by setting the equation for dH_e/dh equal to zero), but it is more important to understand qualitative patterns predicted by the balance model [for example, equation (5.15) predicts that N_e should decrease linearly as h increases, while H_e should vary quadratically with h].

Two mathematical difficulties can arise in equilibrium analysis as described above, especially with step (2), when other than linear relationships are used to describe density dependence in rates. Both appear as difficulties in solving equations like (5.14) for N_e. First, the equation for N_e may be transcendental (may not have an algebraic solution). For example, suppose density dependence in survival is modeled as an exponential (as in Ricker's model of stock and recruitment in fishes) with $s_a = s_o e^{-\beta N_e}$, where β is a density dependence constant, while birth rate is assumed to decrease linearly with density. Then the equilibrium equation for N_e cannot be reduced beyond a form like $N_e + k_1 e^{-\beta N_e} = k_2$, where k_1 and k_2 are constants. Such equations can only be solved numerically.

The second difficulty is conceptually more interesting. The equation for N_e may have more than one nontrivial solution, indicating the presence of multiple equilibria or a critical population size below which the model predicts extinction. I was recently involved in a modeling exercise for barren ground caribou, where one scientist insisted that the number of very young calves killed each year by wolves is nearly independent of caribou population size, and depends mainly on the number of wolves present on the calving grounds. He further argued that the number of wolves involved is largely independent of caribou population, the wolves present being mostly young animals forced to disperse from packs that are maintained by other prey besides caribou. Suppose that the annual total kill averages k young calves per year, that the caribou birth rate is density independent, and that the winter survival rate of subyearlings (after the main period of predation) is linearly dependent on the number of older caribou. These assumptions result in the balance equation

$$N_{t+1} = s_a N_t + (bN_t - k)(s_0 - s_1 N_t) \tag{5.17}$$

where s_0 is the maximum winter calf survival and s_1 is a density-dependence factor. If we set $N_{t+1} = N_t = N_e$ and solve for N_e, the result is

$$k_1 N_e + k_2 N_e^2 = -k_3 \tag{5.18}$$

where $k_1 = 1 - s_a - bs_0 - ks_1$, $k_2 = s_1 b$, and $k_3 = ks_0$. This equation has two solutions for N_e given by $[-k \pm (k_1^2 - 4k_1 k_3)^{0.5}]/2k_2$. The larger of these is a stable "sustainable" population level. The smaller is a critical population size below which the caribou will decline toward extinction (or decline until the wolf kill rate does respond to caribou population). I do not wish in any way to imply that this caribou model is sound; it may just reflect the incompleteness of one scientist's arguments (and is thereby instructive). However, it does illustrate a general point: the most common cause of multiple equilibria in balance equations is the inclusion of "depensatory mortality agents" that exert an increasing relative effect as population size decreases (the specific calf mortality rate due to wolves in the above example is k/bN, which obviously increases as N decreases).

Effects of Population Composition

A common and deceptively incorrect criticism of simple models is that they assume all animals in the population to have equal risks of mortality and/or equal birth rates. Showing why this assertion is wrong gives insight about why simple models often give predictions almost identical to more complex models, and helps to show how simple models can be parameterized from detailed life table data.

Consider partitioning a population into a collection of sex–age–health, etc., classes, where N_{it} is the number of animals in class i at time t. The total population is then $N_t = \Sigma_i N_{it}$ and the proportional representation of each class is $P_{it} = N_{it}/N_t$. Suppose the survival rate through some period for each class is s_{it}. Then the number of animals remaining at the end of the period, say at $t + 1$, is

$$N_{t+1} = \sum_i s_{it} N_{it} = (\sum_i s_{it} P_{it}) N_t = \bar{s} N_t \tag{5.19}$$

Here $\bar{s} = \sum_i s_{it} P_{it}$ is a weighted average survival rate, with each s_{it} weighted by the proportion P_{it} of animals that started the period in class i (the argument is not changed if the animals change class, i.e., age, at the end of the period). Exactly the same argument can be developed for birth rates b_{it}; the male classes would, of course, have $b_{it} = 0$.

Relationships like the above $N_{t+1} = \bar{s}N_t$ are the basis of simple balance models. In no way do we need to claim that all the s_{it} (or b_{it}) involved in $\bar{s}$ are equal to one another, or even that they are constant in time when $\bar{s}$ is assumed to be constant. It is quite possible for the P_{it} to vary in such a way that $\bar{s}$ remains nearly constant while some s_{it} change considerably.

Further, patterned variations in s_{it} and b_{it} with population density should result in patterned changes in the P_{it} and in the aggregated $\bar{s}$ and $\bar{b}$. Balance analysis assumes only that the aggregated patterns can be detected empirically and represented algebraically through simple functions like $\bar{b} = b_0 - b_1 N_t$.

Stable age distributions

It is often reasonable to assume (or assumed anyway in computer models) that some s_{it} and/or b_{it} are constant over time (or, at least, show no patterned variation). The key question is then whether the class proportions P_{it} are stable, or more precisely: following some persistent change in birth/survival rates, will the proportions P_{it} quickly move to values P_i that are independent of time? If the classes i represent age/sex groups, this question is the classic one of stable age distributions, and the answer is generally a resounding yes. Even when density-related changes in s_{it} and/or b_{it} lead to changes in the stable proportions P_i, computer simulations generally show close "tracking" of these stable proportions, unless other mechanisms in the model result in strong population cycles (Botsford, 1979, 1981).

It is easy enough to concoct models with initial age structures and density-dependence mechanisms such that the P_{it} vary in complex or even chaotic fashions over time; a basic requirement in these cases is that the recruitment rate drops very sharply as population size increases. Such drops apparently occur in some fish and marine mammal populations, but are probably uncommon. I find the following rule useful: ignore fluctuations in the P_{it} unless there is evidence of strong, periodic fluctuations in stock size.

An implied assumption in the simple balance model examples above is that average natural survival and birth rates are independent of exploitation rates except through the influence of exploitation on population size. This assumption is obviously false for cases where exploitation has strong effects upon the stable proportions P_i (and therefore on the average rates). However it is easy enough to show mathematically (i.e., Burgoyne, 1981)

Table 5.1. Effects of various harvest policies on the stable age distribution of a hypothetical ungulate population, when no density-dependent rate changes are assumed.

		Age (year of life)										$\bar{b}^a$	$\bar{s}$
		1	2	3	4	5	6	7	8	9	10+		
Per capita birth rate (b_a)		0	0.05	0.3	0.59	0.85	0.86	0.85	0.6	0.25	–	–	–
Annual survival rate (s_a)		0.7	0.8	0.9	0.95	0.95	0.95	0.9	0.8	0.6	0.6	–	–
Harvest policies (h_a) and stable age proportions (P_a)													
(1) No harvest or non-selective rate	h_a[b]	–	–	–	–	–	–	–	–	–	–	–	–
	P_a	0.26	0.17	0.12	0.098	0.084	0.072	0.062	0.05	0.036	0.042	0.27	0.81
(2) Younger animals taken selectively	h_a	0.4	0.2	0.1	0.05	0.95	0.05	0.05	0.05	0.05	0.05	0.05	0.05
	P_a	0.29	0.13	0.089	0.08	0.075	0.07	0.064	0.052	0.082	–	0.29	0.79
(3) Older animals taken selectively	h_a	0.0	0.01	0.05	0.15	0.2	0.25	0.3	0.4	0.04	–	–	–
	P_a	0.26	0.18	0.14	0.12	0.098	0.078	0.059	0.039	0.021	0.012	0.26	0.83
(4) Young and old selectively vulnerable	h_a	0.4	0.2	0.1	0.05	0.05	0.1	0.2	0.3	0.4	0.4	–	–
	P_a	0.30	0.14	0.097	0.086	0.086	0.081	0.064	0.039	0.039	0.026	0.30	0.81

[a] $\bar{b} = \Sigma\, P_a b_a$, $\bar{s} = \Sigma\, P_a b_a$ are average rates appropriate for simple balance models.

[b] Valid for any harvest rate $x < 1$.

that nonselective harvesting (equal h on all classes) has *no effect whatsoever* on the P_i, provided all rates are density independent; changes in survivorship are exactly balanced by changes in population growth rate (which also influences the stable proportions). This very important point has apparently been missed by many authors, particularly those who have advocated use of the P_i as an index of overexploitation. Indeed, even quite age-selective harvesting often has little effect on the stable P_i when no rates are density dependent; Table 5.1 demonstrates this point for a simulated "deer" population.

The stable proportions P_i (and thus the $\bar{s}$, $\bar{b}$ values) can be strongly influenced by harvesting when rates are density dependent. An extreme example is the case where total annual recruitment remains nearly constant across a wide range of parental population densities (juvenile production "bottlenecks;" recruitment per parent is inversely proportional to total parent population). In this case, the stable age proportions are just the total survivorships (l_x), which are obviously influenced directly and cumulatively (across age) by harvesting.

A balance model for lake trout

It is quite possible for a simple model to realistically reflect equilibrium relationships involving harvesting, by representing how equilibrium age structure and average rates are influenced by harvesting, even if this model cannot accurately predict transient population changes. To illustrate this point, the remainder of this section develops a balance model for responses of lake trout (*Salvelinus namaycush*) in the Laurentian Great Lakes to rehabilitation measures (stocking, harvesting, sea lamprey control). The balance model results are compared to a detailed simulation based on the model of Walters et al. (1980). In this problem there would seem to be little hope of learning anything from a simple model: the species is long lived with considerable lags to the age of first harvesting, maturation is even later, lamprey mortality effects are age-dependent and depensatory, and the age structure is initially far from equilibrium since the population must be built up by yearling stocking into a negligible base of natural fish. The following presentation omits references for the various life history and rate estimates used; see Walters et al. (1980), Pycha (1980), and Wells (1980) for further details.

Great Lakes trout become vulnerable to sea lamprey predation and fishing at similar ages, around 4–6 years depending on the lake in question. To develop a balance model, I defined N_t as the number of 5-year-old and older lake trout present in the early summer of year t. Annual natural survival rates (excluding fishing and lamprey mortality) for yearling and older fish appear to average around 0.75-0.8. Lamprey are thought to exert a depensatory mortality influence, and D. Jester (Michigan Department of Natural Resources, personal communication) has used experimental components analysis (Holling, 1965) of their attack behaviors to derive the following approximate survival model:

$$s_{at} = s_0 \exp \frac{-\lambda_t L_t}{N_t} \tag{5.20}$$

where $\lambda_t = P_0 \alpha N_t / (\beta + N_t)$. Here s_0 is the natural survival rate (0.75-0.8), L_t is the number of parasitic lamprey present in the early summer of year t, N_t is the number of vulnerable trout present at that time, P_0 is probability of mortality per lamprey attack, α is the maximum number of attacks per lamprey per year, and β is the prey density needed for the lamprey to achieve 0.5α attacks per year (λ_t is the number of attacks per lamprey in year t). This elegant relationship just says that the trout survival rate will decline if either the lamprey population L_t increases, or the prey population N_t decreases (relative to L_t), yet lamprey have some limit to their searching ability, so λ_t decreases if N_t is sufficiently small.

While instantaneous rate relationships can be used to model the seasonal intermixing of harvest, natural mortality, and lamprey mortality, for the purposes of this discussion it will be sufficient to represent total annual survival rate s_t of 5-year-old and older fish as

$$s_t = s_{at}(1 - h_t) \tag{5.21}$$

where s_{at} is calculated from equation (5.20), and h_t is the exploitation rate in year t. Thus, from N_t we would estimate the number of 6-year-old and older fish in year $t + 1$ as $s_t N_t$.

Females first mature at ages 6-9; their fecundity then increases with age in a roughly linear fashion for at least a few years, with a slope b' of 1000-3000 eggs per year depending on the growth rate. If the population is at equilibrium with $s_t = s_e$ and if the age at maturity is (say) 8 years, we would expect to see s_e^3 8-year-old females with an average fecundity of $b' = $ 1000-3000 eggs for each 5-year-old present in N_t. We would expect to see s_e^4 9-year olds, with an average fecundity of $2b'$; s_e^5 10-year-olds with a fecundity of $3b'$, and so forth. Thus, the total egg production per 5-year-old female present in year t is $b s_e^3 (1 + 2s_e + 3s_e^2 + \cdots)$, while the total number of individuals present per 5-year-old is $1 + s_e + s_e^2 + \cdots$. The ratio of these two series gives an average fecundity $\bar{b}$ *at equilibrium* of all individuals 5 years old and older; after algebraic simplification of the above series, the ratio reduces to

$$\bar{b} = \frac{s_e^3 b'}{1 - s_e} \tag{5.22}$$

We can then predict total equilibrium egg production per year as $0.5\ \bar{b}N_t$, with equation (5.22) accounting for the effect of survival on the age structure and therefore $\bar{b}$ (the 0.5 factor accounts for a 50:50 sex ratio).

While there is no empirical evidence concerning density-dependent survival factors in lake trout, it is reasonable to assume that egg to yearling survival is a decreasing function of total egg deposition, reflecting the physical limits to quality spawning area and/or competition among the juveniles. If E_t is total egg production ($= \bar{b}N_t$ at equilibrium), a survival bottleneck can be represented as

$$s_{jt} = \frac{s_{j0}}{1 + s_{j0}\dfrac{E_t}{Y_m}} \tag{5.23}$$

where s_{j0} is the maximum egg to fry survival (0.005–0.01) and Y_m is the maximum number of natural yearlings produced when egg deposition is very high (bottleneck carrying capacity).

Total yearlings entering the population each year, Y_t, can be predicted by combining equations (5.22) and (5.23), then adding the number of yearlings stocked (S_t):

$$Y_t = S_t + \left[\frac{s_{j0}s_e^3 b' N_{t-1}}{(1 - s_e)}\right]\left[1 + \frac{s_{j0}s_e^3 b' N_{t-1}}{(1 - s_e) Y_m}\right]^{-1} \tag{5.24}$$

Of these Y_t yearlings, only a fraction s_0^4 will survive natural mortality factors to enter the 5-year-old and older stock, N_{t+4}.

Combining the above arguments about survival and recruitment results in the following balance model for lake trout:

$$N_{t+1} = s_t N_t + s_0^4 Y_{t-4} \tag{5.25}$$

where s_t is calculated each year from equations (5.20) and (5.21) (lamprey and harvesting effects), and Y_{t-4} is calculated from equation (5.24); note that Y_{t-4} depends explicitly on N_{t-5}. This model is strictly valid only at equilibrium; in other situations of practical interest (population growth during recovery), it will overestimate the annual egg deposition, since it assumes an equilibrium spawning stock with perhaps substantial numbers of older, more fecund females if s_e is large. However, it remains to be seen whether the equilibrium approximation results in unacceptably large errors compared to an explicit age structure model.

Figure 5.2 compares predictions of equation (5.25) and an age-structured simulation model that contains the same basic assumptions about survival, age at maturity, and fecundity at age a, but does not assume that the age structure of 5-year-olds and older is always at equilibrium. The simulation keeps track of 20 separate age classes. Both models assume that the initial population in all age classes is zero, then try to predict a possible rehabilitation scenario:

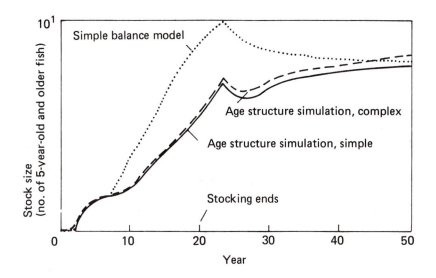

Figure 5.2. Comparison of predicted developments of lake trout populations in a Laurentian Great Lake, using three alternative models. The simple balance model does not account explicitly for age structure effects; the age structure simulations predict slower growth because they account for reduced average fecundity when the stock consists mainly of younger fish.

(1)   2 million yearlings are stocked each year, with genotypes and planting locations chosen so the planted fish should reproduce as well as wild fish; stocking is discontinued at year 20;

(2)   control measures hold the number of lamprey to 50 000, with no control failures;

(3)   the annual harvest rate is initially zero, then grows linearly to 10% between years 5 and 20, and afterwards remains constant.

It is obvious from Figure 5.2 that the simple lake trout model cannot mimic the behavior of its more detailed relative during the early years of rehabilitation. It initiates spawning by stocked fish only 5 years after stocking begins, rather than 8–10 years after as in the simulation. This results in unrealistically high recovery rates until stocking is discontinued. In later years, the model predictions converge toward the same equilibrium stock. Apparently, equation (5.22) gives a reasonable approximation of average fecundity except in the first 25–30 years.

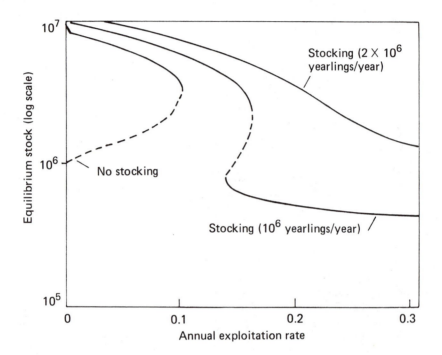

Figure 5.3. Equilibrium stock size of lake trout predicted by the balance model equations (5.20)-(5.21), (5.24), and (5.25), as a function of annual exploitation rate. Note that for low stocking rates, there is an exploitation rate above which the stock is predicted to collapse suddenly. When stocking is present, there is a lower equilibrium stock size consisting mainly of stocked fish.

So we are left without a clear conclusion about the usefulness of a simple balance model for lake trout. The simple model has two features to recommend it for policy analysis: (1) it organizes considerable biological complexity into a few equations that can be examined critically by anyone who can follow the algebraic shorthand; and (2) it captures a key qualitative prediction that is difficult to demonstrate by simulation (Walters et al., 1980), namely, that there can be two stable endpoints (or rehabilitation outcomes) of a given stocking policy, depending on the harvest rate during recovery. This point is illustrated in Figure 5.3, which plots the balance model's estimates of equilibrium N_e in relation to equilibrium harvest rate. At lower harvest rates, the model has only one (high) equilibrium. At intermediate rates, there are both low and high equilibria separated by an unstable critical stock size. The unstable point is generated by depensatory lamprey predation. At high harvest rates, only a low equilibrium remains, maintained primarily by continued stocking. Similar "multiple equilibrium" predictions have arisen in a number of other ecological models used for policy analysis (Ludwig et al., 1978; Holling, 1973, 1980; examples are given in Chapter 4). The detailed simulation exhibits similar qualitative behavior, but the equilibria are reached slowly so it is difficult to see them by doing numerical simulation trials.

The lake trout example as presented above is somewhat deceptive since the simulation used simple "knife-edge" assumptions regarding vulnerability to mortality agents in relation to fish age. Figure 5.2 also compares the balance model to a more realistic simulation with complex age-related characteristics. The parameters for this simulation were chosen to reflect the possibilities that (1) vulnerabilities to harvesting and lamprey attack increase smoothly with fish age; (2) the probability of lamprey mortality per attack decreases with fish age (older fish are more likely to survive each attack); and (3) the natural mortality rate is higher in older fish. Figure 5.2 shows that these complexities do not matter substantially to the simulation model predictions, and should not be considered grounds for not using a simple balance model. Like many biological complexities, they imply a smoothing of dynamic response relative to what would be predicted by simpler "on-off" models; it is not clear that this smoothing property should be of significant concern in policy analysis.

The main conclusion of this section is that single-variable balance models should not be rejected simply because they fail to represent explicitly some aspects of population composition. The rate parameters of balance equations are averages across component rates for composition classes, and as such will be accurate if the composition is either stable or moves in predictable ways as component rates are varied. This optimistic conclusion will be challenged in the next section, which highlights some factors already mentioned in Chapter 4 that can cause serious and persistent prediction errors.

When Balance Models Fail

The key assumption underlying simple models is that there exist stable and immediately repeatable relationships between key rate processes and total population size across some composition classes. Model failure occurs when there is a persistent or periodic change in rates observed at any population levels that are likely to arise under management. Suppose a rate value of, say, s^* is observed while population size is N^*, and the population then changes substantially but finally returns to N^*; the rate may remain persistently different from s^*, and this change may alter even the ranking of

N^* in its desirability relative to other population levels attainable by management. We might tolerate a model whose predictions were randomly too high or low, or even damping in inaccuracy over time; feedback management policies can be designed to deal with such situations (Chapters 7 and 8). But persistent or periodic errors imply that we should try to identify, and perhaps capitalize upon, additional aspects of the system's structure.

It is difficult to provide a simple classification of factors that can cause model failure as defined in the previous paragraph. One obvious class is irreversible change in some key habitat requirement, due either to external forces or to change in population size. For example, a large population increase may cause the effective extinction of some food organisms. As noted by McCullough (1979) and Caughley (1976), interactions between populations and their food supplies are generally expected at least to produce time delays in population responses; large enough delays can result in periodic prediction errors by any model that does not explicitly represent at least the delays, if not food supply itself.

Predation is another factor that can generate irreversible rate changes. The lower stable equilibrium for lake trout in Figure 5.3 is associated with low survival rates of older fish (high lamprey attack rates per fish), while the upper equilibrium is associated with high survival (low lamprey attacks per fish). Lamprey abundance was held constant for those predictions (same number of lamprey present at both equilibria); the presence of two equilibria would not even have been suspected if survival rates had been assumed to vary only with lamprey abundance (as in Pycha, 1980) rather than with the ratio L/N. After seeing a number of examples like this, it is my personal conviction that every applied population analysis should include a very careful evaluation of possible changes in predation mortality rates.

Beyond obvious interactions involving food supply and predation, there may exist persistent effects of more subtle factors like density-dependent selection, loss of genetic variation at small population sizes, and changes in abundance of competing species. Most basic texts contain catalogs of the possibilities. We simply do not have enough long-term case studies at this point to say much more; perhaps it is comforting that existing studies (like the George Reserve deer) show changes that can largely be explained by the obvious factors.

An untestable prediction that would be consistent with our understanding of long-term evolutionary changes is that all quantitative population balance models must eventually fail due to the action of natural selection on the biological factors that result in measurable rates. The key questions become how fast do "parameters" change, and do they change smoothly or abruptly? This we can only discover by experience, and more critically by having clear null hypotheses against which the changes can be detected. That is, the only way we can determine if some parameter has changed is to predict, *using a model*, what should have been seen if no change had occurred. This means we are not going to escape the business of building and testing balance models, and we should welcome their failures as a guide to learning.

When there is no substantial data record against which to test alternative simplifications, the analyst has three basic options: (1) be patient; (2) advocate an "actively adaptive" management policy that will induce informative changes in population size as quickly as possible; or (3) elaborate a complex mechanistic model that tries to build up population dynamics predictions from available understanding of the organisms and processes involved. Only a fool would trust the third approach; it is too easy to miss the key details, and in any case the relationships that were included might be as unrepeatable as the overall density-dependence relationships can be.

When a simple model appears to be failing, there are two possible directions to search for an explanation. First, one may go "downward" in a hierarchical sense, into the details of population composition and rate components. This has been the route taken in most wildlife population modeling, as witness various examples in Fowler and Smith (1981), and also in fisheries, with the recent exceptions cited earlier (Botsford, 1979; Deriso, 1980; Schnute, 1985; Shepherd, 1982). Second, one may go "outward," and try to model the broader factors (food, predation, unregulated harvest components, etc.) that operate on the population of interest. Ultimately the second approach implies ecosystem modeling, which many ecologists suppose *a priori* is hopeless. Certainly, the trophodynamic/biomass dynamic approaches to ecosystem modeling have not shown much promise (Chapter 4), but this may be a matter of poor state-variable selection leading to poor functional assumptions. There is nothing, for example, to prevent us from looking at "linked subsystems" (Overton, 1978) where each subsystem involves a balance model of appropriate type and functional complexity for that subsystem alone, and where variables that link subsystems are chosen for functional convenience in the subsystems where they impact or are used (i.e., number of parasitic lamprey is a useful variable for modeling the lake trout subsystem; entirely different variables might be appropriate for describing the lamprey dynamics as a subsystem).

Appendix 5A: Fitting Balance Models to Time Series Data

This appendix describes a general approach for finding "least squares" estimates of balance model parameters. The idea is to find those estimates that minimize the sum of squared deviations between model predictions and time series data. In what follows it will be assumed that the reader has an introductory knowledge of calculus and matrix algebra. The approach described is based on Bard (1974); further details and options can be found there. Alternative approaches based on the theory of time series analysis are reviewed in Priestly (1982).

Iterative improvement scheme

If one wishes to minimize a sum of squared deviations between observations N_t and predictions $\hat{N}_t$, the following iterative procedure is usually reasonably efficient:

(1)   Assign initial parameter estimates R_{init} to the parameters R_0, R_1, etc., that are to be estimated.

(2)   Calculate the vector of deviations $d = N - \hat{N}$ using the latest available parameter estimates (R_{init} or R_{new} from below). $d_i = N_t - \hat{N}_t$, where $\hat{N}_t$ is calculated from R.

(3)   Calculate the sensitivity (or design or Jacobian) matrix X, where x_{ij} is the derivative of the ith predicted value with respect to the jth uncertain parameter.

(4)   Calculate and invert the cross products (or approximate Hessian) matrix $X'X$, to give $(X'X)^{-1}$.

(5)   Calculate the parameter correction vector $c = \lambda(X'X)^{-1}X'd$, where λ is a "step size correction" that is initially set to 1.0 and reduced for later iterations if the corrections fail to decrease in successive iterations.

(6)   Get new parameter estimates as $R_{new} = R_{init} + c$, and return to step (2) unless all the elements of c are very small.

This procedure is identical to linear regression, and gives best estimates without iteration [without repeating steps (2)-(7)] if the X matrix contains only constants and data (i.e., its elements are independent of the parameter estimates). More elaborate procedures for correcting step size, such as Marquardt's algorithm (see Watt, 1968), sometimes give faster convergence but in my experience are not worth the extra programming effort.

After the estimates have converged, a measure of uncertainty about the correct parameter values can be easily calculated. This measure, the "asymptotic covariance matrix of the parameter," Σ, is given by

$$\Sigma = \left(\frac{\sum_{i=1}^{n} d_i^2}{n - m} \right) (X'X)^{-1}$$

where the d_i and X are as defined above, n is the number of data points (rows of X), and m is the number of parameters (columns of X). Confidence limits for the individual parameters can be estimated as $\hat{\beta}_i \pm t_{\alpha,n-m}\sqrt{\Sigma_{ii}}$ where $t_{\alpha,n-m}$ is the Student's t-statistic for α probability level and $n - m$ degrees of freedom, and Σ_{ii} are the diagonal elements of Σ. The interpretation and use of Σ will be discussed further in Chapter 6.

Calculating sensitivities to parameters

The key step in the above algorithm is calculation of the sensitivities $x_{ij} = d\hat{N}_i/dR_j$. In general, we may express these sensitivities (using the chain rule of basic calculus) as

$$\frac{d\hat{N}_i}{dR_j} = \frac{\partial f_i}{\partial R_j} + \frac{\partial f_i}{\partial \hat{N}_{i-1}}\frac{d\hat{N}_{i-1}}{dR_j} \tag{5A.1}$$

where f_i is the balance model equation used to predict $\hat{N}_i$. Additional terms of the form $(\partial f_i/\partial \hat{N}_{i-k})(d\hat{N}_{i-k})/(dR_j)$ must be added for cases (such as the lake trout example) where $\hat{N}_i$ depends directly on $\hat{N}$ at times $i - k$. When the real system is thought to have mostly "process error," so $\hat{N}_i$ is best predicted by the observed values N_{i-1}, N_{i-k}, etc. (rather than $\hat{N}_{i-1}, \ldots$), then the second term(s) of equation (5A.1) are simply deleted since $\partial f_i/\partial \hat{N}_{i-1} = 0$.

When mostly observation errors are assumed, so $\hat{N}_i$ is predicted from $\hat{N}_{i-1}$, note that equation (5A.1) is recursive in $d\hat{N}_i/dR_j$: one first calculates $d\hat{N}_1/dR_j$, then this is used in calculating $d\hat{N}_2/dR_j$, then this in calculating $d\hat{N}_3/dR_j$, and so forth. The problem is what to do with $d\hat{N}_1/dR_j$, since this should be predicted from $\hat{N}_0$ (and perhaps $\hat{N}_{-k}$). Here two approaches are possible: (1) use the first data point as $\hat{N}_0$, so $d\hat{N}_0/dR_j = 0$; or (2) estimate $\hat{N}_0$ as an additional unknown parameter (say R_m), so $d\hat{N}_0/dR_m = 1.0$. The second of these approaches is usually not worthwhile unless the early population estimates are particularly suspect. In the first approach, the first k observations are not predicted (where k is the largest lag appearing in the model), so the X matrix has $n - k$ rows, where n is the number of years' data.

As an example of the sensitivity calculations, consider the simple example $N_{t+1} = R_0 N_t - R_1 N_t^2$. In this case we have

$$\frac{\partial f_{t+1}}{\partial R_0} = N_t \tag{5A.2a}$$

$$\frac{\partial f_{t+1}}{\partial R_1} = -N_t^2 \tag{5A.2b}$$

$$\frac{\partial f_{t+1}}{\partial N_t} = R_0 - 2R_1 N_t \tag{5A.2c}$$

[but ignore (5A.2c) for process error assumption]. The X matrix elements are calculated by substituting these relationships into equation (5A.1). For the process error assumption, note that only the second, third, etc., data point can be predicted (or, in general, the first k data points are not predicted if the model explicitly contains N_{t-k}); in this case the X matrix for data points $2, \ldots, n$ is just

$$
X = \begin{bmatrix} N_1 & -N_1^2 \\ N_2 & -N_2^2 \\ \vdots & \vdots \\ N_{n-1} & -N_{n-1}^2 \end{bmatrix}
$$

where the first column of X represents R_0 and the second represents R_1.

In programming the above procedures for programmable calculators or computers, note that the rather large X matrix need not be stored explicitly. Only the matrix $X'X$ and vector $X'd$ are used to improve parameter estimates. Except for the calculation of $(X'X)^{-1}$, the above estimation steps (2)–(4) can be accomplished by a simple "loop" over the data points to be predicted. For each point i, first calculate the predicted $\hat{N}_i$ and d_i; then calculate the derivatives with respect to each parameter, storing these in a vector t ($t_1 = \partial\hat{N}_i/\partial R_0$, $t_2 = \partial\hat{N}_i/\partial R_1$, etc.). Then add $t_i \cdot t_j$ to the i, jth element of $X'X$, for all combinations of i and j ($X'X$ is symmetric; some calculations can be saved by noting that $X'X_{ij} = X'X_{ji}$). Finally, calculate $t_1 d_i$ and add it to $(X'd)_1$, add $t_2 d_i$ to $(X'd)_2$, and so forth.

Estimation from population index data

Often N_t is not observed directly, and model parameter estimates must be based on a time series of index values y_t, whose functional relationship to N_t must be assumed *a priori*. If this "observation model" is designated $h(N_t)$, so it is assumed that $y_t = h(N_t)$, then the above procedures can be used to find parameter estimates that most closely predict the index time series. All calculation procedures are as described above, except that (1) deviations d_t are calculated as $y_t - h(\hat{N}_t)$, where $\hat{N}$ is predicted from the balance model as before (using either $\hat{N}_{t-1}$, or N_{t-1} depending on the error assumption), and (2) the elements of X are replaced by sensitivities of the predicted *observations* to the parameters. These modified sensitivities are obtained by the chain rule of calculus:

$$
X_{ij} = \frac{\partial h_i}{\partial R_j} + \frac{\partial h_i}{\partial \hat{N}_i} \frac{d\hat{N}_i}{dR_j} \tag{5A.3}
$$

where $\partial h_i / \partial R_j$ is zero unless R_j is an unknown parameter of the observation model itself (i.e., q in $y_t = qN_t$), and $d\hat{N}_i / dR_j$ is calculated as described above. For the simple case $y_t = qN_t$, the term $\partial h_i / \partial \hat{N}_i$ is just equal to q, that is, all the original sensitivities are "scaled down" by the observation model parameter.

What goes wrong

The above estimation scheme can fail if the data are very noisy or exhibit temporal patterns that the chosen balance model cannot produce. However, the most common cause of failure is reflected in the inability to invert $\mathbf{X'X}$, and/or erratic behavior of the correction vector c. The basic cause of these problems is statistical "confounding" of the balance model parameters, so that some columns of $\mathbf{X}$ are nearly (or exactly) linearly dependent. In simpler terms confounding between two parameters means that they have highly correlated effects on the predictions, at least over the range of past observation. For example, the logistic model leads to an $\mathbf{X}$ matrix with one column of N_t values and one column of $-N_t^2$ values. These columns are linearly dependent (correlation $= 1$) if the N_t values span only a narrow range.

Confounding can also be thought of as the existence of a long, narrow "trough" in the surface of sums of squares plotted as a function of the parameters. Erratic parameter changes usually involve jumping back and forth along such troughs without there being any well defined deepest point. Points along the bottom of the trough represent parameter combinations that are equally good at explaining the observed data.

Poor estimation performance sometimes results from choosing bad initial parameter estimates. This problem can usually be avoided by beginning with the process error assumption, which generally implies less confounding of parameter estimates ("natural" variation generates informative contrasts), applied to a transformed or approximate model that is constructed so that $\mathbf{X}$ does not depend on the parameters (i.e., a model that is linear in its parameters). For example, the balance model $N_{t+1} = R_0 N_t / (1 + R_1 N_t)$ is nonlinear in its parameters; the transformation $(N_t / N_{t+1}) = (1/R_0) + (R_1/R_0) N_t$ allows initial parameter estimates to be obtained by linear regression ($y = N_t / N_{t+1}$, $x = N_t$, intercept $= 1/R_0$, slope $= R_1/R_0$). However, such transformations generally do not result in best estimates relative to the original time series data, and so should not be used to replace the procedures outlined above.

Using time series data to estimate functional relationships can lead to bad bias in the parameter estimates, when process errors are large and influence the future states that are treated as regression "independent variables." Consider the linear-in-parameters case, where we take $\hat{\beta} = (X'X)^{-1}X'Y$. If the model is structurally correct, so $Y = X\beta + w$ where w is the vector of process errors, it follows by substituting the second equation into the first that

$$\hat{\beta} = \beta + (X'X)^{-1}X'w$$

Ordinarily we assume that X and w are uncorrelated, so the expected value of the error vector $(X'X)^{-1}X'w$ is zero. This assumption fails when w_t influences x'_{t+1}, x'_{t+2}, etc. (the future rows of X). For example, the Ricker stock–recruitment model $R_{t+1} = S_t \exp(a - bS_t + w_{t+1})$, leads to the obvious transformed equation $\ln(R_{t+1}/S_t) = a - bS_t + w_{t+1}$, which is a linear regression where $\beta_1 = a$, $\beta_2 = b$, and

$$X = \begin{bmatrix} 1 & -S_1 \\ 1 & -S_2 \\ \vdots & \vdots \\ 1 & -S_{T-1} \end{bmatrix}$$

when there are T years' data. The spawning stocks S_{t+1}, S_{t+2}, etc., depend on w_t whenever the harvest rate has been relatively steady over time, or at least when no deliberate effort has been made to experimentally set S_t to be independent of the recruitments R_t (which obviously depend on w_t). The correlation between X and w in this example leads to a very nasty bias: the productivity parameter a is generally overestimated, and the spawning stock for maximum yield is generally underestimated (by as much as 50% when the w's are realistically large and harvest rates are nearly constant).

Problems

5.1. Construct simple balance models to describe the following "typical" density dependence patterns:

   (a) birth rate decreases linearly with density, survival rates are density independent;

   (b) winter survival rate of juveniles (subyearlings) is inversely related to the abundance of older animals N_t present at the start of winter, i.e., $s_{jt} = s_0/(1 + s_1 N_t)$;

   (c) limited breeding space, so that total births B reach an upper limit B_m and vary with spring population N as $B = b_0 N/(1 + b_0 N/B_m)$; all survival rates are density independent.

5.2. For each of the models in problem 5.1, plot the population size next year (N_{t+1}) as a function of N_t, using parameter values that will give the populations a maximum rate of increase (at low N_t) of about 30% per year and an equilibrium level of $N_e = 1$. Can the behavior of all the models be well approximated by a simple logistic equation $N_{t+1} = R_1 N_t - R_2 N_t^2$?

5.3. Construct a balance model for a population that has limited breeding space, as 5.1.c above, and is subject to depensatory mortality of the juveniles by an efficient predator that does not depend on the population in question for its own well-being (for example, think of feral cats preying on juvenile pheasants, where the cats have many other food sources). Represent juvenile survival rate as $s_{jt} = s_0 B/(B_h + B)$ where B is births, s_0 is the maximum survival rate, and survival goes from zero to this rate as B increases while reaching $s_0/2$ when $B = B_h$. Show that this model can exhibit a stable equilibrium at some high N_e, and a critical population size N_c below which the depensatory predation drives it to extinction. Show how the size of the "stability domain" $N_e - N_c$ varies with the annual survival rate s_a of older animals.

5.4. Construct an age-structured population model with birth and survival rates as in Table 5.1, except make the survival rate for age 1 animals (s_1) depend on total population size

$$N_t = \sum_{a=1}^{10} N_{at}$$

by the relationship $s_{1t} = 0.7/(1 + N_t)$. Using the no-harvest P_a from Table 5.1 and $N_0 = 0.1$, construct an initial age structure $N_{at} = P_a N_0$ to start the simulations. For each simulated year, calculate $P_{at} = N_{at}/N_t$ and use these proportions to estimate time-varying average rates $\bar{b}_t = \sum P_{at} b_a$, $\bar{s}_t = \sum P_{at} s_{at}$ (only the P_{at} and s_{1t} are variable). How do $\bar{s}_t$, $\bar{b}_t$, and N_t vary over time if you use various harvest policies? Is there a repeatable relationship between $\bar{s}_t$ and N_t, provided the population does not grow or decline too rapidly?

5.5. Modify the age-structured model in exercise 5.4 to use the survival relationship $s_{1t} = 0.7 \exp(-0.69 N_t)$. Do your results change? Then multiply all the b_a values by 5; now what happens, especially to the relationship between $\bar{s}_t$ and N_t?

Table 5.2. Estimates of Pacific herring egg deposition and subsequent age 1 recruitment, and the number of predatory cod, in the Hecate Strait, British Columbia. Data assembled by M. Stocker and J. Westerheim, Canadian Department of Fisheries and Oceans, Nanaimo, B.C.

Year	Number of age 1 herring recruits ($\times 10^{-6}$)	Herring egg deposition the previous year ($\times 10^{12}$)	Number of age 2 and older cod in the system ($\times 10^{-6}$)
1955	359	5.3	8.5
1956	586	3.44	5.1
1957	182	3.44	8.5
1958	756	6.24	5
1959	627	4.28	5
1960	514	8.16	5.8
1961	968	8.7	5.1
1962	355	8.4	7.7
1963	128	11.22	11.7
1964	67	8.06	7.3
1965	197	4.64	7.9
1966	214	2.36	12.7
1967	270	2.12	8.3
1968	343	2.94	7.9
1969	329	4.04	3.7
1970	608	5.32	3.8
1971	444	5.9	6.6
1972	550	8.46	8.7
1973	169	8.76	10.9
1974	254	9.8	8.7
1975	210	7.36	8.1

5.6.   Table 5.2 shows crude data developed during an AEA workshop on species interactions among commercial fish species in the Hecate Strait, British Columbia. Estimate a Ricker stock–recruitment model $N_1 = E \exp(a - bE)$ for herring, by linear regression of $Y = \log N_1/E$ against $X = E$, where N_1 = age 1 recruits and E = eggs deposited. How can the parameters a and b be interpreted? Then include abundance of cod (a herring predator) N_c as a second independent variable, so the herring recruitment model becomes $N_1 = E \exp(a - bE - cN_c)$. Does the fit improve? How would you interpret the c parameter, and can you see a way to estimate it by using independent data on cod prey consumption rates? Develop a simple balance model for the herring, using your final stock–recruitment model and an annual adult survival rate $s_a = 0.6$. Use this model to show how the equilibrium relationship between herring yield and harvest rate might be influenced by changes in the cod stock.

Chapter 6

Embracing Uncertainty

*If the outcome of any research project could be safely
predicted, there would be no need for research.*

Cushing (1968)

Previous chapters have stressed that resource managers must learn to
live with some very substantial uncertainties. Modeling helps to clarify and
highlight these uncertainties, but cannot usually resolve them by decomposition (reduction) of relationships into smaller and more understandable
(researchable) pieces. This means in the end that many key management
decisions are essentially *gambles*, no matter how nicely we may try to package the justification for these decisions by presenting reams of data and elaborate calculations. Most people find it rather uncomfortable at first to think
of resource decision making as gambling; somehow we expect governments
and international agencies to act prudently and with at least a modicum of
foresight. Indeed, when uncertainties are revealed in public debates it is
often argued that inaction (wait and see, do more research) is preferable to
the indignity of gambling; such arguments can reflect gross confusion
between personal ethics (gambling as a personal weakness or bad habit) and
public responsibility.

Most prescriptions about how to use model building and prediction in
management have implicitly shied away from treating decisions as gambles.
Typically, it has been argued that one should construct the best possible
model and parameter estimates, and then either act as though this model
were correct or perhaps be more conservative if the estimates are obviously
weak. It has been stressed that point estimates and predictions should be
accompanied by measures of uncertainty, such as confidence limits, but
there is seldom any careful analysis about how to behave differently if the
limits turn out to be quite wide (the usual case). In the language of statisticians, the preoccupation has been with how to construct point estimates and

their policy implications. In the language of control theory, this preoccupation has been called "certainty-equivalent" policy design; it has been shown that for some very special situations (so-called linear-quadratic control problems), it is indeed optimal to act as though the current best estimates were correct.

Statistical decision theorists have developed quite a different notion about how to deal with uncertainty, and much of the remainder of this book is based in one way or another on the approach that they have suggested. Their basic claim is that one should first *embrace uncertainty*, by trying to define not a single best prediction, but rather the set of possible outcomes (models, states of nature) that are consistent with historical experience. Next one should try to assign odds or probabilities to the alternative models, and these odds should be somehow used in decision making. Placing odds is very different, and usually more difficult, than just setting confidence limits.

This chapter will be concerned mainly with how to identify alternative models and place odds on them, but let us first look ahead briefly to how such results can be used. There are two very distinctive uses, and the first of these has not been emphasized in the literature of decision making under uncertainty. This is simply to expose uncertainties in an emphatic manner, so as to stimulate *imaginative* thinking about policy options that may be more robust or informative than the options that would otherwise be evaluated. The second, more formal, use is in ranking or comparison of options that have already been identified; most of decision theory is concerned with how to go about constructing such rankings. The basic idea in ranking alternatives is to construct (at least conceptually) a "decision table," or matrix, which displays the outcome of each policy option for each model or hypothesis that has been identified as plausible. Then the ranking proceeds by looking across the possible outcomes for each policy; it is usually assumed that the best policy is the one with the maximum *expected value*, which is the sum of products of outcomes times the probabilities of occurrence. That is, the expected value of an option is its weighted average outcome, where the weight on each outcome is its probability or odds of occurring. Other ranking criteria are, of course, possible, such as "min–max," in which we seek that policy whose worst possible outcome is highest.

A much oversimplified example from Pacific salmon management will serve to illustrate these ideas. In the early 1960s, it was suggested that artificial spawning channels could be used to increase juvenile survival of sockeye salmon from the Skeena River in British Columbia, and that more juveniles going to sea would result in more harvestable adults returning a few years later. Various experiences elsewhere had shown that juvenile survival could be improved, but there were no good data on the relationship between juvenile production and adult returns. A caricature of this decision problem would be that there were two models of response (adult increase versus no adult increase) and two decisions (build channel, do not build). The following decision table shows a rough estimate (revised from Walters, 1977) of the net economic value of the sockeye fishery (an output measure), in millions of dollars, for each of these models and decisions, as they might have been evaluated before the actual decision:

Option	States of nature	
	No adult increase	Good adult increase
Do not build channel	240	240
Build channel	135	564

According to these predictions, the fishery value would be roughly doubled ($240 to $564 million) if the juvenile to adult survival were favorable, and there would be a loss of around $105 million in construction/operating costs plus lost yields (to allow stock recoveries) if the survival were not favorable. Now suppose it had been decided to place roughly even (50:50) odds on the two states of nature; in this case the expected value of the "do not build" decision would have been 240 = (0.5)(240) + (0.5)(240), and of the "build channel" decision would have been 349.5 = (0.5)(135) + (0.5)(564). Thus, in an expected value sense, building the channel would be a good decision in spite of the great uncertainty. But decision makers are averse to mistakes, and the possible loss of $105 million might weigh more heavily than the expected value calculations would indicate. Then the above decision table would represent a source of some considerable frustration (how to balance the 135 against the 564, and these against the "sure" conservative decision giving 240), and we would hope that this frustration would lead to some imaginative thinking about other policy possibilities. An obvious option would be to construct a much cheaper temporary channel, with an expected outcome of, say, $200 million in the "no increase" case and a lower payoff of, say, $500 million in the "good increase" case. The expected value of this "experimental" option would be 350 = (0.5)(200) + (0.5)(500), and we begin to see the importance of setting the odds carefully based on past data and judgment. In an expected value sense, the experimental option is barely better than the "build" option, and this advantage would be lost if we assess slightly higher odds for the "good increase" model.

The salmon example shows that by embracing uncertainty we may come to see decision problems quite differently than if we insisted on trying to provide single, best models and predictions in the first place. For example, we would not agonize very long in the face of inadequate data about whether "good increase" is the best model, and we would not suggest expensive research programs to try and resolve the uncertainty prior to the decision. We would instead be quite concerned about the issue of risk aversion, and whether to count a \$105 million loss as somehow different from a \$224 million gain (see Chapter 2). We would worry about how to set odds on these outcomes, and on the identification of still other outcome possibilities. And, perhaps most important, we would try to eliminate entirely the original problem of a difficult choice between extreme alternatives, by seeking adaptive policy options.

The following sections will discuss some ideas about how we recognize and measure uncertainty, when we are seeking to identify alternative models and place reasonable odds on them. Most of this discussion is really just a review of material that can be found in various textbooks on statistics and decision theory, and for further reading I particularly recommend Raiffa (1968), Behn and Vaupel (1982), and Bard (1974). My particular concern here will be to try to bring together some concepts and approaches that are usually presented as scattered and distinctive, and to show how they are applied in resource systems. The final section returns to questions raised in Chapters 4 and 5 about how complex the alternative models should be.

Levels of Uncertainty

It is usual to distinguish between three types of uncertainty about natural systems. First, we must admit that certain inputs or disturbances that occur rather regularly or frequently over time will generate unpredictable and uncontrollable changes. This background variation or *noise* can be of large magnitude without having any profound effects on management decision making except (1) to make it imperative to have "feedback policies" involving monitoring and adjustment to the changes, and (2) to retard efforts to learn about underlying average patterns of response.

Second, there is statistical or "parametric" uncertainty about the forms and parameter values of various functional responses, such as production rates as a function of stock size. With this type of uncertainty we worry about what equations to use, how to estimate parameters from noisy data, and how to assign probabilities to various hypotheses expressed as alternative equations and/or parameter values.

Third, there is always basic structural uncertainty about even what variables to consider (in the words of Chapter 3, how to "bound the problem"). Incomplete structural representation implies that we can expect not only time-varying parameters in functional responses, which can sometimes be dealt with as statistical patterns without too much difficulty, but also some surprises involving large and unexpected changes in response patterns.

In most of the following discussion I will deal mainly with the first two types of "measurable" uncertainty. I doubt that there can, in principle, be any consensus about how to plan for the inevitable structural uncertainties that haunt us, any more than we can expect all human beings to agree on matters of risk taking in general.

It is important to note that the distinction between noise, parameter error, and structural uncertainty does not stand up under close logical inspection. It is just a simple way of modeling a continuous spectrum of things that can go wrong at increasing temporal and spatial scales moving away from any particular point of analysis. Noises are usually treated as random and uncorrelated over time; when we see strong correlations (space/time patterns) in noises, we seek to explain and model these by (uncertain) functional relationships. As we gain greater experience, sources of structural surprises become familiar and can also be modeled through functional relationships. All this is just another way of saying that the "natural systems" seen by managers are not really natural at all, but rather are the result of arbitrary problem boundings and selections of state variables.

Another way of categorizing situations involving uncertainty is to break up the sequence of decisions that are taken during resource development into phases based on the availability of management experience. The earliest phase we might call the *preadaptive phase*, in the sense that no direct data on the system's response is available so decisions must be based on earlier experience in "similar" situations. In this phase many decision options may appear to be equally advantageous (or risky), and practically anything that is done will yield valuable information for future decision makers. The key policy issue in this preadaptive phase is how much to invest in monitoring systems that will give a consistent picture of responses during the development.

The next or *adaptive phase* begins when there is enough experience at hand to begin sorting out clear hypotheses (alternative models) about responses to further action. The key policy issue becomes whether to act informatively with respect to hypotheses that imply opportunities for improved performance by moving outside the range of experience available (a good example is in Figure 1.1; should spawning stocks be increased to test whether larger recruitments would result?).

The final phase, which may never be attained in practice, might best be called the *certainty-equivalent phase*. In this phase the system's responses and limits have been well tested through experience, so that there is no probable advantage in acting any other way than suggested by the best available models. I personally have seen only one resource that I felt has entered this phase, through a very long and painful pattern of development, collapse, and rebuilding. This is the fishery for Pacific herring off the British Columbia coast. In this case a reduction fishery built up and slowly depleted a rich complex of substocks until alarming signs of recruitment failure were evident. The fishery was then shut down, and many substocks recovered rapidly. A second fishery for roe (a prized delicacy in Japan) then developed, and this fishery, which takes place near spawning areas, has been closely regulated to allow adequate spawning stocks in most locations. There are still major uncertainties about the sustainability of harvests taken from spawning areas and about how to regulate the various local fisheries efficiently and safely, but there is at least consensus on the basic limits and potential for further development. Only further experience will resolve the sustainability issue, and there is little point in either inviting a repeat of the early disasters or managing so conservatively as to avoid all risk.

Again it should be obvious that there is no really clear distinction between the preadaptive, adaptive, and certainty-equivalent phases of resource development. In particular, there is a large element of luck involved in whether early development decisions result in a solid basis for more thoughtful and experimental progress later. There is a large element of creativity (which also involves luck) in the imaginative discovery of untested opportunities and ways to pursue them with reasonable safety. And, finally, there is the danger that strong aversion to further change, or simple complacency, will result too early in the adoption of stabilizing policies that prevent further informative variation. In a sense, it would be reasonable to say that the major role of formal decision analysis all along the way is to help maintain an open and balanced picture of what has been learned versus what remains open to question.

Judging the Credibility of Alternative Models

For the remainder of this chapter we will examine what can be done about assigning odds to alternative models, in situations where analysis has proceeded to the point where various alternatives have been clearly defined. To simplify the discussion a bit, let us concentrate on what can be said at a single decision point in time, on the basis of information available at that time; in Chapter 7 we will turn to the question of how to represent (model) the propagation of uncertainties dynamically over time in relation to management policies viewed as decision sequences.

At any decision point, the analyst has three ingredients to use as the basis for inference:

(1)   a *set of models* M, where each element m_i of this set represents one specific hypothesis about how the managed system responds;

(2)   a *set of "prior probabilities"* $P_0(m_i)$ that would be placed on the alternative models in the absence of any specific data on the system under consideration, on the basis of previous experience with similar systems;

(3)   a set of historical observations **Y** on the system under consideration, where each measurement y_j is drawn from some larger (and usually ill-defined) "database" about the system.

We seek to somehow combine these ingredients so as to make a stronger statement about the *posterior probabilities* $P_t(m_i)$ that should be placed at decision point t on each alternative hypothesis m_i, using **Y** as well as P_0. Keep in mind that we wish to explicitly avoid the old business of finding a single best model or estimator m^*, though it would be natural to define m^* as that hypothesis (if it is unique) with the maximum P_t. In the following discussion I shall use the terms "hypothesis" and "model" interchangeably; instead of thinking of alternative models in terms of alternative equations, think of each alternative model m_i as being *any quantitatively distinct rule* for prediction. Then, for example, model m_1 might be a logistic equation $N_{t+1} = 1.2\,N_t - 0.1\,N_t$, model m_2 might be another logistic equation $N_{t+1} = 1.3\,N_t - 0.1\,N_t$, while model m_3 might be the equation $N_{t+1} = 1.3\,N_t/(1 + N_t/10)$, and so forth.

Readers with some statistical background will recognize that I have just laid out the basic problem of *Bayesian inference*. In our context, Bayes' theorem implies that, in principle, there is a well defined solution to the problem; we should compute $P_t(m_i)$ for each model m_i as

$$P_t(m_i) = \frac{1}{\alpha}\,L(Y\,|\,m_i)\,P_0(m_i) \tag{6.1a}$$

where $L(Y\,|\,m_i)$ is the likelihood (probability) of obtaining Y given that m_i is the correct model, and α is the total probability of getting the data Y (α is simply the sum or integral of the products $L(Y\,|\,m_i)\,P_0(m_i)$ across all models i). Each of the $L \cdot P_0$ products can be viewed as a measure of "relative credibility" for one model. If we call $L(Y\,|\,m_i) \cdot P(m_i)$ just T_i, then (6.1a) can be written more simply as

$$P_t(m_i) = \frac{T_i}{\displaystyle\sum_{j=1}^{N} T_j} \tag{6.1b}$$

Most modern statistics texts have some discussion about Bayes' theorem and equation (6.1a), usually in the context where the model equation structure is fixed and the m_i are generated by varying some unknown parameters continuously [so the sum in (6.1b) is replaced by an integral]. They note that in practice it is usually not a simple matter to compute or represent $P_t(m_i)$ across large sets **M**, and warn that there are some basic theoretical difficulties to be aware of as well. An obvious warning is that it is dangerous to use equation (6.1a) as the basis for "scientific" inference, since the scientist will almost inevitably inject personal biases through arbitrary choices of

$P_0(m_i)$ or by restricting the set $\mathbf{M}$. The decision theorist's answer to this warning is that scientific inference should not be confused with decision making, which *inevitably* involves some subjective weighting of possible outcomes [$P_0(m_i)$ present and used even if not made explicit], even when the decision maker tries to be scientifically objective. More vociferous advocates of Bayesian inference would go further, and argue that even scientists do not (and cannot) avoid various subjective weighting schemes. For our purposes, it is enough to assert that Bayes' theorem provides the logical machinery for including prior information in assessments where appropriate or necessary. Another, perhaps more serious warning, is that the likelihood of Y given any model cannot be assessed without making some strong assumptions about probability distributions of observed outcomes. We shall discuss this difficulty in more detail below.

There is essentially no way to avoid some version of equation (6.1a) if we wish to assign odds to alternative models based on a data set Y. Arguments about prior probability distributions can be avoided by setting all the $P_0(m_i)$ equal (so-called "uniform prior" distribution), and it can usually be shown that the $P_t(m_i)$ are insensitive to details about the assumed data distribution, $L(Y|m_i)$. But it remains necessary to undertake a series of analytical steps that can be quite difficult but exciting. The following subsections outline these steps in more detail.

Establishing the data set Y

The data set Y for a problem will consist of a collection of measurements of variables such as stock sizes, catches, and harvesting effort. It is usually convenient to partition this set into time series of "input" or control variables, such as harvesting effort, and "output" or state response indices, such as catches or catches per effort. The set may contain detailed statistics, such as age composition of catches, and also aggregates or averages of these. Generally, in what follows, we will assume that at least the detailed observations are statistically independent of one another, in the sense that they are gathered as independent measurements. Nonindependence between two statistics usually arises when they are both computed from the same detailed measurements. So, for example, the statistic $Y_t = Z_t - Z_{t-1}$, where Z is a more primitive measurement, is obviously not independent of Y_{t-1}, since both depend on the measurement Z_{t-1}.

It would be foolish to make any pretense that Y comprises *all* the data that have ever been gathered about a system of interest, or even that it represents a complete and unbiased statistical summary of available information. Instead, we must take Y to be an arbitrary set of statistics that has been assembled for dealing with whatever decision problem is at hand from some larger "database." This is an important point: it is seldom practical or necessary to assemble all of the raw data available about a managed system, just as it is impossible to develop any complete model of that system. If you doubt this point, recall the earlier discussion (Chapters 2 and 4) about how managed systems are defined by placing arbitrary boundaries with respect to space, time, and disciplines of concern. Right from the start, data-gathering programs reflect these arbitrary boundaries: we cannot even decide what data to collect without some sort of vague model about what processes or phenomena are important. Various practical constraints, such as funding for sampling and lack of understanding about sampling design and procedures, take a further toll on the set of potentially useful measurements.

One requirement that must be placed on Y is that it does not represent a deliberate selection of data (from some larger set) intended to support a particular model or hypothesis. This seems like an obvious requirement, but it is regularly and blatantly violated by resource analysts; the literature is full of examples where data have been carefully selected or massaged (i.e., by running averages) so as to show good fits to particular models.

In practice, the set Y need not be considered as fixed and immutable at any particular decision point. Preliminary analysis may show that a first set, say Y_1, selected from historical records is inadequate to say much about some models that have been proposed. By examining why this "failure" has occurred, it may become apparent that other existing data should be included in an extended set Y_2, and so forth. The "final" analysis may be based on a set Y that is an aggregation or selection from Y_1, Y_2, provided that Y does not deliberately support any particular model.

Let me inject a final word of caution about Y. Analysts are human beings, and most of us find it impossible to be completely objective about various models. Especially when we have been involved in developing them, or have worked hard to understand them from papers or university lectures, we tend to develop affection for particular formulations. Further, we may unconsciously feel that professional reputation or credibility depends on success in defending one formulation or other. These psychological considerations will almost inevitably creep into the necessarily intuitive step of sifting out firm sets Y from the databases available for most resources. Perhaps the only practical means to assure greater objectivity in this interplay between models and data is by asking several analysts who likely have different biases to work independently on the same database. Standard scientific review procedures are a step in this direction, but they are seldom structured so as to result in a synthesis that objectively embraces uncertainties.

Defining the model set M

We noted in Chapter 4 that dynamic models for resource systems should be thought of as having two basic components: a "state dynamics" submodel for changes in the actual system state, and an "observation" submodel for how observable quantities are related to the actual states. This distinction becomes critical when we wish to determine the credibility of a model in relation to some data set Y. One general way to represent most state dynamics/observation models m_i is by partly dropping the distinction between variables and parameters, so all uncertain quantities are placed in a vector x. The models are then written as

$$\text{State:} \qquad x_{t+1} = f(x_t, u_t) + w_t \qquad\qquad (6.2)$$

$$\text{Observation:} \quad y_t = h(x_t) + v_t$$

where the u_t are controls, the w_t are random "process errors," and the v_t are random "measurement errors." In this formulation, we represent parameters as those x's such that $f = x_t$, i.e., $x_{t+1} = x_t$. "Parameters" that may drift or move in unpredictable ways over time are modeled as random walk processes by allowing $x_{t+1} = x_t + w_t$, where the variance of w_t is chosen to reflect prior belief about how fast the parameter x may change. The error terms need not be assumed additive, though this considerably simplifies some calculations such as likelihood functions. In this representation, Y is the set of observed u_t and y_t values.

It is essential that each model m_i specify probability distributions (usually as density functions) for the w_t and v_t. If these variables are continuous, the usual assumption is that w and v are normally distributed with mean 0 and covariance matrices Σ_w and Σ_v; a simple justification is that the normal distribution minimizes the information content (in a Shannon–Weiner sense) assumed about the distribution (Bard, 1974). Covariance matrices Σ_w and Σ_v can be considered unknown along with some or all of the x's, but, as we shall see below, this can greatly complicate the analysis. A reasonable starting point is to assume that Σ_w and Σ_v are known in advance, and to proceed by assigning conservatively large values to them (so that none of the models m_i is assumed to hold very precisely).

The set of alternative models may be generated by varying the uncertain initial states/parameters x_0, by changing the state and observation functions f and h, and by combinations of these. Thus, the set M can contain infinitely many elements, but this can create serious practical difficulties in computing α and $P_t(m_i)$ in equation (6.1a). The computation is simplified in some special cases where P_t is a known and simple probability density (i.e., P_t is a normal distribution when P_0 and the likelihood function are normal). While any collection of varying x´s, f´s, and h´s can, in principle, be used to form the set M, the set should, in practice, include only those m_i which

(1)   make distinctive predictions over some range of control or state variable values, and

(2)   if correct, imply different optimal management policies.

So, for example, in the analysis of alternative stock–recruitment models (see Figure 1.1), it is not worthwhile to include alternative functions f that behave in the same way over the observed range of data, and also make similar predictions outside that range. Likewise, obviously nonidentifiable (redundant) parameters should be omitted from the formulations. If, say, x is a population size of older animals that is measured annually, one can of course describe the dynamics of x as $x_{t+1} = s_s s_w x_t$, where s_s and s_w are summer and winter survival rates. But the data on x will allow identification of only a single, lumped parameter $\beta = s_s s_w$ since there are infinitely many s_s and s_w values that will give the same prediction of the x_t time series; in Bayesian terms, there are infinitely many combinations s_s and s_w that will have the same posterior density $P_t(s_s, s_w)$, since they all predict the data equally well and thus have equal likelihood $L(x \mid s_s, s_w)$. Examples of "lumped" parameters appropriate for Bayesian analysis were presented in the previous chapter, where they were called R_1 and R_2.

It is often not obvious whether several alternative functions f and h will, in fact, give essentially the same predictions, and which unknown parameters can be lumped without changing the predictions. Thus, the safest tactic is to start with a bit of overkill, by including more alternatives than really seem necessary. Then later examine those alternatives that turn out to have equal P_t and retain from this set only a representative subset whose members imply different best policy choices.

The key point to keep in mind while constructing the model set M is that the objective is to identify alternative hypotheses that *matter in terms of policy choice*. It is easy to "waste" a great deal of time comparing models whose differences (for example, in detail of representation) are certainly scientifically interesting, but that all imply practically the same best management strategy. To avoid this, it is important to keep iterating back and forth between the statistical analysis and the broader decision analysis for which the $P_t(m_i)$ are only one input. To be understandable and therefore useful, the final decision table (see first section) should be as simple as possible; it should contain no more "states of nature" than necessary.

Establishing prior probabilities $P_0(m_i)$

According to Bayesian decision theorists, $P_t(m_i)$ is to be interpreted as a measure of relative credibility for the outcome (model) m_i, rather than as a "frequency of occurrence" in the sense of classical probability theory. If so, there is nothing wrong with injecting purely subjective judgments of credibility, in the form of $P_0(m_i)$, into the calculation of P_t.

Through many hard experiences in situations like AEA workshops, I have come to distrust strongly the subjective judgments made by most resource analysts (including myself). Too often they are based not on real physical constraints (i.e., survivals must be less than 1.0) or past experience with other systems, but instead on accumulated folklore (wishful thinking) and earlier application of inappropriate estimation methods to Y itself. Assignment of P_0 based on the data themselves is, of course, a nasty version of circular reasoning. In view of these difficulties, I advise always starting with a uniform prior (all P's equal) and concentrate on making the range of models included in the analysis wide enough to include some alternatives that may initially seem physically infeasible. Then apparent inconsistencies, such as high probabilities assigned to survival rates greater than 1.0, may be revealed and help to identify errors in model structure and biases due to difficulties with the data.

For some ecological parameters it is possible to construct reasonable prior distributions on the basis of a large experience with similar systems. This is the case, for example, with stock–recruitment parameters for Pacific salmon, where many populations have been studied for several decades. Mortality rates for most fish can at least be bounded from simple data on growth, using the extensive summaries by Pauly (1979). There are clear bounds for reproduction parameters in most mammal species. However, again let me emphasize that such bounds must be used with great caution; animals can perform in surprising ways in populations disturbed through harvesting, and the data may contain effects of hidden (unmodeled) processes, such as immigration from outside the system defined for analysis.

In analyses of how uncertainties propagate over time, it is often more convenient to assume a normal rather than a uniform prior distribution for uncertain parameters. There is no practical difference between these assumptions provided the variances for the normal prior distributions are made very large, except that known physical bounds must be ignored. When computed posterior distributions place high probabilities on values outside such bounds, the analyst's response, as noted above, should be to reexamine the model structures and data used (rather than worrying about the very formidable computational difficulties of estimating P_t for a mixture of normal and bounded uniform distributions).

Computing the likelihood of **Y** given a model m_i

In principle, the likelihood function for a data set Y given a particular model (all parameters fixed) is simple enough to define; it is the probability of obtaining Y given that the model is correct. If all the elements y_i of Y are independent of one another, and if the probability of getting y_j given m_i is $d_j(y_j|m_i)$, then the likelihood is just the product of the probabilities of the individual "events" y_j:

$$L(Y|m_i) = \prod_{j=1}^{n} d_j(y_j|m_i) \tag{6.3}$$

if there are n observations. Note that there is no presumption here that all the observations are drawn from the same distribution d, or even that the observations have the same form of distribution. In this section, we shall assume that Y has been constructed so that all the y_j are independent of one another (see above), so that equation (6.3) can be used as a basic building block for L. The problem then becomes to define the set of distributions $d_j(y_j|m_i)$ for the individual observations.

Generally, it is easier to describe uncertainties about dynamic behavior and observations for a given model m_i in terms of probability distributions for deviations like w_t and v_t in equation (6.2), rather than directly in terms of distributions for y_j. This means that to find the distributions $d_j(y_j|m_i)$, we must make a change of variables from the assumed distributions for variables w and v. This can be tricky, especially when the deviations are not assumed to be additive. In general, if the model is written as $y = h(x, z)$, where z is a random variable that we assume to be distributed as $p(z) = g(z)\,dz$, then in order to find $P(y)$ we must be able to solve for z as a function h^{-1} of y so as to give $P(y) = g[h^{-1}(y)]\,|(\partial y/\partial z)|$. For the special case $y = h(x) + z$, this rule reduces to $P(y) = g[y - h(x)]\,dy$. In the following brief discussion, we will examine only the additive deviations case, i.e., models in the form of equations (6.2).

In Chapter 5 we noted that parameter estimation for dynamic models should be done differently depending on whether we assume only process errors [$\Sigma_v = 0$ in equation (6.2)] or only observation errors ($\Sigma_w = 0$). The mixed, or "errors-in-variables," situation is considerably more difficult to handle, and we must usually assume that Σ_v and/or Σ_w are known *a priori* in order for unique estimates to exist at all. The construction of a likelihood function is most easily discussed in terms of these three situations.

(1) *Assuming only process deviations* w_t. In this situation we assume that the observations y_j are exactly equal to some functions $h_j(x_t)$ of states x, and that we know the probability distributions $d_t(w_t)$ *a priori*. Equation (6.2) then reduces to

$$x_{t+1} = f(x_t, u_t) + w_t$$

$$y_t = h(x_t)$$

Two cases are possible, depending on whether or not we can uniquely "reconstruct" the states x_t by solving h for x_t given y $[x_t = h^{-1}(y_t)]$. For example, if the state x_t is a scalar population size and y_t is an abundance index such that $y_t = qx_t$, then for any particular hypothesis about q (i.e., given q) we can reconstruct $\hat{x}_t = y_t/q$. Here, $h^{-1} = (1/q) y_t$.

In the case where x_t can be uniquely reconstructed from y_t, we first do so and then substitute the estimates $\hat{x}_t$ back into equation (6.2) so as to solve for $\hat{w}_t = \hat{x}_{t+1} - f(\hat{x}_t, u_t)$. The likelihood for the data set is then simply the product

$$\prod_{t=1}^{T-1} d_t(\hat{w}_t) J_t$$

if there are T years' data. Here J_t is the absolute value of the determinant of the matrix $\{\partial h_j^{-1}/\partial y_i\}$ evaluated at time t (a way of defining "uniquely reconstructible" is that this determinant be different from zero).

If x_t cannot be uniquely reconstructed from Y, given the parameters of h, the likelihood function for Y becomes much more difficult to calculate since Y, given x, does not uniquely determine the set of deviations w_t. To proceed, we must in essence first solve a "partial realization problem" (Kalman et al., 1969) of finding the set of all values x_t^* that are consistent with Y. We then find the w_t^* associated with each feasible combination (x_{t-1}^*, x_t^*), and integrate probabilities of w^* across all combinations to find the marginal distribution for each y_t. I am not aware of any practical example involving nonlinear f and h functions where these formidable steps have been accomplished fully.

(2) *Assuming only observation deviations* v_t. This is the situation usually treated in textbook discussions on fitting dynamic models to time series data. Equation (6.2) reduces to

$$x_{t+1} = f(x_t, u_t)$$

$$y_{t+1} = h(x_t) + v_t$$

Notice that since the state dynamics are assumed to be deterministic, every predicted x_t can be computed if we are willing to assign a value to x_0 (the initial state). It must either be known, or included among the set of unknowns that are varied to generate the model set M. Then the likelihood computation is quite easy. We simply simulate the trajectory x_t for m_i and calculate the predicted observations $h(x_t)$. The deviations v_t for which we assume probability distributions, say $d_t(v_t)$, are calculated as $v_t = y_t - h(x_t)$.

The likelihood of Y is just

$$L(Y|m_i) = \prod_{t=1}^{T} d_t(v_t)$$

If the reader is confused at this point, recall that in order to specify m_i, we must specify x_0 exactly, and the values (u_t, y_t) are part of the set Y. Without all of these specifications, we cannot do a simulation and compute the v_t values from which we compute $L(Y|m_i)$.

(3) *Assuming mixed process and observation errors* w_t *and* v_t. In this situation, each model m_i is specified partly by a particular parameter combination x in equation (6.2), and we must in addition assign some prior probability distribution, say $m_0(x_0)$ to the initial system state. We begin by noting that each observation y_t is the sum of two independent random variables: (1) $h(x_t)$ which is random because of w_t, and (2) v_t. Thus, if we denote the density function for x_t as $m_t(x_t)$ and the density function for v_t as $d_t(v_t)$, and note that $v_t = y_t - h(x_t)$, it is easy enough to write the marginal distribution for y_t, by integrating across the set X_t of possible values of x_t:

$$P(y_t) = \int_{X_t} m_t(x_t)\, d[y_t - h(x_t)]\, dX_t \tag{6.4a}$$

To solve for such elements of the likelihood function $(L = \Pi_t\, P_t)$, we must also find $m_t(x_t)$. This can be done recursively given the initial density $m_0(x_0)$, which is analogous to assuming x_0 known in the observation error case, as

$$m_t(x_t) = \int_{X_{t-1}} m_{t-1}(x_{t-1})$$

$$\tag{6.4b}$$

$$\times\, g[x_t - f(x_{t-1}, u_{t-1})]\, dX_{t-1}$$

where g is the density function of $w_t = x_t - f(x_{t-1}, u_{t-1})$. Equations (6.4a,b) are obviously not simple to solve for most nonlinear functions f and h; in general, numerical integration is required, and the density function $m_t(x_t)$ must be approximated either numerically on a grid (impractical if x has many dimensions) or by some simple density function.

To avoid the integrations in (6.4a,b) it is tempting to use the approximation

$$P_t(y_t) \approx d[y_t - h(\tilde{x}_t)]$$

where the estimate $\tilde{x}_t = f'(\tilde{x}_{t-1}, u_{t-1})$ is computed using a model f' that accounts for the average effects of process errors. In other words, use the pure observation error procedure [situation (2) above], but with a modified deterministic model f'. [As we shall see by example in the next section, it is *not* good to use $\hat{x}_t = f(\hat{x}_{t-1})$ when process errors are actually present.] In principle, the integration in equation (6.4a) can always be replaced exactly by $g[y_t - h(x_t^*)]$ for some value x^*; the question is whether x_t^* can be computed as some "nominal trajectory" or state reconstruction $\tilde{x}, \ldots, \tilde{x}_T$ from the stochastic dynamics $f(x) + w$. There are as yet no clean theoretical answers to this question, and little numerical experience using obvious state estimation procedures such as the extended Kalman filter (see Chapter 7) to obtain $\hat{x}_T$.

The one thing we know for sure about mixed-error situations involving dynamic production models is that you should not try to approximate the likelihood function by assuming only process errors [assumption (1) above]. The result can be a badly biased posterior distribution $P_t(\beta)$, favoring parameter values β that imply the stock is much more productive at low stock sizes than it actually is (see Chapter 4; Uhler, 1979; Walters and Ludwig, 1981). Most of the fisheries literature on estimation methods for production models makes exactly this mistake, by assuming that abundance indices, such as catches per effort, can be substituted directly back into the production models, using (stock) $= h^{-1}$ (index) $= y_t/q$. In the next section we shall see just how bad the biases can be, using a very simple example.

Computing the posterior distribution $P_t(m_i)$

Unless you can find an exceptionally patient and clever mathematician to help with the calculus required to find analytical expressions for α and $L(Y|m_i)$ in equation (6.1a), the practical approach to finding $P_t(m_i)$ for nonlinear f and h models involves using the deceptively simple formula (6.1b). For each model m_i, you just calculate $T_i = L(Y|m_i) P_0(m_i)$, then add these up and divide each by the total. But we noted in the last subsection that the computation of each $L(Y|m_i)$ involves at least one simulation of the period of historical record, and some side calculations involving the probability density functions of the deviations. Thus, it is not practical to compute $L(Y|m_i)$ for a very large number of model equations, or very many parameter values for a single-equation system. When the models m_i are generated by changing values of some parameter vector β, there are difficulties in even visualizing how P_t varies with the parameter values if there are more than two or three uncertain parameters. We address some approaches to dealing with complex (many-parameter/many-equation) models in the next section; for now, let us examine what can be done by "brute force" with simple models, using logistic population growth as an example.

Let us suppose that nature has generated a population dynamics that follows the equation $x_{t+1} = x_t + 0.1 x_t - 0.1 x_t^2 - u_t + w_t$, where x_t is population size ($x_0 = 1$), u_t is a time-varying harvest (assumed known exactly), and w_t is a normally distributed environmental effect (process error) with mean zero and variance σ_w^2. Let us assume that this population has been estimated each year $t = 1, \ldots, T$ by a procedure that gives $y_t = x_t + v_t$, where v_t is a normally distributed observation error with mean zero and variance σ_v^2. Let me emphasize that situations this simple are uncommon at best, but we need to have a known and transparent starting point against which to see what can go wrong in more realistic cases.

Now suppose we are confronted with a data set $Y = \{u_1, \ldots, u_T, y_1, \ldots, y_T\}$ from this population, and that the Almighty has told us the correct model structure ($f = x_t + \beta_1 x_t - \beta_2 x_t^2 - u_t + w_t$, $h = x_t + v_t$), but not the parameters β_1, β_2, σ_v, and σ_w. Suppose we do know that β_1 and β_2 should lie between 0 and 0.3, so the set of models m_i should involve combinations (β_1, β_2) from this range. We might then take M to be all combinations of (β_1, β_2) values between 0 and 0.3, with each parameter varying in steps of 0.025 ($\beta_1 = 0, 0.025, 0.05$, etc.; $\beta_2 = 0, 0.025, 0.05$, etc.). Then we can easily visualize $P_t(\beta_1, \beta_2 | Y)$ as the vertical dimension in a 3-D plot with β_1 and β_2 as the horizontal axes. Likelihood functions for the process and measurement error extreme situations are easily seen to be of the form

$$L(Y | \beta_1, \beta_2) = K e^{-\Upsilon/2\sigma^2} \tag{6.5}$$

where K is the same for all (β_1, β_2) for both situations, and $\Upsilon = \Sigma_t E_t^2$. For the process error only assumption, the prediction errors E_t are calculated as

$$E_t = y_{t+1} - y_t - \beta_1 y_t + \beta_2 y_t^2 + u_t$$

$$t = 1, \ldots, T - 1$$

(i.e., E_t is the error in predicting y_{t+1} using β and y_t). For the measurement error only assumption, $E_t = y_t - \hat{x}_t$, where $\hat{x}_0 = x_0 = 1.0$ and $\hat{x}_t = \hat{x}_{t-1} + \beta_1 \hat{x}_{t-1} - \beta_2 \hat{x}_{t-1}^2 - u_t$ (i.e., the simulated population sequence). If we assume a uniform prior distribution for the β's over 0-0.3, the posterior probabilities are given by calculating L using equation (6.5) at each (β_1, β_2) grid point, summing the L's across grid points, and then dividing each L by the sum over all grid points tested.

Figure 6.1 shows some numerical results obtained with an Apple microcomputer using a harvest sequence u_t that increased for a few steps to drive x_t down to around 0.5, then were chosen so as to hold x_t near 0.5. $P_{10}(\beta | Y)$ forms a long hill, and I roughly sketched the contours of this hill for several situations. Figure 6.1 is intended to show what you would conclude about P_t by assuming either all process or all measurement error, when the data are in fact generated from other assumptions. The first thing to notice is that P_t looks like a bivariate normal distribution in all cases, with a high correlation between the two "random variables" β_1 and β_2. Also

Assumed errors

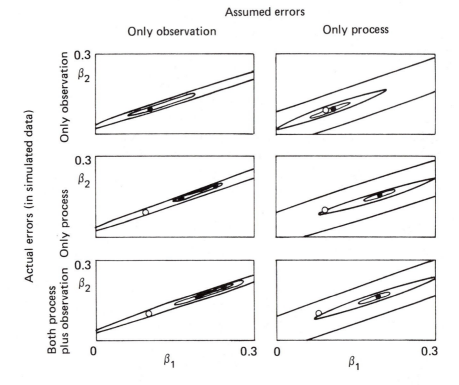

Figure 6.1. Contours of equal posterior probability for various parameter combinations β_1, β_2 in a logistic model, using 10 years' simulated data. Actual error pattern in data shown on left; assumed error (in probability calculation) shown above the panels. o marks the true parameter values used to generate the data: ● marks the parameter estimate with highest posterior quality.

notice that the distributions do not peak up sharply at a single point after 10 years, but instead just form a narrower ridge. This is because harvesting holds the population almost steady around 0.5 after year 5, so that, roughly speaking, only the linear combination $0.5 = 0.5 + \beta_1 0.5 - \beta_2 0.5^2 - u_t + w_t$ is observed after this time; the ridge of P_t values shows combinations of β_1 and β_2 that satisfy this "constraint equation." Notice that the ridge is correctly placed for all combinations of assumed and actual error structure. Finally, notice that the distributions have peaks (maximum likelihood estimates of β_1, β_2) that are badly biased upward when the wrong error structure is assumed, even though the deviation variances used were quite small by ecological standards. That is a major lesson to be learned from this example: be careful to test alternative assumptions about error structure when attempting to measure the relative credibility of alternative hypotheses.

Compressed Representations of Uncertainty

As the logistic growth example of the previous section demonstrates, it is not a trivial matter to evaluate posterior probability distributions even for parameters of simple models. This section discusses possible ways to get compressed, understandable representations of uncertainty in situations where the alternative models are complex and/or have too many unknown parameters to allow brute force representation with a grid of parameter combinations. The approach suggested here involves three steps. First, find an approximate representation of P_t as a function of the unknown parameters $\boldsymbol{\beta}$, generally as a normal distribution centered on the maximum likelihood (or maximum a *posteriori* density) estimates of $\boldsymbol{\beta}$. Second, use the covariance matrix of this distribution to guide the construction of a reduced parameter set with elements that are not well determined by Y. Third, from the models defined by this reduced parameter set, extract a few representative examples that highlight qualitative and quantitative differences in predictions.

As a preview to these steps, consider Figure 6.1. Here it is obvious that $P_t(\beta_1, \beta_2)$ can be well approximated by a normal distribution, even when only a few observations are available. Further, we could find the mean and approximate covariance matrix of this distribution quickly by standard estimation methods, without searching the entire grid. The covariance matrix would clearly show the β_1, β_2 correlation (high probability ridge) and our second step would be to express one or another parameter as a function of the other. Ridges of high probability in this example define a straight line, $\beta_2 = k_1 + k_2 \beta_1$; treating β_1 as the remaining unknown, the state dynamic model becomes $x_{t+1} = x_t + \beta_1 x_t - (k_1 + k_2 \beta_1) x_t^2 - u_t + w_t$, which has only one unknown (β_1) provided we accept that k_1 and k_2 are well defined by the available data Y. Finally, from this reduced model we select reasonably probable cases, such as $\beta_1 = 0.05$, $\beta_1 = 0.1$, $\beta_1 = 0.2$ that are consistent with Y but imply quite different sustainable harvest rates.

Approximation of the *a posteriori* density function

Suppose that instead of calculating $P_t(\boldsymbol{\beta} \mid Y)$ at many $\boldsymbol{\beta}$ points, we instead just try to find the maximum value P^* or at least some point along

the ridge of highest probability. This is a standard problem in statistical estimation, and various techniques to make the search for P^* more efficient are
well reviewed in Bard (1974). Usually it is easiest to search for the maximum of ln P_t. The maximum of P_t occurs at the maximum likelihood estimates $\hat{\beta}$ if we assume a uniform prior $P_0(\beta)$. Considering that we will be
using the estimation results to help in model compression rather than to
define a single "best" model, note that we need not be concerned about
whether the final estimates $\hat{\beta}$ are even unique; to find a ridge of equally
probable combinations is enough to help guide the search for a compressed
model structure.

Provided the observation set Y is not too small (5–10 observations per
state variable x_t), ln P_t will generally behave as a quadratic function of β
near any peak or ridge point $\hat{\beta}$. This means that P_t will be shaped like a normal distribution near $\hat{\beta}$, with covariance matrix $\Sigma_\beta = -(H^*)^{-1}$, where $H^*_{ij} =$
$\partial\varphi/\partial\beta_i\,\partial\beta_j$ and $\varphi = -\ln[P_t(\beta)]$. For details of this argument and approximate methods for calculating Σ_β, see Bard (1974) and Appendix 5A. The
first things to look for in Σ_β are strong correlations π_{ij}, where $\pi_{ij} =$
$\Sigma_{ij}/\sqrt{\Sigma_{ii}\Sigma_{jj}}$. High correlations ($> 0.9$) indicate that points along linear
combinations $\beta_i = k_1 + k_2\beta_j$ have nearly equal probability P_t. It is not particularly critical that Σ_β be calculated accurately, since the informative correlations π_{ij} stand out even in the crudest approximations; the so-called
"asymptotic covariance matrix" estimated from $\partial\hat{y}_t/\partial\beta_i$ (see Appendix 5A) is
adequate for the purpose.

Given estimates $\hat{\beta}$ and Σ_β, it is a good idea to calculate $L(Y|\beta)\,P_0(\beta)$
for a variety of β values near $\hat{\beta}$ (preferably on a grid), to confirm the
existence and narrowness of P_t ridges along the dimensions β_i and β_j that
show high correlations π_{ij}. Here it is especially valuable to have an interactive computer program that calculates $L \cdot P_0$ for any β you input. Such programs help you determine quickly whether the search for $\hat{\beta}$ has been successful, and whether the Σ_β being used is really giving a good approximation to
the shape of P_t.

Identification of reduced parameter sets

Here we seek to reduce the dimension of β by finding combinations
(functions) of parameter values that are well determined by the existing data
Y. One formal procedure that can be applied here is *principal components
analysis* of Σ_β. The principal components of Σ_β are the new parameters γ
defined by the transformation $\gamma = D\beta$, where D is the matrix of normalized
eigenvectors of Σ_β. The parameters γ are statistically uncorrelated and have
variances $\sigma^2_\gamma = \pi_i$, where π_i is the ith eigenvalue of Σ_β (see Bard, 1974, p
183). In seeking reduced parameter sets, we seek those principal

components γ_i with lowest variance π_i. These components define linear equations

$$\gamma_i = \sum_j D_{ij} \beta_j$$

from which we can solve for some of the β's in terms of others, leaving only those others as generators of alternative hypotheses that are consistent with Y. When the data define a ridge of equally probable combinations β_i, β_j, standard parameter estimation procedures fail and will not provide estimates of Σ_β. These procedures work by approximating Σ_β^{-1}, usually (as noted in Appendix 5A) by:

$$\Sigma_{ij}^{-1} = \frac{1}{s^2} \sum_{t=1}^{T} \left(\frac{\partial \hat{y}_t}{\partial \beta_i} \right) \left(\frac{\partial \hat{y}_t}{\partial \beta_j} \right)$$

or related information measures. In such cases, the required π_i can still be found, since $\pi_i = 1/\lambda_i$, where λ_i is the ith eigenvalue of Σ_β^{-1}; matrix inversion fails when some $\lambda_i = 0$, implying that combination γ_i has infinite variance (points β defined by such γ_i are points along the ridge).

However, principal components analysis as a mechanical procedure does not give much insight about why certain parameter combinations are well determined by the existing data Y, nor is the linear transformation $\gamma = D\beta$ always an appropriate functional form for estimating some parameters from others. Thus, the principal components, along with parameter correlations π_{ij}, should be used at first only as a guide in looking back at the model structure and data to see which functions of the parameters have actually been well defined by experience. There are three common conditions to look for as starting points in this examination:

(1) *Observations giving only a narrow range of states.* Such situations are typical in the early development of resources, or where there has been a strong policy to stabilize the system. In this case Y defines only the average response, i.e., mean harvest $\bar{u}$ in the logistic model of the previous section, as a function of the average state $\bar{x}$. Then we have "seen" only $\bar{x} = f(\bar{x}, \bar{u}, \beta)$, $\bar{y} = h(\bar{x}, \beta)$, and it is obvious how to solve for some of the β's given others while taking $\bar{y}$, $\bar{u}$ as determined exactly (see example in previous section).

(2) *Masking of a relationship due to errors-in-variables effects.* A good example of this condition arises in the estimation of natural and harvesting mortality rates from age composition data. When gathered over many years, such data permit accurate assessment of the average total mortality rate $\bar{z}$, and the average harvesting effort $\bar{E}$. It is usually assumed (see Paloheimo, 1980) that $z_t = M + qE_t$, where M is a natural mortality rate

parameter and q is the "catchability" parameter (see Chapter 4). But the effective effort E_t is usually measured with considerable error (equivalently, q_t is time-varying around an average q to be determined), and this destroys the z versus E correlation. The data then determine only the linear combination $\bar{z} = M + q\bar{E}$ without large bias. One can calculate q given M [$q = (\bar{z} - M)/\bar{E}$, or by cohort analysis], and M becomes the hypothesis-generating parameter.

(3) *System remaining near equilibrium with respect to inputs.* Sometimes a wide range of states is sampled, but this is done by varying input controls (i.e., harvest rates) slowly so that the state is always near a time-varying equilibrium set by the controls. This condition has been actively promoted by analysts seeking to estimate parameters of surplus production models. For example, suppose the harvest is modeled as $H_t = qE_t x_t$, where q is the catchability parameter, E is effort, and x_t is stock size. Suppose the dynamics of x are modeled by $x_{t+1} = x_t + R_1 x_t - R_2 x_t^2 - H_t$. Then, provided E_t changes slowly, the model will predict $x_{t+1} \approx x_t$ and the analyst will see only $H = R_1 x - R_2 x^2 = qEx$. If $y_t = H_t/E_t = qx_t$ is taken as an abundance index, we will see only the relationship $E = (R_1/q) - (R_2/q^2) y$, which can be used to establish equilibrium yields in relation to E, but not dynamic transients. If the parameters $\alpha_1 = R_1/q$ and $\alpha_2 = R_2/q^2$ are determined precisely, then we may pick q as the hypothesis-generating unknown (i.e., total stock size unknown), while assuming that $R_1 = q\alpha_1$ and $R_2 = q^2\alpha_2$.

Let us generalize a little from these conditions and examples. For any data set Y, we can always compute a smaller set of statistics Z (mean values, covariances, etc.) that are well determined if Y is reasonably large. The model functions f and h predict how the elements of Z should vary as a function of the parameters β. Thus Z, along with f and h, establish a series of constraints on the values of β that are consistent with Y. When Z does not imply β uniquely (the usual case), the constraints are used to solve for some β's in terms of a reduced set α of hypothesis-generating unknowns. While the choice of α is in principle arbitrary, there are usually "natural" choices (like M and q in the examples above) that simplify the algebra and make the results simpler to explain to scientists and policymakers.

Using singular value decomposition to find reduced parameter sets

The main reason for introducing principal components analysis in the previous section is that it is a familiar statistical tool to most biologists. A preferable procedure for calculating which combinations of parameter values

β are equally consistent with the data set Y is to find the "singular value decomposition" of the Jacobian matrix X, where $X_{ij} = \partial \hat{y}_i / \partial \beta_j$ evaluated at the "best" parameter estimate $\hat{\beta}$ (which may be a point on a ridge of equally likely estimates). The singular value decomposition breaks X into three matrices

$$X = U S V'$$

where U and V are orthogonal (i.e., $V V' = I$) and S is the diagonal matrix of "singular values" σ_i ($S_{ii} = \sigma_i$). Efficient routines for computing U, S, and V are generally available; for small problems (< 10 parameters) they do not take much more effort to compute than the $X'X$ and $(X'X)^{-1}$ matrices needed in parameter estimation.

To find equally likely parameter combinations, we first form the orthogonal (uncorrelated) deviations $z = V'b$, where $b_i = \beta_i - \hat{\beta}_i$. Confidence limits for the z_i are computed as $z_i \pm \sqrt{d/\sigma_i}$, where $d = ns^2 F_{\alpha, m-n}$ [n = number of parameters, m = number of observations, F_α = Fisher's F-statistic for probability level α, $s^2 = \Sigma (y - \hat{y})^2 / (m - n)$ is the residual error variance]. For two-parameter problems, confidence regions can be plotted by rotating an angle Θ from 0 to 2π and calculating $z_1 = \sqrt{d/\sigma_1} \cos \Theta$, $z_2 = \sqrt{d/\sigma_2} \sin \Theta$ at each angle. Note that a zero singular value ($\sigma_i = 0$) implies that z_i can be made arbitrarily large without leaving the confidence region; i.e., the data do not set any limits at all on z_i for which $\sigma_i = 0$. Next, we transform back to find $b = \beta - \hat{\beta}$, by the matrix operation $b = V z$. For two-parameter problems, finding $V z$ means rotating the confidence ellipse. A graphical example of the confidence ellipse is shown in Figure 6.2.

Generally, we can write the deviations b as

$$b = z_1 V_1 + z_2 V_2 + \cdots + z_n V_n$$

where V_i is the ith column of the V matrix. For any z_i that is very poorly determined (σ_i zero or very small), we can increase its value greatly (and thus generate $b = z_i V_i$) without leaving the confidence region for β. Thus, a practical prescription is to (1) find the smallest σ_i, and (2) generate parameter combinations β that are equally (or almost equally) likely as $\beta = \hat{\beta} + z_i V_i$ by increasing z_i.

When singular value decomposition is used in parameter estimation, a simple parameter correction scheme for nonlinear iteration j is given by

$$\beta^{(j)} = \beta^{(j-1)} - \lambda V S^* U' (\hat{y} - y)$$

where $0 \le \lambda \le 1$ is a step size correction and V, S^*, U', and the deviations $\hat{y} - y$ are calculated using $\beta^{(j-1)}$. S^* is a diagonal matrix with $S_{ii}^* = 1/\sigma_i$ if $\sigma_i \ne 0$, and $S_{ii}^* = 0$ if $\sigma_i = 0$. Note that this iterative scheme will simply not move once it reaches a ridge of equally likely parameter combinations where

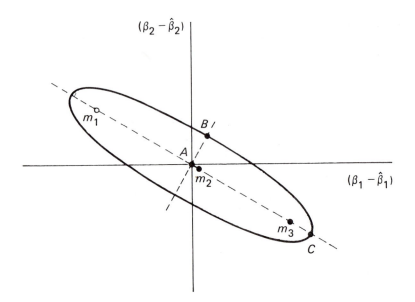

Figure 6.2. Confidence regions for uncertain parameters can be calculated by singular value decomposition of the regression design or Jacobian matrix $\mathbf{X} = \{\partial \hat{y}_i / \partial \beta_j\}$. Using symbols as defined in the text, the length of line AB is given by $\sqrt{d/\sigma_i}$, and points along it are found by varying z in the scalar–vector product $z \, \mathbf{V}_1$. The length of line AC is $\sqrt{d/\sigma_2}$, and points along it are given by $z \, \mathbf{V}_2$. The parameter combinations m_1, m_2, and m_3 define a reduced model set for use in decision analyses.

some $\sigma_i \approx 0$ and values $\hat{\beta}^{(j+1)} = \hat{\beta}^{(j)} + z_i \mathbf{V}_i$ would represent moves along the ridge.

Consider the following simple example. Suppose we wish to find out as much as possible about β_1, β_2, β_3 in the "experimental design model" $y_i = \beta_1 + \beta_2 X_{i1} + \beta_3 X_{i2}$, where $X_{ij} = 0$ or 1, depending on whether treatment j was present or absent. Suppose no "control" observations were made with both treatments absent, and that there are two observations for each treatment level. Letting the first two observations be for treatment 1 present, and noting that $\partial \hat{y}_i / \partial \beta_1 = 1$ for all observations, we get

$$\mathbf{X} = \begin{bmatrix} 1 & 1 & 0 \\ 1 & 1 & 0 \\ 1 & 0 & 1 \\ 1 & 0 & 1 \end{bmatrix}$$

The singular value decomposition of this matrix gives

$$U = \begin{bmatrix} 0.5 & -0.5 & 0 \\ 0.5 & -0.5 & 0 \\ 0.5 & 0 & 0 \\ 0.5 & 0 & 0 \end{bmatrix}$$

$$S = \begin{bmatrix} \sigma_1 & 0 & 0 \\ 0 & \sigma_2 & 0 \\ 0 & 0 & \sigma_3 \end{bmatrix} \qquad \text{where} \qquad \begin{aligned} \sigma_1 &= 2.44949, \\ \sigma_2 &= 1.41421, \\ \sigma_3 &= 0 \end{aligned}$$

$$V' = \begin{bmatrix} 0.8165 & 0.4082 & 0.4082 \\ 0 & -0.7071 & 0.7071 \\ 0.5773 & -0.5773 & -0.5773 \end{bmatrix}$$

The fact that $\sigma_3 = 0$ implies that not all the parameters can be estimated; we can add any multiple of the last column of V to the estimates, without changing the predictions at all. Suppose we have the data

$$y = \begin{bmatrix} 0.9 \\ 1.1 \\ 1.9 \\ 2.1 \end{bmatrix}$$

and we take an initial estimate

$$\beta^{(0)} = \begin{bmatrix} 0 \\ 0 \\ 0 \end{bmatrix}$$

Then the iterative scheme above will converge in one iteration if $\lambda = 1$, since the model is linear in its parameters (X independent of β). The scheme gives

$$\hat{\beta} = \begin{bmatrix} 1.0 \\ 0 \\ 1.0 \end{bmatrix}$$

Further, we know that any estimate

$$\beta^* = \hat{\beta} + z_3 V_3$$

$$= \begin{bmatrix} 1.0 \\ 0 \\ 1.0 \end{bmatrix} + z_3 \begin{bmatrix} 0.5773 \\ -0.5773 \\ -0.5773 \end{bmatrix}$$

will fit the data equally well. Intuitively, the interpretation of this example is that the absolute values of β_2 and β_3 cannot be determined uniquely, since no observations were made with both treatments absent (i.e., of the β_1 effect only). We can only determine the effects of β_2 and β_3 relative to the background mean β_1; provided we subtract the same amount (for example, 0.5773) from β_2 and β_3 as we add to β_1, the resulting model will predict the original data exactly as well as the nominal model $y_i = 1 + (0)\,X_{i1} + (1)\,X_{i2}$.

Definition of representative models

In the end we seek to identify a set of models $\boldsymbol{M}$ that can be used in decision analysis. To review, this means that we want to be able to compute the expected value V_j of any policy u_j from a set of choices U, as

$$V_j = \sum_i P_t(m_i)\, V(u_j | m_i)$$

where $V(u_j | m_i)$ is the performance expected if policy u_j is used and the system turns out to respond as predicted by m_i. Then we seek to provide a ranking of the u_j according to expected performances V_j. The final analysis should contain no more models m_i than are really necessary to do this ranking. The obvious implication of this argument is that the reduced model set should be defined by looking back at the decision choices U available, using some iterative procedure. By recomputing V_j using progressively "coarser" summations across fewer models m_i, it can often be shown that the ordering of V_j values remains stable until the decision table contains far fewer models than policy choices. When there is a continuous spectrum of policy choices (as generated by "policy variables" like fishing efforts or quotas) and models generated by varying some parameters β, the iterative procedure may be replaced by a search for some smooth (and simple) relationship between the optimum choice and the parameters of a probability distribution used to approximate $P_t(\beta)$. However, as we shall see in Chapter 9, the optimal choice depends on Σ_β as well as $\hat{\beta}$ when that choice affects Σ_β (so-called dual effect of control), and this relationship is very difficult to estimate.

I have found that practically any portrayal of uncertainty in terms of a few alternative models or hypotheses that are consistent with past data (for example, m_1–m_3 in Figure 6.2) will stimulate some fruitful discussions about new policy choices to include in the formal decision analysis. Often choices are identified that force a basic rethinking about what models m_i and data Y should be included. The analyst must be careful not to inhibit this discussion by initially presenting a numbing barrage of quantitative analyses about precisely what policy would be best in relation to the original definition of choices. This leads to a simple rule for defining the reduced model set:

initially, present the best guess ($\hat{\beta}$) and two extremes (low and high) with assessments of P_t for these choices. Then "fill in the gaps" between these extremes later, after adequate discussion is allowed to really clarify the policy choices.

Too Many Assumptions?

In this chapter I have deliberately avoided discussion of some traditional statistical methods for measuring uncertainty, except as they relate to Bayesian inference. I have not, for example, even mentioned topics like least squares estimation and residuals analysis. If you look back through the chapter, you will see that the word "assume" appears far too many times for comfort, and that the construction of probabilities for alternative models requires far more acts of faith than any sane scientist would dare make. Let me close with a simple reminder about why it is defensible to construct such houses of cards: management decision making and science are not the same thing. We cannot avoid making decisions, and in some way or other these will be based on the data available. The decision theorist's argument, with which I wholeheartedly agree, is that *any* inference from data to decision must be based somehow on prior beliefs, likelihoods, and models; to proceed without admitting and articulating these is the really unscientific thing to do. The issue is not whether one should adopt a Bayesian approach, but rather how to deal in a practical and efficient way with the formidable technical (computational) difficulties that it entails without restricting analysis to the most simplistic of model and policy choices.

In Search of Optimum Model Complexity

To readers with experience in renewable resource modeling, it may seem that I have chosen in this and the previous chapter to emphasize overly simplistic models (such as the logistic) as examples. There is more behind this choice than a desire to illustrate basic concepts clearly; in many practical situations it is unwise or even deceptive to try and penetrate more deeply into the biological and economic details, though we know full well that these details may be important in some sense. Two things go wrong as we try to articulate more and more detailed models; it becomes more difficult to specify functionally how the detailed components interact with one another, and each additional model parameter becomes less well specified (flatter probability distribution, wider confidence limits) by the available historical data set Y. It is a reductionist myth that models become monotonically better as the details of natural processes are more fully articulated. Practical problems of structural specification and estimation imply that there is some balance point or "optimum complexity" *beyond which model performance will actually deteriorate.* Indeed, much of the art in model building consists of being able to recognize, usually by trial and error, when a reasonable balance has been achieved. This is a key point: the "best" model complexity is a function of the skill and experience *of the model builder,* as well as of the "objective" data Y and functional understanding available in the disciplines of concern.

Costanza and Sklar (1983) have developed an example of how "descriptive accuracy" (ability to fit historical data) is likely to vary with model complexity (Figure 6.3). They reviewed 87 mathematical models for freshwater wetlands (swamps, marshes, etc.) and shallow water bodies. For about half of these they were able to measure descriptive accuracy quantitatively in terms of the correlations between model predictions and historical data. They measured model complexity in terms of an "articulation index" based on the number of model components, temporal resolution of predictions (number of time steps in simulation, etc.), and spatial resolution (number of spatial areas modeled). They took a product of their accuracy and articulation indices as a rough measure of overall model "effectiveness." They found that effectiveness defined this way was a dome-shaped function of the model complexity: it is possible to predict a few things very well, or a lot of things very poorly; between these extremes there is a balance point where reasonable accuracy is maintained without too much loss of biological "realism."

The concept of optimal complexity can be made somewhat more precise for management situations when it is possible to define (1) a fixed data set Y from which model parameters are to be estimated; (2) a series of alternative models f_1, f_2, ... , where the models are ordered in terms of the number of unknown parameters to be estimated from Y (f_1 has just β_1, f_2 has β_1 and β_2, etc.); and (3) a well defined management policy variable $U(\beta)$ to be computed from a subset of the parameters. An example of a sequence of population models would be (N_t = stock, h_t = harvest, Z_t = an environmental variable, such as water temperature)

$$f_1: \quad N_{t+1} = \beta_1 N_t - h_t$$

$$f_2: \quad N_{t+1} = \beta_1 N_t - \beta_2 N_t^2 - h_t$$

$$f_3: \quad N_{t+1} = \beta_1 N_t - \beta_2 N_t^2 + \beta_3 N_t Z_t - h_t$$

$$f_4: \quad N_{t+1} = \beta_1 N_t - \beta_2 N_t^2 + \beta_3 N_t Z_t + \beta_4 N_t^2 Z_t - h_t$$

An example of a policy variable would be the stock size producing maximum equilibrium yield; in this case U is not defined for f_1, and $U \approx (\beta_1 - 1)/2\beta_2$ for $f_2 - f_4$ provided Z varies randomly over time.

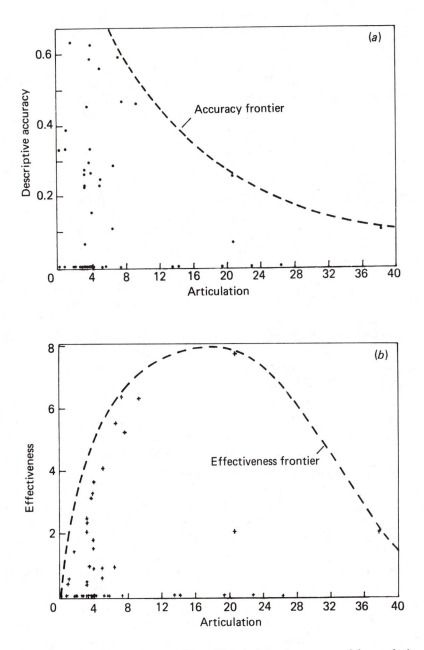

Figure 6.3. Descriptive accuracy (fit to historical data) versus model complexity (articulation) measured across a series of models for freshwater wetlands and water bodies. Overall "effectiveness" is a product of accuracy and articulation. *Source:* Costanza and Sklar (1983).

A basic management concern would be about the accuracy of estimation of U. This accuracy can be measured by the approximate variance formula

$$\sigma_U^2 = \boldsymbol{h}' \Sigma_\beta \, \boldsymbol{h} \tag{6.6}$$

where $\boldsymbol{h}_i = \partial U / \partial \beta_i$, and Σ_β is the covariance matrix for the parameters β. Let us examine how σ_U^2 will vary in relation to model complexity measured by the number of parameters in β. Recall that Σ_β can be approximated by

$$\Sigma_\beta = s^2 (\boldsymbol{X}' \boldsymbol{X})^{-1} \tag{6.7}$$

where s^2 is a measure of the residual error variance $[s^2 = \Sigma (y - \hat{y})^2 / d,$ where $d =$ degrees of freedom] and $\boldsymbol{X}_{ij} = \partial \hat{y}_i / \partial \beta_j$ measured at the best parameter estimates $\hat{\beta}$. Substituting equation (6.7) into equation (6.6), we get

$$\sigma_U^2 \approx s^2 \boldsymbol{h}' (\boldsymbol{X}' \boldsymbol{X})^{-1} \boldsymbol{h} \tag{6.8}$$

Our concern is with what happens to s^2 and the elements of $(\boldsymbol{X}' \boldsymbol{X})^{-1}$ as the number of parameters (number of columns in the $\boldsymbol{X}$ matrix) increases (the best estimates $\hat{\beta}$ will also change, but the main issue is uncertainty in U).

For ordered model sets, s^2 is expected to decrease as the number of parameters increases (Figure 6.4). A simple way to think about this effect is to imagine that additional model components "absorb" or "explain" (perhaps spuriously) more of the variation in observations, making the deviations $y - \hat{y}$ decrease monotonically as the number of parameters increases. Notice that this rule seems to be violated in Figure 6.3; there descriptive accuracy is measured roughly by $R^2 = 1 - s^2/v^2$, where v^2 is the total variance in observations, so s^2 must be increasing across the model set if R^2 is decreasing as the figure seems to imply. But the models used in Figure 6.3 are not an ordered set, and increasing complexity (articulation) is represented there by models that try to explain more kinds of variables, rather than the same variables in more detail. When s^2 does, indeed, increase with model complexity, the argument presented in Figure 6.4 is made even stronger.

On the other hand, the elements of $(\boldsymbol{X}' \boldsymbol{X})^{-1}$ generally increase monotonically as more columns (parameters) are added to $\boldsymbol{X}$ (Figure 6.4). This effect is enhanced if the elements of the new columns are correlated with columns already present. (For example, the $\boldsymbol{X}$ matrix for f_1 above has a single column consisting of N_t values; f_2 has this column, plus a column of N_t^2 values which are correlated with N_t especially if the variation in N_t is not large.) Thus the variances of individual parameters β_i, measured by $s^2 (\boldsymbol{X}' \boldsymbol{X})_{ii}^{-1}$, will increase whenever a column is added that increases $(\boldsymbol{X}' \boldsymbol{X})_{ii}^{-1}$ more than it decreases s^2; new columns that are strongly correlated with old ones will have little effect on s^2 (explain little more than the old ones already do), yet will cause $(\boldsymbol{X}' \boldsymbol{X})_{ii}^{-1}$ to increase.

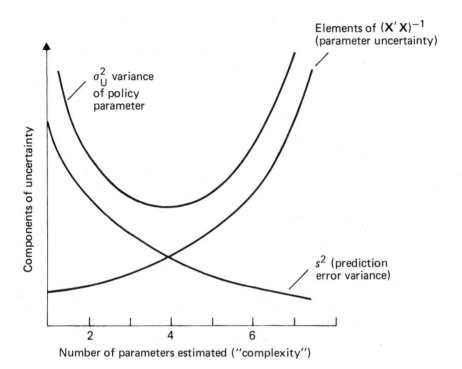

Figure 6.4. Uncertainty about a policy parameter U is likely to be minimized for a fixed data set Y by basing the calculation of U on parameter estimates from a model of intermediate complexity, in this case four parameters (for explanation see text). The minimum for σ_U^2 corresponds roughly to the maximum in "effectiveness" of Figure 6.3.

Taking the product of decreasing s^2 times increasing $(\mathbf{X}'\mathbf{X})^{-1}$, and then calculating the quadratic form for σ_U^2 [equation (6.8)], we see that σ_U^2 is likely to decrease at first as parameters are added, then later increase pathologically as confidence decreases in each of the parameters from which U is calculated. There are exceptions to this rule (level sets of U coincide with principal axis of Σ_β, so U is well determined, though the β's from which it is calculated are not), but they unfortunately are not that common in practice.

So if we take σ_U^2, the variance of a key policy variable, as a measure of the best model to use, we should seek an intermediate level of complexity where σ_U^2 is minimized (Figure 6.4). This conclusion is in striking contrast with a common practice in the renewable resource literature, which is to measure the "quality" of models only in terms of descriptive ability as measured by R^2 or s^2. As we see in the figure, it is quite possible for a very good "predictive" model (low s^2) to give very poor (highly uncertain) estimates for key variables of policy interest.

This conclusion applies to model structure as well: it can be best to deliberately use a model structure that is known to be incorrect, if that structure leads to better performance in estimation of important policy measures $U(\beta)$. A good example is provided in Ludwig and Walters (1985). We simulated fish population "histories" using a Deriso model (Chapter 4), and then tried to recover the (*known*) optimum equilibrium harvest effort using estimation schemes for the "data" based both on the Deriso model and on a simpler production model that did not correctly represent time delays and growth effects. Over many simulation trials, we found that the simpler model often outperformed the Deriso model (better effort estimates). The estimation procedure for the correct model involved trying to sort out, often unsuccessfully, the effects of recruitment versus growth on net production. These effects were "lumped" in the simpler model, and the estimation scheme for it was more often able to capture the lumped effects accurately.

The above arguments do not in any way represent a coherent theory of optimum model complexity. But they do suggest that such a theory can be developed, and will likely become an important research area in the near future. Based on experience with a variety of models for renewable resources, I predict that the theory will show that quite compact models (2–4 parameters per measured state variable) are optimum for most practical situations.

Problems

6.1.   For the two alternative salmon models in Figure 1.1, we estimated (Walters, 1977) the dollar value of future catches, discounted at 1% per year, for four alternative escapement policies. Three of these policies represented "experiments," with increased escapements over periods we felt would be necessary to see which model is correct. The policies and value estimates were:

		Correct model	
Policy option		η_1 (domed curve)	η_2 (saturating curve)
(1)	Hold escapement at 1 million	\$232 million	277
(2)	1.5 million spawners for 15 years, then optimum	233	325
(3)	2 million spawners for 5 years, then optimum	226	332
(4)	3.0 million spawners for 3 years, then optimum	215	332

Here the decreasing values associated with increasingly severe experiments, if η_1 proves correct, are due to low harvests in the experimental period and also to lower recruitments produced by the experimental escapements. Assuming equal prior probabilities $P(\eta_1) = P(\eta_2) = 0.5$, which option has the higher expected value? How low would $P(\eta_2)$ have to be for you to conclude that option 1 (no experiment) has the highest expected value? What would happen to your assessment if a higher discount rate were assumed, so the low catches during the experimental period (due to allowing higher experimental escapements) were given more weight in the total value calculation?

6.2. Bayes' theorem says the probability $P(m_i | Y)$ that you should place on a model m_i, given a data set Y, is proportional to a relative credibility measure $P_i L_i$ for model i. Here P_i is the prior probability you would place on model i before looking at the data, and L_i is the probability (or likelihood) of getting the observed data given that m_i is actually correct (roughly, L_i measures how well the data fit m_i, in light of prior expectations about how much variability there should be around the correct model). Further, Bayes' theorem says that $P(m_i | Y)$ is inversely proportional to the sum of credibility measures $(\Sigma P_j L_j)$ over all models j included in the analysis. In view of these comments, indicate at least three qualitative ways to construct misleading estimates $P(m_i | Y)$ for any model i that you might personally favor or dislike.

6.3. The methods introduced in this chapter seem to rely heavily on the analyst's ability to define a closed model set $m_1, m_2, \ldots$, that includes the "true" model from which the historical data Y were generated. Such an approach may seem to you like taking an examination with only multiple-choice questions, where the correct answer is *always* "none of the above." What is wrong with this analogy? (Hint: should you think of a model for management as "true" or "false" in the first place?)

6.4. Consider again the George Reserve deer population (problem 1.3, Figure 5.1) and a logistic model for it, $N_{t+1} = R_0 P_t - R_1 P_t^2$ [equation (5.11); R_0 and R_1 are parameters, $P_t = N_t - H_t$ is the population after harvest in year t]. Assuming that P_t (Table 1.1) was measured without error, the deviations $w_t = N_{t+1} - R_0 P_t + R_1 P_t^2$ must be due to effects (biotic and environmental) not included in the logistic model. Assume further that these effects were drawn by nature from a normal distribution with mean zero and standard deviation σ_w

$$p(w) = \frac{1}{\sqrt{2\pi\sigma_w^2}}\, e^{-w^2/2\sigma_w^2}$$

The likelihood of the data given any particular possibility (R_0, R_1) is then $p(w_1)p(w_2) \cdots p(w_t)$, where you use R_0, R_1 and the N, P data to calculate each w_t. Write a computer program to calculate such products, and use it to show how the likelihood varies with R_0 and R_1 as you depart from the "best" estimates $(1.93, 0.00494)$, for various assumed values of σ_w between 5 and 50. Are your qualitative conclusions (about which R_0, R_1 combinations are most credible) very sensitive to your choice of σ_w?

$$L = f(N,s) = \left[(2\pi\sigma^2)^{-\frac{1}{2}}\right]^N \exp\left(-\frac{N}{2}\right)$$

6.5.   For the George Reserve data and the model in problem 6.4, a singular value decomposition procedure gave

$$\hat{R}_0 = 1.892 \qquad\qquad \hat{R}_1 = 0.00458$$

$$\sigma_1 = 63344.47 \qquad\qquad \sigma_2 = 178.80$$

$$V = \begin{bmatrix} -0.0086 & -1.0 \\ 1.0 & -0.0086 \end{bmatrix}$$

$$s^2 = \sum (N - \hat{N})^2/(42 - 2) = 271.57$$

$$(\text{so } \sigma_w \approx 16.48)$$

Using $F_{0.95,2,40} = 3.23$, we see that the 95% confidence limits for $z_2 = (-1)(R_0 - \hat{R}_0) + (-0.0086)(R_1 - \hat{R}_1)$ will be ± 3.13, indicating that $\hat{R}_0$ and $\hat{R}_1$ can be increased jointly (somewhat) without causing a much poorer fit to the historical data. How does this conclusion compare with your findings from problem 6.4? With the above information, construct a 95% confidence region for R_0 and R_1, and use it to define a reduced set of three models that are consistent with the data but imply different management strategies. In particular, consider the (unlikely) simple model $R_0 = 1.36$, $R_1 = 0$; how poor is the fit to this model, and what are its management implications?

6.6.   To illustrate the ideas of optimum model complexity, write a program to generate fake data sets for the 4-parameter production model

$$N_{t+1} = N_t + rN_t^m \left(1 - \frac{N_t}{k}\right)^n - h_t N_t + w_t$$

Use $N_0 = 0.9$, $r = 0.5$, $k = 1$, $m = 1.2$, $n = 0.8$, and $h_t = 0.05\,t$, $t = 1, \ldots, 10$. Use w_t normally distributed with mean zero and standard deviation σ_w, where $0 < \sigma_w < 0.03$. For this model, the policy parameter "stock size for maximum surplus production" is given by

$N^* = mk/(m + n)$. Assuming that the time series N_t, h_t, $t = 0, \ldots,$ 10 has been observed exactly (only process errors), develop parameter estimation procedures (Appendix 5A, or above section on singular value decomposition) for each of the following models of your fake data:

$$f_1: \quad N_{t+1} = N_t + rN_t \left(1 - \frac{N_t}{k} \right) - h_t N_t$$

("logistic", $m = 1$, $n = 1$)

$$f_2: \quad N_{t+1} = N_t + rN_t \left(1 - \frac{N_t}{k} \right)^n - h_t N_t$$

("general", $m = 1$)

$f_3:$ the four-parameter correct model

Include in your program a procedure for estimating the variance of N^*, using the formulas for σ_U^2 in the final section of this chapter. By generating data sets with increasing σ_w and fitting these data to all three models, show that $\sigma_{N^*}^2$ is minimized by assuming f_3 if σ_w is small, but by assuming f_1 if σ_w is large. Compare this result to Figure 6.3. What can you say about the problem of bias in N^* when f_1 or f_2 is assumed? What happens to your conclusions if the data are generated from a more informative harvest sequence, i.e., $h_t = 0.1$, $0.1, 0.9, 0, 0, 0, 0, 0, 0.5, 0$?

6.7.   In exercise 6.6, you needed to use an approximation $s^2(\mathbf{X}'\mathbf{X})^{-1}$ for the covariance matrix of the parameters, and a further approximation for the variance of N^*. To test the validity of these approximations, generate 20 fake data sets with $\sigma_w = 0.001$ (an optimistic situation), fit the correct model f_3 to each, and calculate the sample covariance matrix

$$\frac{1}{20} \sum_{k=1}^{20} (\hat{N}_k^* - N^*)^2$$

You should find that $s^2(\mathbf{X}'\mathbf{X})^{-1}$ usually underestimates the variation $\hat{\beta}$ among trials, while the approximation for $\sigma_{N^*}^2$ is not too bad. Why does $\hat{N}^*$ not have large variance, considering that it depends on several quite uncertain parameters?

Chapter 7

The Dynamics of Uncertainty

*Aristotle's experimentations were confined
to catching nature in the act, without attempting,
after the modern fashion, to put her to the torture.*

John Gillies (1797) *Ethics and Politics*

This chapter examines how uncertainty, measured by odds placed on alternative models, is likely to propagate over time in relation to management decisions. In Chapter 1, I stressed that responses of resource systems to management can, in the end, only be learned through experience, so the common prescription to "wait until we understand the system better" is based on a false presumption that extrapolation (to the conditions created by *not* waiting) will become possible even without experience. Then, in Chapter 4, I suggested that the need to gain experience should not be treated as an excuse for blind management by trial and error; we can, at least, construct alternative models as a guide to possible responses. Chapter 6 then looked into how we can, on the basis of historical data sets **Y**, place odds on these alternatives. I turn now to the question of how to model (predict) rates of learning when accidents or directed trials ("management experiments") send a managed system into states for which responses are uncertain. Perhaps the most important lesson from this analysis is that, in the face of natural variation and noisy measurements, we should expect learning rates to be discouragingly slow except when rather drastic experiments are undertaken.

As a prelude to the examination of how probabilities placed on alternative models propagate over time, the first section below looks at how bounds can be placed on the importance of learning by calculating the expected value of a magical experiment that instantly resolves all uncertainties, then comparing this value to the best that could be expected in the complete absence of learning. This calculation shows that, surprisingly often, learning will not be as valuable as we might intuitively expect. Then subsequent sections examine simple statistical models for propagation of probabilities over time, for cases where the value of learning is expected to be significant. The models presented in these sections are intended to provide (1) a theoretical basis for understanding the factors that determine how fast uncertainties can be reduced, and (2) practical calculation procedures that can be used in "Monte Carlo" simulation studies of learning rates under particular policy options.

Bounding the Importance of Learning: The Expected Value of Perfect Information

This section returns to the simple idea, introduced in Chapter 6, of constructing a decision table that lays out the possible outcomes of alternative policies. We will first see how to construct a table that defines the best policy to take if no learning is possible. The expected value of using this policy is a lower bound on future performance. Then for each possible "state of nature" or model of outcomes, we pretend for a moment that the model is known to be correct, and calculate the best policy to use and the value that would be obtained. This results in a set of most optimistic (perfect knowledge) estimates of value, and we compare these to the values expected if the best "no learning" policy were used. To see intuitively how the calculations go, let us use the simple decision example from Chapter 6, involving whether or not to build a salmon spawning channel:

	Options	
	Do not build	Build
Models	channel	channel
No response	240	135
Good response	240	564

For this example, we assumed 50:50 odds on the two models, and estimated expected values for the two policies as

Value of "do not build" $= (0.5)(240) + (0.5)(240) = 240$
Value of "build" $= (0.5)(135) + (0.5)(564) = 349.5$

Clearly the best policy, in the absence of further information, is to build. But suppose we could make a magical study that would resolve beforehand which model is correct. Then for the "no response" model, we could obtain the value $240 million, and for the "good response" model, $564 million. But before we do the magical study, we should place 50:50 odds on these outcomes, for an expected value of $(0.5)(240) + (0.5)(564) = 402 million.

This is the expected future value from the system, measured today, of doing the magical study then acting correctly for whichever model turns out to be right. If we compare this value to the best that can be expected without the magical study (349.5), we get the "expected value of perfect information," $52.5 million $= 402 - 349.5$. Notice that this value of information can also be calculated as the average, using today's odds, of the improvements in value associated with learning that each model is correct:

$$52.5 = (0.5)(240 - 135) + (0.5)(564 - 564)$$

Notice that the gains are measured relative to the values (135 and 564) expected under the best choice without magic, which in this example was to build a channel.

Definition of the expected value of perfect information (EVPI)

Let us now try to make these notions more general and precise. Suppose that at some moment in time, statistical analyses such as those suggested in Chapter 6 have led to probabilities $P_t(m_i)$ on a series of models m_i in a set M. Suppose, further, that there has been defined a set U of policy options, where each option u_j specifies a whole course of future control actions. u_j might, for example, be a feedback policy, such as "harvest $u_t = x_t - \hat{u}_j$ animals if the number of animals x_t exceeds $\hat{u}_j$, but harvest none if x_t is less than $\hat{u}_j$." We then define the value $V(u_j|m_i)$ as some measure of average or total future returns if the policy u_j is followed and if model m_i turns out to be correct.

Now suppose we assume that no learning will take place in the future. Then we can calculate a (pessimistic) expected value for each policy option u_j, as

$$V(u_j) = \sum_i P_t(m_i) V(u_j|m_i) \tag{7.1}$$

This expected value is just a weighted average of the possible outcomes of applying u_j, with each outcome model m_i given a weight $P_t(m_i)$. Then suppose we calculate $V(u_j)$ for every policy u_j, and find that policy u^* which has the maximum expected value $V(u^*)$. Keep in mind that in order to obtain this maximum, we have to calculate the conditional values $V(u^*|m_i)$ for every model m_i; for some models, u^* may be a very good policy [high $V(u|m)$], yet it may be very bad for others (i.e., if others turn out to be correct).

Next, suppose we consider each of the models m_i separately, and find the policies u_i^{**} that maximize $V(u_j|m_i)$. That is, u_i^{**} is the best policy (from U) to follow if m_i is known to be correct. Then for each model m_i, we will have defined two extreme outcomes:

$V(u^*|m_i)$: best value given m_i but no learning (m_i not known); *value of policy u^* given m_i*

$V(u_i^{**}|m_i)$: best value given m_i, and m_i known with certainty.

Notice that the second quantity must be greater than or equal to the first; otherwise we have incorrectly estimated the optimum u_i^{**} for model i. The values will coincide only when $u_i^{**} = u^*$ or when m_i predicts the same outcome by following u_i^{**} as u^*.

Finally, the expected value of perfect information (EVPI) is defined as the average gain associated with using u_i^{**} instead of u^*, where the average is weighted by the odds currently placed (before perfect information becomes available) on the models:

$$\text{EVPI} = \sum_i P_t(m_i)[V(u_i^{**}|m_i) - V(u^*|m_i)] \qquad (7.2a)$$

This can also be written as

$$\text{EVPI} = \sum_i P_t(m_i) V(u_i^{**}|m_i) - V(u^*) \qquad (7.2b)$$

where $V(u^*)$ is the value associated with the policy u^* that is best if we never expect to learn which model is correct. The simplest interpretation of EVPI is that it is our best estimate, given the odds currently placed on the alternative models m_i, of the expected or average gain to be obtained if we could suddenly resolve all uncertainty about which model is correct. Another interpretation is that EVPI is the amount we should be *willing to pay*, in units of V, for a magical study or measurement that would resolve which m_i is correct.

There are three reasons why the value of learning, measured as an upper bound by EVPI, is often not as large as we would intuitively expect. First, the optimal policies u_i^{**} for the various models need not differ greatly from u^*. Second, production models tend to predict nearly the same yield across a fairly wide range of harvest policies, i.e., the differences $V(u_i^{**}|m_i) - V(u^*|m_i)$ may be relatively small for all m_i that are assigned high probability $P_t(m_i)$. Finally, u^* itself tends to be "close" to the u_i^{**} for those models m_i that are assigned high probability $P_t(m_i)$; for these models, $V(u_i^{**}|m_i) - V(u^*|m_i)$ is small unless the models are very sensitive to u. These points will become a matter of critical concern in Chapter 9, where we will examine methods to estimate optimum adaptive policies.

EVPI in a stock–recruitment example

A simple example will serve to illustrate how EVPI can be estimated, and why it is often quite small. A common situation in salmon management is to have a fairly long series of spawning stock and recruitment observations, but all gathered over a rather narrow range of spawning stocks. Suppose we have such a record, and believe that average recruitment can be approximated by the Ricker model $R_{t+1} = S_t \exp [\bar{a} - b(S_t - \bar{s})]$, where $\bar{a}$ is the average value of $\ln (R_{t+1}/S_t)$, and $\bar{s}$ is the average historical value of S_t (see Ludwig and Walters, 1981). Following the arguments of Chapter 6, let us assume that the data are good enough to precisely estimate $\bar{a}$ and $\bar{s}$, so we can generate a range of models (hypotheses) by varying just the single uncertain parameter b. Let us assume that, based on the data, we would assign b a normal distribution with mean b_0 and variance Σ_b. That is, each model m_i is the Ricker curve with an assumed value b and M is the set of all b values; the probability placed on each model is

$$P_t(m_i) = P_t(b) = \frac{1}{\sqrt{2\pi \Sigma_b}} \exp \left[\frac{-(b - b_0)^2}{2\Sigma_b} \right]$$

Let us define a control policy u_j by the number of spawners S_j to allow each year, where S_j is fixed over time. The annual catch is then 0 if $R_t \leq S_j$, and $R_t - S_j$ when R_t exceeds S_j. U is the set of all possible values of S_j. Finally, let us use the average annual catch $\bar{R}(S_j) - S_j$ as our measure of value V. For a fixed S_j, the average recruitment $\bar{R}$ is given approximately by $\bar{R} = S_j \exp [\bar{a}' - b(S_j - \bar{s})]$, where $\bar{a}' = \bar{a} + \sigma^2/2$, and σ^2 is the variance of environmental effects on recruitment. According to these definitions

$$V(u_j|m_i) = V(S_j|b) = \bar{R} - S_j$$

$$= S_j \exp [\bar{a}' - b(S_j - \bar{s})] - S_j \tag{7.3}$$

Using this definition, it is possible to find the best policy S_j given no learning: it is the value of S_j that maximizes

$$V(S_j) = \int_{-\infty}^{+\infty} P_t(b) \, V(S_j|b) \, db \tag{7.4}$$

This integral represents the analogue of equation (7.1) for m_i generated by continuous variation in b; for small Σ_b it has an analytical solution which is a function of S. The S for which the integral is maximum [$V(S)$ maximum] can be found by a simple numerical procedure, and this value S^* is the optimum escapement if no learning is expected. The same numerical procedure can be used to find the escapement S_b^{**} that is optimum for any fixed, known b [i.e., S_b^{**} maximizes $V(S|b)$]. The expected value given perfect information is then given by

$$\text{EVPI} = \int_{-\infty}^{+\infty} P_t(b) \, V(S_b^{**}|b) \, db - V(S^*) \tag{7.5}$$

[This is the continuous version of equation (7.2b).] The integral in (7.5) can be calculated numerically using various computer algorithms.

Table 7.1. Expected values of perfect information for a stock–recruitment system, measured as a percentage of the average annual harvest expected if there were no future learning. Parameters $\bar{a}'$ and b_0 are determinants of productivity, while the ratio σ_b/b_0 is a measure of relative uncertainty.

		Initial recruitment parameters			
		$\bar{a}' = 0.1$		$\bar{a}' = 1.0$	
		$b_0 = 0.5$	$b_0 = 1.0$	$b_0 = 0.5$	$b_0 = 1.0$
Initial	$\sigma_b/b_0 = 0.1$	0.13%	0.12%	0.11%	0.10%
uncertainty	$\sigma_b/b_0 = 0.5$	0.11%	0.32%	0.69%	2.26%

The key variables affecting EVPI in this example are $\bar{a}'$, b_0, and Σ_b. $\bar{a}'$ is a measure of average productivity per spawner, b_0 is a prior measure of expected sensitivity of recruitment rate to changes in S, and Σ_b, of course, measures uncertainty in b. Table 7.1 summarizes how EVPI varies with these parameters, as a percentage increase in potential performance (average annual catch) relative to the no learning maximum $V(S^*)$. From the table, we see immediately that when Σ_b is large, the value of information increases with $\bar{a}'$ and b_0, which measure how sensitive $V(S|b)$ is to the choice of S [i.e., how big the average difference $V(S_b^{**}|b) - V(S^*)$ is expected to be; to see this, plot the average catch as a function of S, for different values of $\bar{a}'$ and b_0]. Also, EVPI increases with Σ_b, since by increasing this variance we admit higher probabilities for more extreme b values such that S_b^{**} differ greatly from S^*. But we see from Table 7.1 that the value of perfect information is only a small percentage of the average annual catch unless the stock has been very productive (large $\bar{a}$ historically); for unproductive stocks, the Ricker model predicts low sensitivity of $V(S|b)$ to changes in S. We could, of course, construct recruitment models such that average catch would be predicted to fall off more sharply as S is varied away from S^{**}, but there is little evidence of such sensitivity in the case of Pacific salmon.

Let us close this example by looking ahead briefly at questions to be addressed in the remainder of this chapter and in Chapter 9. First, notice that EVPI is a well defined function of Σ_b; it increases as this variance increases. Next, consider the Ricker model as written above, with only b unknown: $R = S \exp [\bar{a} - b(S - \bar{s})]$; in this form, it is obvious that we can only learn about b (reduce Σ_b) by choosing S away from the historical average $\bar{s}$. (Otherwise, we "see" only $R = \bar{s} e^{\bar{a}}$.) In subsequent sections, we will examine methods for estimating the effect of actions like S_t on uncertainty measures like Σ_b. If we can estimate changes in Σ_b, then we can also predict how EVPI will change in relation to policy choices like S_t. The change in EVPI is a simple measure of the "information value" associated with any choice (though it is incomplete since it ignores the way *future* decision makers may also use information values in selecting decisions). Now, suppose we calculate this information value for every choice S_t, and act as though it is a real "benefit" of S_t by adding it to the nominal expected value $V(S_t)$ that we calculate while assuming no learning. If the initial EVPI is large and some extreme decisions will significantly reduce Σ_b, we are likely to find that the combined information plus nominal value is maximized at some $\hat{S}_t$ quite removed from S^* (best S if no learning), if S^* is near $\bar{s}$. When this happens, $\hat{S}$ is in some sense an optimal "probing decision." In Chapter 9, we will look at more precise methods for estimating the combined values of alternative decisions.

Information States and Sufficient Statistics

As we saw in the previous chapter, precise computation of the odds $P_t(m_i)$ that should be placed on alternative models given a data set Y is usually a tedious and time-consuming task. When we turn to the study of how $P_t(m_i)$ propagates over time, it becomes important to have compressed equations (simple models of P_t) that can be used to give analytical insight and permit practical Monte Carlo (simulation) studies. Such compressions have been a topic of much study in the field of control system theory, where the subject is called "adaptive filtering." Most of the following discussion is borrowed from the literature on that subject; for further reading, I particularly recommend Young (1974), Gelb (1974), Meinhold and Singpurwalla (1983), and Detchmendy and Sridhar (1966).

Recall that $P_t(m_i)$ is computed from the likelihood function $L(Y|m_i)$ and from prior probabilities $P_0(m_i)$. Since Y is growing in time, there is potentially a growing "information state" needed to compute P_t (and therefore any decisions that depend on P_t). As a first step toward avoiding this growth in models of P_t, we usually assume that the observation set y_t (a vector) is statistically independent of $Y_{t-1} = \{y_1, \ldots, y_{t-1}\}$. Then the likelihood function can be written as $L(Y_t|m_i) = L(y_t|m_i)L(Y_{t-1}|m_i)$, and the general Bayesian formula for P_t can be written recursively as

$$P_t(m_i) = \frac{1}{\alpha_t} L(y_t|m_i) P_{t-1}(m_i) \tag{7.6}$$

where $\alpha_t = \Sigma_j L(y_t|m_j) P_{t-1}(m_j)$. That is, when the observations are independent over time we can compute P_t knowing only y_t and P_{t-1}.

If the models form a discrete set $m_1, \ldots, m_N$, with N alternatives, the N numbers $P_{t-1}(m_1), \ldots, P_{t-1}(m_N)$ form an information state of fixed dimension, which is fully sufficient to permit computation of P_t after y_t is observed. We can "forget" about the original data Y_{t-1} from which P_{t-1} was computed, since P_{t-1} is a sufficient "memory" (contains all the information needed) to compute P_t. In statistical terms, we say that P_{t-1} is a *sufficient statistic* for Y_{t-1}. Since the P_{t-1} must sum to 1.0, there are in effect only $N - 1$ distinct elements in the information state P_{t-1}. These $N - 1$ numbers are the minimum we can get away with storing, and using as the basis for feedback policies (action as a function of information state) when there are N discrete model alternatives.

The situation becomes more interesting when the model set is generated by continuously varying a set of unknown parameters β, so we describe uncertainty in terms of the probability density function $p_t(\beta)$. This function is generally very complicated (cannot be represented exactly by an analytical equation) for nonlinear models. However, as noted in Chapter 6, p_t can be approximated in the neighborhood of the most probable estimates $\hat{\beta}_t$ as a normal distribution, with mean $\hat{\beta}_t$ and covariance matrix $\Sigma_{\hat{\beta}_t}$. If we then use $\{\hat{\beta}_t, \Sigma_{\hat{\beta}_t}\}$ as an approximate information state for the calculation of p_t from p_{t-1}, and if there are M unknown parameters, the information state has roughly $M + M^2/2$ dimensions (M estimates $\hat{\beta}_t$, plus roughly half the elements of $\Sigma_{\hat{\beta}_t}$ since $\Sigma_{\hat{\beta}_t}$ is symmetric).

There is a special class of dynamic models for which $\hat{\beta}_t$ and $\Sigma_{\hat{\beta}_t}$ do constitute a complete set of sufficient statistics for p_t. This is the class of models for which (1) the state dynamics $x_{t+1} = f(x_t, u_t, \beta)$ are "linear in parameters," i.e., can be written as

$$x_{t+1} = \sum_{i=1}^{M} \beta_i f_i'(x_t, u_t) + w_t$$

where w_t is normal and the f_i' do not depend on β; and (2) the observation dynamics $y_t = h(x_t)$ are linear and invertible with no observation noise, i.e., $y_t = A x_t$, where A is a known matrix having inverse A^{-1}. It is often possible to convert other nonlinear models into the linear-in-parameters form by a change of variables, though the condition of no observation noise is more difficult to meet or justify. When the dynamic model can be written as above, $\hat{\beta}$ and its covariance matrix $\Sigma_{\hat{\beta}}$ depend only on sums of products among the f_i' and $x_t = A^{-1}y_t$; these sums, or equivalently $\hat{\beta}$ and $\Sigma_{\hat{\beta}}$, are thus sufficient statistics for Y. For models that are not linear-in-parameters, it is not possible to calculate even $\hat{\beta}_t$ exactly without looking back at all the past data. (See Appendix 5A and note that the sensitivity $\partial \hat{y}_i/\partial \beta_j$ and cross products matrices involved in updating the estimates $\hat{\beta}$ usually involve a complicated interdependence of all the y's and β's.)

To make these ideas clearer, let us examine two stock–recruitment models. The Ricker model $R_t = S_{t-1} \exp [a - bS_{t-1} + w_t]$ can be transformed into $\ln R_t = \ln S_{t-1} + a - bS_{t-1} + w_t$, which is linear in the parameters a and b when we treat $\{\ln R_t, S_{t-1}\}$ as the basic observation set $Y(y_t = \ln R_t, A = 1, u_t = S_{t-1})$. To update parameter estimates, we need only keep track of *sums* of cross products involving the $\ln R_t$ and S_t. The Beverton-Holt model $R_t = \alpha S_{t-1}e^{w_t}/(\beta + S_{t-1})$ does not admit such a transformation. To get a form that is additive in the errors w_t, we take $\ln R_t = \ln \alpha + \ln S_{t-1} - \ln (\beta + S_{t-1}) + w_t$. In this case we can treat $\ln \alpha$ as one unknown parameter, but the term $\ln (\beta + S_{t-1})$ causes trouble. To estimate parameters, we must calculate $\partial y_t/\partial \beta = -1/(\beta + S_{t-1})$, where $y_t \equiv \ln R_t$, for each y_t. These derivatives involve all the past data points S_t individually, and change over time as we update $\hat{\beta}$.

The precise definition of the sufficient statistics for a parameter set β given a data set Y is a set of functions $l_1(Y), \ldots, l_m(Y)$ of the data, such that the probability distribution of Y given these functions is independent of β. Then, in some sense, the functions contain all the information about β, contained by the original set Y. In practice, we search for these functions (statistics) by examining the likelihood function $L(Y|\beta)$, to see if it can be expressed as functions of Y (such as $\Sigma_t y_t$, $\Sigma_t y_t^2$) that do not contain β, which operate on (add to, multiply, etc.) the β. These functions are sufficient statistics, and it is usually obvious how they will propagate over time as new information is gathered. Such sufficient statistics form a natural information state description for p_t. Unfortunately, they cannot be found easily, or do not exist, for most nonlinear models. Also there is no way to determine how well they are approximated by the first and second moment statistics $\hat{\beta}$, $\Sigma_{\hat{\beta}}$ of p_t, except by comparing computations based on them to some more tedious and accurate evaluation of p_t, on a case by case basis. Thus, in practice the discovery of reasonable information states to describe learning about parameters of nonlinear systems is very much a trial and error modeling process, akin to the search for reasonable dynamic models (we need both m_i that approximate Y, and statistical models that approximate p_t).

Effects of Control Sequence on Uncertainty

In this section we examine some models that can be used to predict how the odds placed on alternative dynamic hypotheses will change over time in relation to the control (harvest, other disturbance) sequence employed in management. These models help us to understand why learning rates are often discouragingly slow, and give general guidance about how to design more informative policies. As in earlier sections, we will focus on the Bayesian description of uncertainty

$$P_t(m_i) = \frac{1}{\alpha_t} L(y_t|m_i) P_{t-1}(m_i)$$

As a preamble, let us discuss briefly the idea of a single "informative decision" u_t for models m_i expressed in the format $x_{t+1} = f(x_t, u_t, \beta) + w_t$, $y_t = h(x_t, \beta) + v_t$. In this format, notice that all the information we get about the effects of any u_t will be through the observations $y_{t+1}, y_{t+2}, \ldots,$ that measure *in part* the state responses $x_{t+1}, x_{t+2}, \ldots,$ to u_t. Let us now define u_t as *locally informative* with respect to any pair of models m_i and m_j if $L(y_{t+k}|m_i, u_t) \neq L(y_{t+k}|m_j, u_t)$ for some $k \geq 1$. In other words, u_t is informative if the models m_i, m_j assign different likelihoods to observations y_{t+k} for at least one future time k. This is not the same as saying that m_i and m_j assign different likelihoods to every possible outcome y; there generally will be certain, hopefully few, outcomes that are equally likely under both models. (Murphy's law states that such outcomes are bound to happen.) Extending the notion a bit, an obvious step is to define u_t as *globally informative* if it is locally informative with respect to every model pair m_i, m_j in the set M under consideration. The purpose behind these definitions is to highlight the fact that analysis of uncertainty about dynamic models is not the same as classical experimental design in statistics, where we generally think only about taking a single output measurement as a response to each well defined (designed) combination of inputs. Then we seek to replicate at least some combinations. In dynamical systems, single decisions may generate whole sequences of responses (not replicates!), and we must look carefully at how the alternative models make different predictions about these sequences. The distinction between locally and globally informative has obvious importance when the analysis involves only a small set of alternative models; it may be practical to take small steps (incremental decisions) that give information about part of the model set, but not to make the large changes usually needed to be globally informative.

Learning rates about discrete alternative hypotheses

Perhaps the simplest possible learning situation in renewable resource management is shown in Figure 1.1. In such situations, we seek only to understand the stock-recruitment (or production) response to the stock size left after harvesting, and we suppose that stock and production can be measured exactly. Then the alternative hypotheses m_i reduce to models of the form $x_{t+1} = f_i(u_t) + w_t$, $y_t = x_t$, since we can view the stock after harvest as the control u_t. Here, f_i is any alternative production equation or curve of mean response to varying u. Figure 1.1 shows two such functions (hypotheses!) labeled η_1 and η_2. If we take x_t to be the logarithm of recruitment in year t, the Ricker stock-recruitment model (curve η_1 in Figure 1.1) would then be written as

$$x_{t+1} = \ln (u_t) + a - bu_t + w_t \tag{7.7}$$

(a and b are parameters, w_t is a normally distributed environmental effect) and the Beverton–Holt model (curve η_2 in Figure 1.1) would be

$$x_{t+1} = \alpha' + \ln \frac{u_t}{\beta + u_t} + w_t \tag{7.8}$$

(α' and β are parameters, w_t is normally distributed). To fix the two hypotheses in Figure 1.1, I chose $a = 1.96$, $b = 0.44$, $\alpha' = 2.2$, $\beta = 1.3$, and I estimated the variance of environmental effects $\sigma_w^2 \approx 0.1$. (An alternative parameterization of the two curves is given in problem 1.5.) With w as an additive error, both models predict (conservatively and realistically) that recruitment (e^x) would be more variable around η_1 and η_2 if higher spawning stocks were allowed.

Now suppose for this example we choose a spawning stock u_t, and later (four years in the sockeye salmon case) observe x_{t+1}. Then for each of the two models we can calculate what $_iw_t = x_{t+1} - f_i(u_t)$ must have occurred if the model were correct. The likelihood of y_{t+1} ($= x_{t+1}$) given model m_i is just the probability of obtaining this w_t. Since we have hypothesized that w_t is normally distributed with mean zero and variance σ_w^2, its probability is given by

$$g(w) = \frac{1}{\sqrt{2\pi \sigma_w^2}} e^{-w^2/2\sigma_w^2} dw$$

Thus, $L(y_{t+1}|m_i)$ is $g(_iw_t)$, where $_iw_t$ is the value of w for model i. Notice that the models η_1 and η_2 predict nearly the same x_{t+1} for $u_t < 1$ million spawners, so they will imply nearly the same likelihood for any x_{t+1} that arises. But the models diverge more and more for $u_t > 1$ million spawners, and only recruitments that fall halfway between them will be assigned equal likelihoods (see the above comment about Murphy's law).

If we call the likelihood given the Ricker model L_1, and the likelihood given the Beverton–Holt model L_2, then the probability that we will place on the Ricker model η_1 after observing a new x_{t+1} is given very simply by Bayes' theorem:

$$P_{t+1}(m_1) = \frac{L_1 P_t(m_1)}{L_1 P_t(m_1) + L_2 P_t(m_2)} \tag{7.9}$$

With just two models, $P_t(m_2) = 1 - P_t(m_1)$, and there is no need to even keep track of $P_t(m_2)$. Using this, and calling $P_{t+1}(m_1)$ just p_{t+1}, we can rewrite the dynamics of p_t even more simply:

$$p_{t+1} = \frac{p_t}{p_t + \varrho_t(1 - p_t)} \tag{7.10}$$

where ϱ_t is the "likelihood ratio" L_2/L_1. In this form we see very clearly that the learning rate, measured by $p_{t+1} - p_t$, depends very much on choosing u_t

values such that the likelihoods (and their ratio ϱ) are expected to be as different as possible. Using the normal distribution equation for L_i, we see that ϱ can be written as

$$\varrho = \exp\left[\frac{1}{2\sigma_w^2}\left({}_1w_t^2 - {}_2w_t^2\right)\right] \tag{7.11}$$

Large values of ϱ result in lower p_{t+1}, and these will occur if either ${}_1w_t^2$ (${}_1w_t^2$ is calculated by assuming that m_1 is correct) is large, or ${}_2w_t^2$ is small. Notice that the ratio becomes "less responsive" to the w_t values as σ_w^2 is increased; in other words, individual prediction errors w_t have less effect on ϱ (and hence p_{t+1}) when we assume that they were sampled from a distribution with larger variance.

The easiest way to discover how p_t will propagate over time when there are just a few alternative models is by Monte Carlo simulation. The procedure is essentially trivial. First, select one of the hypotheses as true, and generate a set of "data" from it by solving the model equations over time with random effects included. For each of the simulated data points y_t, calculate $L_{it} = L(y_t|m_i)$ for every model m_i (see the general procedures for likelihood evaluation in previous chapter). Then set initial odds $p_0(m_i)$ on all the models, and calculate $p_t(m_i)$ recursively over time from the general rule

$$p_t = \frac{L_{it}\,p_{t-1}}{\sum_j L_{jt}\,p_{t-1}}$$

Notice that the p_t values generated in this way are random variables, since they are calculated from the random variables y_t that determine the likelihoods L_{it}. Thus, you need to repeat the simulation several times, with different random error sequences, in order to get a good feeling for the range of learning patterns that might occur.

Figure 7.1 shows three sample learning trajectories for the Fraser River two-model example. In each case the spawning stock was fixed at 1 million fish until time (generation) 10, then changed to a more informative level for the next 10 generations. The correct model was the Ricker curve η_1 of Figure 1.1. For the first trajectory, the informative spawning stock was 3 million fish, for which the two models make very different predictions of average recruitment. Note that even in this case, it takes six generations to be reasonably sure of the correct model. In the other two cases (2 million and 1.5 million spawners after the tenth generation), learning rates are discouragingly slow and the correct model is not obvious even after 10 generations. For the "mild experiment" (1.5 million spawners), which has actually been recommended as "practical," higher odds are placed on the wrong model even after 10 generations.

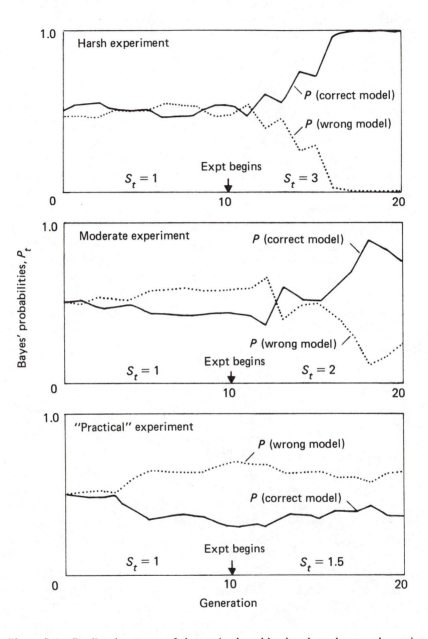

Figure 7.1. Predicted patterns of change in the odds placed on the two alternative recruitment models of Figure 1.1, for different escapement experiments (S_t in millions of spawners) after generation 10. A typical Monte Carlo trial is shown for each experiment; exact outcomes cannot be predicted since they depend on the future sequence of random environmental effects.

The predicted learning rates in Figure 7.1 are very typical for stock-recruitment and surplus production models when realistic assumptions about random variation are used in the calculations. That is, learning rates tend to be slow relative to the generation times of managed populations (and resource managers), even when management involves rather drastic experimental action. With hindsight this conclusion is rather obvious, but it came as quite a shock to us and to some Canadian government salmon biologists in the mid-1970s. At that time, we did a series of simulations like Figure 7.1 (and problem 1.5) to estimate how long it would take to evaluate the performance of some salmon enhancement projects (such as hatcheries and spawning channels) that were planned as part of a \$300 million enhancement program. Various uncertainties had been well recognized in planning for that program, and it had been divided into two phases, each about 10 years long. The hope was that results from projects in the first phase would be used to do a better job in the second phase. Well, it takes 2-5 years just to get each project started (design, construction, etc.), then at least two more years before the first results (returning adult fish) are seen. This means there would be 4-6 observations at most from each project in phase I, to use as the basis for phase II planning, even if all phase I projects were initiated in the first year of that phase (a budgetary and technical impracticability). Needless to say, our gloomy warnings about the need to delay phase II were not received very well by the people responsible for project development. As of this writing, the fate of phase II had not yet been decided.

In one sense the learning rates shown in Figure 7.1 are even a bit too optimistic, since only two alternative models are considered. By including more models that give predictions similar to the correct one, it is, of course, possible to increase the denominator in $p_{t+1} = L\, p_t / \Sigma\, Lp_t$ so that the odds with respect to any one alternative will change more slowly. But this is not so serious a problem as it may initially appear, when you consider how the results will be used in decision analysis. Models that make similar predictions generally imply similar prescriptions about the best policy to use, and hence act almost as a single model in weighting the outcomes of any policy choice. In fact, the probability of even including *the* exactly correct model in the first place is vanishingly small; the best we can usually hope for is to include one that implies roughly the correct policy choice.

Learning about unknown parameters

Suppose now that, instead of placing odds on a few alternative hypotheses, we choose to represent uncertainty in terms of a probability density function $p_t(\beta)$ for the unknown parameters in a simple functional model. We seek simple expressions that will give some insight about how $p_t(\beta)$

propagates over time; in practice this usually involves finding expressions for how the mean or mode $\hat{\beta}_t$ and covariance matrix $\Sigma_{\hat{\beta}_t}$ of $p_t(\beta)$ are likely to change. Such expressions are called "filtering equations," or simply "filters" for β. Quite elaborate filtering schemes can be developed for nonlinear parameter estimation (see Gelb, 1974; Soeda and Yoshimura, 1973), but a general feeling for the factors that affect learning can be obtained by looking at two special cases: recursive linear regression, and its simplest nonlinear extension, the "extended Kalman filter."

Recursive Linear Regression

Linear regressions are models of the form

$$y_i = \sum_j \beta_j Z_{ij} + w_i$$

where y_i is the ith measured response (dependent variable), the Z_{ij} are independent variables or inputs that are assumed to influence y_i through the unknown parameters β_j, and w_i is a random error that reflects the combined effects of measurement error and model inadequacy. Usually the model is written in shorthand matrix form $y = Z\beta + w$, where y, β, and w are vectors and $Z = \{Z_{ij}\}$ is called the input or design matrix. Least squares estimates of β are given by $\hat{\beta} = (Z'Z)^{-1} Z'y$, with covariance matrix $\Sigma_{\hat{\beta}} = s^2 (Z'Z)^{-1}$, where s^2 is an estimate of the variance of the random effects w. When w is assumed to be normally distributed, these estimates are also the maximum likelihood and describe the Bayes posterior distribution $p(\beta)$ completely (it is also normal) when the prior parameter distribution $p_0(\beta)$ is diffuse or normal with very large variances and zero covariances.

Dynamic models that are linear in parameters

$$x_{t+1} = \sum_j \beta_j f_j(x_t, u_t) + w_t$$

can be treated as linear regressions, provided the x_t can be reconstructed *exactly* (without measurement error) from the observation model $y_t = h(x_t)$. When the observation model is linear in a matrix β_0 of unknown parameters, so that $y_t = \beta_0 x_t$, and when β_0 has an inverse, so that $x_t = \beta_0^{-1} y_t$, we can sometimes substitute this expression for x_t back into the linear-in-parameters dynamic model to obtain a new model that contains only y's, w's, and a new set of parameters β' that are functions of β and β_0. Unfortunately, the covariance matrix of β and β_0 can usually be obtained only approximately from the covariance matrix of β' estimated by linear regression.

The estimation equations for linear regression can be written exactly as a recursive relationship which updates the estimates and their covariance matrix after each new observation is obtained. Suppose the regression based on $t - 1$ observations has resulted in $\hat{\beta}_{t-1}$ and $\Sigma_{\hat{\beta}_{t-1}}$, and we make a new observation y_t, Z_t, where Z_t is the vector of independent variable values at time t (i.e., Z_t' is the tth row of Z). Then the recursive (learning) equations for $\hat{\beta}_t$ and $\Sigma_{\hat{\beta}_t}$ are very simply (see Young, 1974)

$$\hat{\beta}_t = \hat{\beta}_{t-1} + k_t E_t \tag{7.12}$$

$$\Sigma_{\hat{\beta}_t} = \Sigma_{\hat{\beta}_{t-1}} - k_t Z_t' \Sigma_{\hat{\beta}_{t-1}} \tag{7.13}$$

where E_t is the "innovation" or error in predicting y_t from previous estimates $\hat{\beta}$ and current inputs Z_t:

$$E_t = y_t - Z_t' \hat{\beta}_{t-1} \tag{7.14}$$

(think of $Z_t' \hat{\beta}_{t-1}$ as $\hat{y}_t$, so $E_t = y_t - \hat{y}_t$), and k_t is the "filter gain" or "Kalman gain" vector defined by

$$k_t = \frac{\Sigma_{\hat{\beta}_{t-1}} Z_t}{s^2 + Z_t' \Sigma_{\hat{\beta}_{t-1}} Z_t} \tag{7.15}$$

Notice that this filter gain is an increasing function of the prior uncertainty $\Sigma_{\hat{\beta}_{t-1}}$, and a decreasing function of the variance of w, measured by s^2.

Equations (7.12)–(7.15) define a remarkably simple model for learning about the unknown parameters of dynamic models that can be rewritten to look like linear regressions. First, equation (7.12) says that the amount by which we should change the parameter estimates $\hat{\beta}$ at each step is just proportional to the prediction error E_t, where the proportionality constants k_t depend on how uncertain we already were ($\Sigma_{\hat{\beta}_{t-1}}$) and on how much random error (s^2) that we expect to see. Equation (7.13) says that uncertainty measured by $\Sigma_{\hat{\beta}}$ will decrease over time at a rate determined by its current level, by the error variance s^2, and by the input choice Z_t.

To see how these equations work, let us consider again the Ricker stock–recruitment model with only one unknown parameter b (see section above on bounding the value of learning), which we can express as $x_{t+1} = \bar{a} + b(S_t - \bar{s}) + w_t$, where x_{t+1} is ln (recruits/spawners), and we presume $\bar{a}$ and $\bar{s}$ are known. We can rewrite this by defining $y_t = x_{t+1} - \bar{a}$, $Z_t = S_t - \bar{s}$, and $\beta = b$, to give the simplest possible linear regression

$$y_t = \beta Z_t + w_t \tag{7.16}$$

where y_t is the observed deviation at time $t+1$ of the "productivity index" x_{t+1} from its historical average $\bar{a}$, Z_t is the deviation of spawning stock S_t from its historical average $\bar{s}$, and β is the sensitivity of productivity to the spawning stock allowed. The recursive regression equations for β reduce to just

$$k_t = \frac{\Sigma_{\hat{\beta}_{t-1}} Z_t}{s^2 + Z_t^2 \Sigma_{\hat{\beta}_{t-1}}} \tag{7.17a}$$

$$\hat{\beta}_t = \hat{\beta}_{t-1} + k_t(y_t - \hat{y}_t) \qquad (7.17b)$$

$$\Sigma_{\hat{\beta}_t} = \Sigma_{\hat{\beta}_{t-1}}(1 - k_t Z_t) \qquad (7.17c)$$

where $\hat{y}_t = \hat{\beta}_{t-1} Z_t$. It is quite obvious that the key variable in these equations is the spawning stock "disturbance" $Z_t = S_t - \bar{s}$; learning rates measured by k_t and $\Sigma_{\hat{\beta}_t} - \Sigma_{\hat{\beta}_{t-1}}$ are roughly proportional to this disturbance when it is small, and reach a maximum set by $\Sigma_{\hat{\beta}_{t-1}}$ when the disturbance is large. The learning rate is zero when $S_t = \bar{s}$.

Similar predictions are obtained for examples involving more than one parameter; the Z_{it} must differ from $\bar{Z}_i$ in order to provide information about β_i. However, an additional complication enters the picture; we must also worry about correlations among the Z_i. Such correlations are almost always large when the Z_i represent functions of the state variables x and controls u in dynamic models. High correlation between any pair Z_1 and Z_2 implies that the effects of β_1 and β_2 cannot be distinguished from one another; in other words, many combinations of β_1 and β_2 could, with almost equal probability, have given rise to the observed response. A good example of this difficulty is with Schnute's (1977) method for surplus production (see Chapter 4). His method in its simplest form uses the linear regression

$$y_t = \beta_1 Z_{1,t-1} + \beta_2 Z_{2,t-1} + \beta_3 Z_{3,t-1} + w_t$$

where y_t is an index of population change (log of catch per effort at t divided by catch per effort at $t-1$), $Z_1 \equiv 1.0$, Z_2 is the average catch per effort over $t-1$ to t, and Z_3 is the average effort over $t-1$ to t. Generally, we find Z_2 and Z_3 to be highly and inversely correlated, since abundance as measured by Z_2 decreases as effort Z_3 increases, and it is hard to break up this correlation even by deliberately varying the effort level.

The Extended Kalman Filter

The extended Kalman filter (EKF) can be used to study uncertainty in general state/observation models of the form $x_t = f(x_{t-1}, u_{t-1}, \beta) + w_t$ and $y_t = h(x_t, \beta) + v_t$. To simplify the discussion, as we did in parts of Chapter 6, let us drop the distinction between state variables and parameters, and simply call x_t a parameter if $f_i = x_t$, i.e., $x_{t+1} = x_t$. Then the model is written as $x_t = f(x_{t-1}, u_{t-1}) + w_t$ and $y_t = h(x_t) + v_t$, and the EKF propagates uncertainty about all state variables (and parameters) simultaneously. Its general form is very much like equations (7.12)-(7.15), except that the state/parameter covariance matrix $\Sigma_{\hat{x}}$ is seen as changing dynamically *between* observation times $t-1$ and t, and the independent variables Z_t are replaced by sensitivities $\partial f/\partial x$ and $\partial h/\partial x$ of state dynamics and observations to the unknown x's. The EKF equations are (for details see Jazwinski, 1970)

$$\hat{x}_t = f(\hat{x}_{t-1}, u_{t-1}) + k_t(y_t - \hat{y}_t) \tag{7.18}$$

$$\Sigma_{\hat{x}_t} = \Sigma_t^* - k_t D_{ht}' \Sigma_t^* \tag{7.19}$$

Here the prediction $\hat{y}$ is $h[f(\hat{x}_{t-1}, u_{t-1})]$ and the filter gain k_t is given by

$$k_t = \Sigma_t^* D_{ht} [D_{ht}' \Sigma_t^* D_{ht} + V_v]^{-1} \tag{7.20}$$

where D_{ht} is the observation sensitivity matrix $\partial h_i / \partial x_j$ calculated at $x = f(\hat{x}_{t-1}, u_{t-1})$ and V_v is the covariance matrix of observation errors v (usually assumed to be known in advance). Σ^* is a predicted covariance matrix of x, based on uncertainty in x after observation at time $t - 1$ and on process effects occurring over $t - 1$ to t; it is given (*very approximately*) by

$$\Sigma^* = D_{ft} \Sigma_{\hat{x}_{t-1}} D_{ft}' + V_w \tag{7.21}$$

where D_{ft} is the process sensitivity matrix $\partial f_i / \partial x_j$ evaluated at $x = \hat{x}_{t-1}$, and V_w is the covariance matrix of process errors w (again usually assumed known in advance). If observation errors are ignored (V_v assumed 0) and $V_w = 0$ for a subset of x's that have $f(x) = x$, i.e., a subset of "parameters," the EKF equations for this subset reduce to looking like the linear regression equations with $Z = \partial h / \partial x$.

Let me inject a word of warning before proceeding. The EKF equations are attractively simple, and it is tempting to use them for practical parameter estimation (instead of messier, iterative, nonlinear estimation algorithms or posterior density function evaluations). In my experience, the EKF usually performs rather poorly for the highly nonlinear functions and small data sets usually encountered in resource problems. It must be carefully tested for each case by using Monte Carlo simulations, which in the end often requires more effort than using messier algorithms in the first place. I introduce the EKF here strictly as a model from which we may obtain some qualitative conclusions about learning rates.

As with recursive linear regression, the EKF model [equations (7.18)-(7.21)] implies that learning rates are proportional to prediction errors $(y - \hat{y})$ and to current uncertainty (Σ^*), and are inversely related to the expected magnitude of measurement errors V_v. Process errors have a more subtle effect, which we will explore further in a later section on "parameters that aren't." As we assume larger values for V_w, we increase $\Sigma_{\hat{x}_t}$ correspondingly. But we also increase k_t and thereby make the estimation of $\hat{x}$ more sensitive (responsive, adaptive) to prediction errors as signals of possible changes in the system state.

A lesson from the EKF that is perhaps not so clear in linear regression is that reduction in uncertainty (as measured by $\Sigma_{\hat{x}_t}$) is very much dependent on how sensitive the predictions $\hat{y}_t$ are to the unknown states x_t. Our ability to "see" the x_t through prediction errors $y - \hat{y}$ is measured through the observation sensitivity matrix D_{ht}. Those state variables x_i that are not directly observed [i.e., parameters, etc., that do not appear in $h(x)$] enter the k_t calculation only through their covariances $\Sigma_{x_i x_j}$ with those variables x_j that are observed through $h(x)$. These covariances are in turn determined [equation (7.21)] by the sensitivities of state variables to one another, measured by $\boldsymbol{D}_{ft}$. Thus, direct and indirect linkages among observations and state variables can be quite complex in nonlinear filtering, and it is difficult to give simple rules of thumb about how to reduce correlations among the variables by providing contrast in inputs.

Tracking Parameters that Aren't

Chapters 4 and 5 discussed various reasons why it is practically impossible to construct models of renewable resource systems such that we may confidently assume complete constancy in those numbers treated as parameters. Almost always, the state variables we include in resource models are aggregate indices or measures whose statistical behavior we hope will be reasonably stationary. Parameters are usually defined by seeking biological and economic processes (such as birth and investment) having average outcomes that should be in some degree repeatable over time independently of the individuals (organisms, boats, etc.) that are engaged in them. To the extent that mixes of individuals change over time due to processes that are not modeled (evolution, undetected environmental changes, or even pure chance!), we should at least expect "drift" in parameter values. Worse, we can be confident that there will occasionally occur abrupt and irreversible parameter changes due to human activities and/or natural events that can be neither anticipated nor controlled, that is, have origins outside the arbitrarily defined "system" upon which management efforts and monitoring activities have been concentrated.

In this section we will first look at how models for recursive state/parameter estimation can be modified to permit tracking of parameter changes. Then we will use the results to examine a very disturbing question: under what conditions will learning rates even exceed rates of parameter change? Considering the slow rates of learning that we must often live with because of large process and measurement errors, we shall see that it is quite possible to become more, rather than less, uncertain over time unless variation in management inputs is kept above some critical "uncertainty threshold."

Approaches to parameter tracking

Time-varying parameters are a problem that must be faced even in engineered control systems, and researchers in that field have developed a number of practical approaches to parameter tracking and detection of abrupt changes. Three of these approaches are reviewed here: exponential forgetting of data, assumption of random drift by parameters, and innovation tests on the sequence of prediction errors.

A very crude way to be adaptive in the face of parameter changes would be to just base all calculations of $p_t, \hat{\beta}$, etc., only on a window of data extending back T time steps. A bit more elegant approach is exponential past weighting of data. The idea here is that if parameters are changing, the observations y_{t-1} should be assigned a weight, say λ ($\lambda < 1.0$), times the weight placed on observations at time t. Then y_{t-2} gets weight λ^2, and so forth. This is equivalent to saying that the observations should be assigned increasing error variances σ^2/λ, σ^2/λ^2, etc., moving back in time from the present, where we take the variances as a measure of information content with respect to the parameter values prevailing at time t. For estimation problems representable by linear regression, exponential past weighting is trivially easy to include in the recursive estimation equations; we simply include λ in the filter gain equation (7.15), as

$$k_t = \frac{1}{\lambda s^2 + Z_t' \Sigma_{\hat{\beta}_{t-1}} Z_t} \, \Sigma_{\hat{\beta}_{t-1}} Z_t \qquad (7.22)$$

Appearing in the denominator, $\lambda < 1$ has the effect of increasing k_t which makes each prediction error $E_t = y_t - \hat{y}_t$ have a larger effect on $\hat{\beta}_t - \hat{\beta}_{t-1}$. It is difficult to say theoretically what value of λ should be assumed in relation to expected rates of parameter change; good values of λ are usually found by Monte Carlo simulations of estimation performance, where the "data" presented to the estimation are generated by systematically varying key parameters at reasonable rates.

I have already introduced the more elaborate approach of assuming parameter drift, in the above discussion on the extended Kalman filter. There we dropped the distinction between parameters and variables, and treated parameters as variables assumed to follow a random walk model $x_{t+1} = x_t + w_t$. In the EKF, the diagonal elements of the process error covariance matrix V_w can be viewed as measures of how fast those x's treated as parameters are likely to change. The standard error of w_t for x_i, $\sqrt{V_{w_{ii}}}$, measures roughly the average parameter change expected per time step. As V_w is increased, the filtering equations are "freed up" (larger k_t values) to track changes in x_t that may be systematic (directional) instead of random. The linear regression equations (7.13)–(7.14) can be similarly modified to permit selective (differential) parameter movement ($V_{w_{ii}} \neq V_{w_{jj}}$), by replacing $\Sigma_{\hat{\beta}_{t-1}}$ wherever it appears by $\Sigma_{\hat{\beta}_t^*} = \Sigma_{\hat{\beta}_{t-1}} + V_w$ where the diagonal elements of V_w are again the squares of the average rates of parameter change expected.

Figure 7.2 shows some simulations of tracking performance for the Schnute method of estimating surplus production parameters. Here "data"

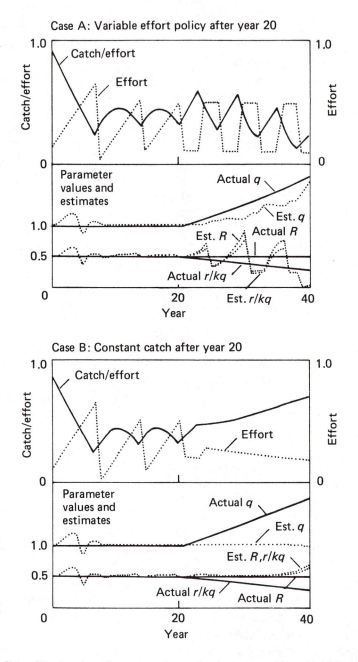

Figure 7.2. Simulated performance of an adaptive parameter estimation scheme for the logistic surplus production model. In case A, strong variations in harvesting effort help to recognize that the catchability parameter q is increasing after year 20. In case B, the catch is held constant after year 20 (so effort decreases), and the change in q is not estimated correctly.

were generated with the logistic model structure assumed by Schnute, but with the catchability coefficient increasing over time. Stable parameters were assumed for the early development (the first 20 years) of the simulated fishery (a bad assumption!), and an informative fishing effort sequence was used. Then catchability was assumed to increase by 3% each year after year 20, and two possible scenarios for regulatory response were simulated. Schnute's linear regression estimates of $\beta_1 = r$, $\beta_2 = r/kq$, and $b_3 = q$ were calculated recursively, but with variance components (V_w) added to the second and third (β_2 and β_3) diagonal elements of $\Sigma_{\beta_{t-1}}$ at each time step. Only small random errors w_t were assumed ($V_w = 0.001$), so we see that the estimates converge rapidly. The estimation "recognizes" that q is not changing until year 20. But when q begins to grow, the recursive estimation is at first "confused," and its later performance depends very much on the effort sequence that is followed. When effort is adjusted to hold only the catch constant (implied correction for growing q), there is little informative variation and the change in the q parameter is not detected even after 20 years. On the other hand, when effort is pushed up and down every three years after q begins to change, the filter tracks the changes fairly well but becomes confused about the r and r/qk parameters. Notice in both effort scenarios that r and r/qk are badly correlated, since it is difficult to generate a sequence of stock sizes varying widely enough to separate their effects.

Figure 7.3 shows a more realistic (and perhaps pessimistic) scenario, in which the catchability is changing rapidly during the early phase of resource development, while effort and stock size are also changing. We see that the estimation remains confused until there is a sharp effort change, in spite of low random errors. Effort disturbances help to sort out the biases induced by correlated early changes in both state variables and parameters, and there is little learning after efforts are later stabilized.

Abrupt parameter changes are in a sense easier to deal with than slow, progressive changes that might be confused initially with state changes or minor inadequacies in model structure. When there is a reasonable history of experience with one set of parameter values, abrupt changes show up as persistent prediction errors when the data are analyzed forward in time using filtering equations for parameter estimation. A simple tactic for dealing with abrupt changes has been suggested by Yoshimura et al. (1979). They advise setting up a two-level estimation system, with the first level being a recursive filter that assumes constant parameters. Then at each time step, a second level of analysis (and potential adaptation) is provided by examining the most recent prediction error. If the most recent $E_t = y_t - \hat{y}_t$ is large (i.e., unlikely) compared with the historical average (measured by the residual error variance s^2), then the parameter covariance matrix Σ_β is arbitrarily increased (for example, by doubling all the diagonal elements). This increases the filter gain k_t, and the estimation is freed up to seek new $\hat{\beta}$

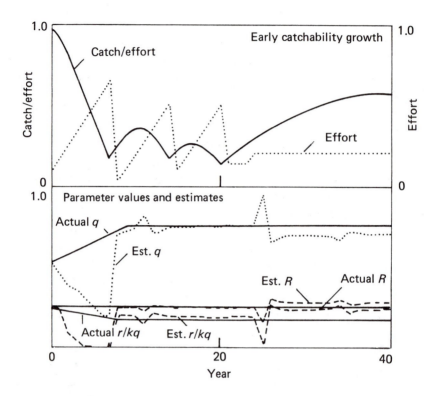

Figure 7.3. Simulated performance of adaptive parameter estimation for logistic production, for a situation in which the catchability parameter increases over years 1–10. The parameter estimates get progressively worse until there is a sudden change in harvesting effort at year 8.

values. If the large innovation is not repeated, the estimates $\hat{\beta}$ will not move too much, because the moves are also proportional to E_{t+1}, E_{t+2}, etc., even though k_t has been made larger than it theoretically should be. Elaborate statistical tests based on likelihood ratios can be developed to decide exactly when an innovation E_t is large enough to warrant changing Σ_{β_t}, but these tests are unnecessary when the analyst has adequate time to reanalyze the data repeatedly while trying different assumptions about what changes actually took place at each point where the prediction errors change abruptly. In other words, the analyst should act as an intelligent and intuitive "second-level controller" of the adaptive estimation scheme.

Figure 4.1 shows my favorite example of an abrupt parameter change in a stock–recruitment system. Chinook salmon in the Columbia River showed one rather consistent pattern of recruitment until the mid-1950s, when recruitment rates per spawner dropped quickly and have remained lower ever since. This example illustrates nicely that big changes may occur due to factors that are not immediately obvious. Anyone looking at Figure 4.1 would first say "which dam did that?," and pin the blame on the water developments for which the Columbia River is so famous. The trouble is, *none* of the developments coincides very closely with the change, and in fact it is remarkable how little effect the various developments seem to have had on recruitment rates per chinook spawner. My suspicion is that the change was due to something else entirely—a rapid increase during the early 1950s of commercial salmon troll fishing off the Pacific coast of Oregon, Washington, and British Columbia. The recruitment rates in Figure 4.1 are measured as the number of fish surviving their ocean life to return to the Columbia River mouth, and troll harvests of them during ocean residence cannot be measured accurately because the harvests are taken in areas where many stocks are mixed. Even a small change in troll effort could have resulted in large changes in ocean mortality, because many of the fish are exposed to trolling for at least two years of ocean residence. In fact, total catch statistics indicate that the effort increase was large.

An uncertainty threshold principle

We have seen that learning rates about uncertain parameters can be quite slow, and that these rates are dependent on how much the managed system is disturbed through changing harvest policies. If so-called parameters are also changing over time, it is natural to ask whether, or under what policy conditions, learning rates can even keep up with parameter changes at all. To examine this question, let us take the parameter covariance matrix Σ_{β_t} as a basic measure of uncertainty, and examine how it propagates over time. To simplify the discussion a bit, let us look only at the case where estimation can be reduced to a problem in linear regression; the conclusions are equally valid for the more complicated algebra of nonlinear filtering.

According to the standard linear regression model, Σ_{β_t} [equation (7.13)] should decrease monotonically over time. But if we assume that parameters may change between observations by random movements, with covariance matrix V_w (diagonal elements of V_w measure average movements, off-diagonal elements measure correlations between moves by different parameters), equation (7.13) should be expanded as

$$\Sigma_{\beta_t} = \Sigma_{\beta_{t-1}} + V_w - \frac{(\Sigma_{\beta_{t-1}} + V_w) Z_t Z_t' (\Sigma_{\beta_{t-1}} + V_w)}{\sigma^2 + Z_t' (\Sigma_{\beta_{t-1}} + V_w) Z_t} \qquad (7.23)$$

Let us examine this rather messy equation simply as a dynamic model, forgetting for a moment that the "state variables" Σ_{β_t} have anything to do with uncertainty. What we see immediately is a model with a positive growth term, V_w, and a negative feedback term that is a complicated function of the state, the growth term, and an external input (the independent control variables Z_t). Obviously the growth term can exceed the negative feedback term; the simplest condition for this would be $Z_t = 0$.

Now, any linear regression of the form $Y = \Sigma \beta x + w$ (where the w's are independent) is exactly equivalent to the regression $y - \bar{y} = \Sigma \beta (x - \bar{x}) + w$, since we can write $y = \Sigma \beta (x - \bar{x}) + \beta \bar{x} + w$ and $\bar{y} = \beta \bar{x}$. This means that the inputs Z_t in (7.23) can be measured as disturbances from their historical means $\bar{Z}_t$, provided we are careful to keep track of how each new Z_t alters $\bar{Z}_t = [(t - 1) \bar{Z}_{t-1} + Z_t]/t$; recursively, the disturbance $Z_t - \bar{Z}_t$ is equal to $[(t - 1)/t] (Z_t - \bar{Z}_{t-1})$. If we propagate equation (7.23) forward in terms of these disturbances, we see that constant Z_t values will stop having any effect on Σ_{β_t} once $\bar{Z}_{t-1}$ becomes stable at the constant value chosen. So if the mean historical inputs $\bar{Z}$ and response $\bar{y}$ are well fixed by the data, the negative feedback term in (7.23) will vanish, and Σ_{β_t} will grow by increments V_w, unless there is disturbance in inputs away from $\bar{Z}$. Indeed, *input disturbances must exceed some threshold values in order to prevent Σ_{β_t} from growing.*

We see an idea here analogous to the notion of multiple equilibria and "domains of stability" in ecological dynamics. *There is a domain of input (state, control) combinations Z_t, whose size is determined by current uncertainty Σ_{β_t}, random errors σ^2, and expected rates of parameter change V_w, such that "choices" of an input combination from within this domain will result in uncertainty actually increasing over time.*

To illustrate this very disturbing idea, let us return again to the almost trivial example of a Ricker stock–recruitment curve with only one unknown parameter {recall that $R_t = S_{t-1} \exp [\hat{a} - b(S_{t-1} - \bar{s}) + w_t]$, $\hat{a}$ and $\bar{s}$ are known}. In this case we have the regression model $y_t = b Z_t + w_t$, where $y_t = \ln (R_t/S_{t-1}) - \hat{a}$, and $Z_t = S_{t-1} - \bar{s}$ (a deviation!). The recurrence relationship (7.23) reduces to a scalar dynamic model for Σ_{b_t}:

$$\Sigma_{b_t} = \Sigma_{b_{t-1}} + Q - \frac{Z_t^2 (\Sigma_{b_{t-1}} + Q)^2}{\sigma^2 + Z_t^2 (\Sigma_{b_{t-1}} + Q)} \tag{7.24}$$

where $\sqrt{Q}$ is a rough measure of the expected or average change $|b_t - b_{t-1}|$ per generation in the actual responsiveness of productivity to spawning stock. If we plot Σ_{b_t} for this model as a function of the absolute spawning stock disturbance $|Z_t|$ (Figure 7.4), we see that Σ_{b_t} decreases monotonically as $|Z_t|$ is increased. At $Z = 0$, it has the value $\Sigma_{b_{t-1}} + Q$, and at some nonzero Z^* it will just equal $\Sigma_{b_{t-1}}$ (i.e., no change in uncertainty). Thus Z^*

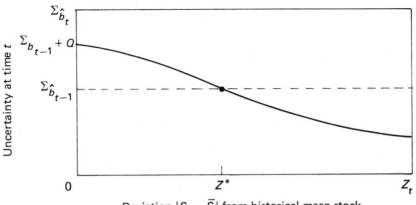

Figure 7.4. When a supposed model parameter b is actually expected to change slowly over time, uncertainty about its current value as measured by the variance $\Sigma_{\hat{b}_t}$ will increase over time ($\Sigma_{\hat{b}_t} > \Sigma_{\hat{b}_{t-1}}$) unless the system state is disturbed by an amount greater than Z^*.

divides the disturbance into two domains: for $|Z_t| < Z^*$, uncertainty will increase from $t - 1$ to t; for $|Z_t| > Z^*$, uncertainty will decrease from $t - 1$ to t. Z^* can be calculated easily; it is

$$Z^* = \left[\frac{\bar{Q}\sigma^2}{\Sigma_{\hat{b}_{t-1}}(\Sigma_{\hat{b}_{t-1}} + Q)} \right]^{0.5} \tag{7.25}$$

From this equation we see the rather obvious fact that the level of disturbance Z^* needed to prevent increasing uncertainty gets bigger as the expected rate of parameter change (Q) gets larger and/or the environmental variation (σ^2) gets larger. Perhaps a little less obvious is that uncertainty is easier to decrease if it is larger in the first place (Z^* gets smaller as $\Sigma_{\hat{b}_{t-1}}$ gets larger); however, this fact is hardly comforting, since it only ensures that $\Sigma_{\hat{b}_t}$ will not just keep increasing toward infinity if there is at least some disturbance in S_{t-1}. To get a little better feeling for the practical implications of equation (7.24), let us plug in a few representative estimates for Pacific salmon. Letting $\bar{s} = 1$ (i.e., be the unit of measurement of S_t), and assuming $\sigma^2 \approx 0.1$, $\Sigma_{\hat{b}_{t-1}} = 0.01$, we find that the disturbance required to hold uncertainty steady is around 0.5 (i.e., 0.5 $\bar{s}$) if $\sqrt{Q} = 0.05$, and is around 0.3 if $\sqrt{Q} = 0.01$. That is, for typical uncertainty about Pacific salmon, we would need to introduce (or allow) disturbances of $\pm 30\%$ in S_t just to hold uncertainty steady if only a 1% change in b is expected per year! In practice, the statistical situation is usually much worse because of measurement errors in

S_t (see Ludwig and Walters, 1981). Also, b is at least partly a measure of habitat "capacity" for producing more recruits when spawners are increased; considering just human activities associated with salmon spawning streams, I would be surprised if actual habitat capacities are changing by so little as 1% per year.

A more complicated example of uncertainty domains is provided by the problem of assessing impacts of fishing and lamprey control on the survival of lake trout in the Laurentian Great Lakes (see Chapter 5). Using various statistical rituals on samples of the age distribution of the stock, it is possible to obtain reasonable estimates of the total instantaneous mortality rate z_t suffered each year by the trout (Pycha, 1980). Provided the stock size is not changing too rapidly, we expect this total mortality rate to be roughly proportional to lamprey abundance (say L_t) and fishing effort (say E_t):

$$z_t = M + q_1 L_t + q_2 E_t + w_t \tag{7.26}$$

where the intercept M is an estimate of "natural" mortality rate, q_1 and q_2 are catchability coefficients (fraction of stock taken by one lamprey and one unit of fishing effort, respectively), and w_t represents all sorts of measurement and process errors that we need not discuss for the purpose of this example. L_t and E_t have been inversely correlated in recent years (lamprey decreasing slowly, fishing effort growing), but there has been enough contrast in their effects to provide at least reasonable estimates of M, q_1, and q_2. However, these coefficients will almost surely change over time. The lamprey coefficient q_1 depends on the abundance of lake trout and other prey fish, and on the distribution of all prey species relative to the lamprey's spawning rivers; these dependences are complex and involve at least some variables that are not adequately monitored. Likewise, the catchability coefficient q_2 will change as fishermen develop new tackle and acquire more mobile vessels; again the detailed factors cannot be predicted or monitored completely. So let us ask what level of variation in lamprey abundance L_t and/or fishing effort E_t over time will be necessary at least to maintain current levels of uncertainty about q_1 and q_2. To simplify the presentation a little, let us assume that $\bar{z} = M + q_1 \bar{L} + q_2 \bar{E}$ is well determined from historical experience, so the model (7.26) can be compressed to

$$y_t = q_1 Z_{1t} + q_1 Z_{2t} \tag{7.27}$$

where $y_t = z_t - \bar{z}$, $Z_{1t} = L_t - \bar{L}$ (deviation in lamprey abundance from historical average), and $Z_{2t} = E_t - \bar{E}$ (deviation in effort from historical average). Using this compression, we will obtain somewhat overly optimistic estimates of the variation in L_t and E_t needed to reduce uncertainty about q_1 and q_2.

Since in this example we have some control of two input variables (L_t and E_t) with uncertain effects measured by q_1 and q_2, we must be more careful to say what is meant by "growing uncertainty." Some control choices (combined changes in L and E) may permit growth in the variance of q_1, while actually reducing uncertainty about q_2, and vice versa. The obvious approach here is to find regions of choices that result in increasing variance for each parameter, and from these identify the region of overlap where both parameters would become more uncertain. Equation (7.23) permits us to do this in a straightforward fashion; the region of increasing uncertainty for q_1 is the set of all Z_{it} $(L_t - \bar{L}, E_t - \bar{E})$ combinations such that Σ_{11} (variance of $\hat{q}_1 = \hat{\beta}_1$) is larger at t than at $t - 1$, and the region for q_2 is the set of Z_{it} combinations such that Σ_{22} is larger. Looking at the negative feedback term in (7.23), it is an exercise in algebra to show that these two regions are defined by quadratic forms, and are therefore elliptical in shape. The orientation of the ellipses is set by Σ_{12} which measures correlation between $\hat{b}_1$ and $\hat{b}_2$ due to correlation in past input values. In the example we know that this correlation will be negative.

Figure 7.5 shows how the domains of increasing uncertainty for q_1 and q_2 look, for the hypothetical parameters $\sigma^2 = 0.1$,

$$
\Sigma_{\hat{\beta}_t} = \begin{bmatrix} 0.13 & -0.12 \\ -0.12 & 0.13 \end{bmatrix} \qquad V_w = \begin{bmatrix} 0.002 & 0 \\ 0 & 0.001 \end{bmatrix}
$$

(i.e., roughly 30% coefficients of variation for measurement error and for the current parameter estimates, 4% expected change per year in q_1, and 3% change per year in q_2). The domain of joint increases in uncertainty is long and narrow, indicating that uncertainty is likely to increase from $t - 1$ to t only if effort increase is accompanied by a lamprey decrease, or effort decrease by a lamprey increase. Unfortunately, precisely such combinations of input changes are likely to arise through management, since it has been a policy to encourage more fishing as lamprey abundance has been reduced, and local increases in lamprey abundance have been followed by restrictions to fishing.

Management donuts

When I showed the lamprey/fishing example above to my dear friend and colleague Joe Koonce, he said immediately "good grief, that means they have to manage in a donut!" Koonce is given to inexplicable outbursts at times, but in this case I think he made a profoundly important point that Figure 7.6 tries to show as a crude diagram. The argument is quite simple: ecological systems tend to show stability domains, bounded by combinations of policy and state variable values that result in abrupt and perhaps irreversible changes. Chapter 5 presented a model of effects of lamprey control and fishing on lake trout, and it predicted a sharp drop in equilibrium stocks if

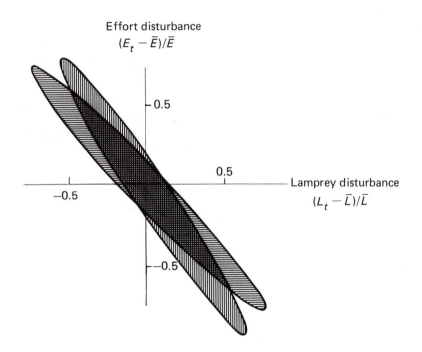

 Uncertainty in fishing effort effect (q_2) increasing over time
Uncertainty in lamprey effect (q_1) increasing over time
Uncertainty in both effects increasing over time

Figure 7.5. Domains of change in two control variables, harvesting effort and lamprey abundance, such that changes within the domains will result in increasing uncertainty over time about the effect of each control variable on the total mortality rate of lake trout.

either of those controlled variables increase too much. Such obviously dangerous policy combinations define the outside of Koonce's donut. But on the other hand, we see from the previous section and Figure 7.5 that it is necessary to induce or permit at least some changes in order to even measure the control effects and thereby hope to better predict and avoid the outside of the donut. The hole in the donut represents policy combinations that are undesirable in view of this need for continuing adaptive learning.

Points within the management donut represent levels of disturbance or variation in management actions that strike some reasonable balance between avoidance of danger, versus probing that is necessary to even estimate where the danger lies. The size of this domain for balanced action is a

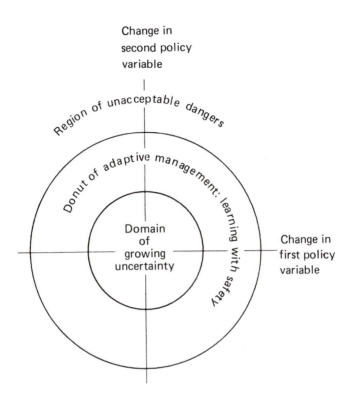

Figure 7.6. Koonce's donut. Changes in policy variables must be reasonably large to allow learning about policy effects, but very large changes imply unacceptable risks.

function of the state of the art in modeling and monitoring, which determines the magnitude of measurement and process errors in estimation schemes, as well as the resilience of the managed system itself. Improved monitoring and estimation schemes cannot reduce the outside diameter of the donut, but they can increase the donut's size by making the hole smaller.

Design of Informative Input Sequences

Let us close this chapter with a brief discussion about the design of input (disturbance, control, harvest) sequences that maximize learning rates about uncertain parameters. A policy that is designed solely to gain information would rarely be optimum in any broad management sense, but it can provide a useful backdrop against which to compare policies designed with other objectives in mind. Also, when uncertainty is very high (preadaptive phase of management), an optimum experimental policy may be a better initial management plan than policies that emphasize slow development and avoidance of risks.

Measures of learning performance

If we want to talk about an "optimum" input sequence for learning, we must first decide on a reasonable criterion (objective function) for ranking alternative sequences against one another. It is quite possible, for example, to design a harvesting policy that will allow precise determination of the intrinsic rate of increase of a population (r of logistic, Ricker a, etc.), yet give no information about other parameters (just drive the population way down, then watch it grow). Once a criterion is selected, it is usually necessary to search for a maximum (or minimum) value by numerical optimization procedures, so it is important that the criterion not be too difficult to compute. Four possibilities that are relatively easy to compute are outlined here.

If it is not clear how uncertainty about particular parameters will affect later policy decisions, we may take some general measure of uncertainty based on $p_t(\beta)$ as the criterion to be minimized. Here a natural choice would be the Shannon information statistic $I = \Sigma_i \, p_i \ln p_i$, which for continuous variables is written as an integral

$$I = \int_\beta p_t(\beta) \ln p_t(\beta) \, d\beta$$

Obviously this statistic can be very hard to compute for general distributions $p_t(\beta)$, but if we approximate p_t as a normal distribution with mean $\hat{\beta}_t$ and variance $\Sigma_{\hat{\beta}_t}$, then the Shannon statistic is proportional to the log of the determinant of $\Sigma_{\hat{\beta}_t}$:

$$I = k_1 + k_2 \ln \left| \Sigma_{\hat{\beta}_t} \right|$$

(see Bard, 1974, Chapter 10). k_1 and k_2 can be ignored. For an "experiment" involving control choices $u_1, \ldots, u_\tau$, the sequence of τ values that is predicted to minimize $\ln \left| \Sigma_{\hat{\beta}_\tau} \right|$ is then taken as the optimal input design. For computational purposes, it is usually simpler to maximize an equivalent measure, $\left| \Sigma_{\hat{\beta}_t}^{-1} \right|$, since the elements of Σ^{-1} can be approximated by

$$\Sigma_{ij}^{-1} = \sum_{t=1}^{\tau} \left(\frac{\partial \hat{y}_t}{\partial \beta_i} \right) \left(\frac{\partial \hat{y}_t}{\partial \beta_j} \right) \frac{1}{\sigma_{y_t}^2}$$

i.e., Σ_{ij}^{-1} is the sum of cross products of the sensitivities of predicted outputs $\hat{y}_t$ to the parameters β_i and β_j, weighted inversely by the expected variance of the observation y_t. Notice that the evaluation of Σ_{ij}^{-1} usually requires some initial or prior assumption about β, since the predicted sensitivities depend on β.

An even simpler criterion to minimize is the approximate variance of the prediction error $y_\tau - \hat{y}_\tau$ at the end of the τ-year experimental period. The variance of this prediction is roughly approximated by $\sigma_{y_\tau}^2 + s_\tau' \Sigma_{\beta_t} s_\tau$, where $s_{i\tau} = \partial \hat{y}_\tau / \partial \beta_i$. As with the Shannon information criterion, we must presume to predict Σ_{β_t} for any input sequence $u_1, \ldots, u_\tau$; then we seek the sequence that will minimize $s' \Sigma s$.

A third, related criterion can be defined if there are particularly important policy variables H, such as optimum equilibrium harvesting effort or optimum escapement, that can be expressed as functions of the original model parameter β. Suppose there are K such variables, and $H_k = f_k(\beta)$, where $k = 1, \ldots, K$. Then the covariance matrix of the vector H is given approximately by $\Sigma_{\hat{H}} = S' \Sigma_{\beta_\tau} S$, where the matrix S is defined by $S_{ij} = \partial H_i / \partial \beta_j$. The obvious criterion to minimize is then the information statistic for H, $\ln |\Sigma_{\hat{H}}|$. This criterion has the advantage that there are usually fewer key policy variables H than uncertain parameters β ($|\Sigma_{\hat{H}}| = 0$ if there are more H's than β's). Parameters that do not directly affect H will enter the analysis only through their effects on uncertainty about the parameters that do affect H (these effects are measured in Σ_β).

Notice that the above three criteria all represent weighted sums and cross products of the elements of the predicted covariance matrix Σ_{β_τ}. They depend on prior estimates $\hat{\beta}_0$ of $\hat{\beta}_\tau$ only in so far as the elements of Σ_{β_τ} depend on the parameters; this dependence can be both direct (β appearing in $\partial \hat{y} / \partial \beta$) and indirect ($\beta$ affects $\hat{x}_t$, $\hat{y}$ depends on $\hat{x}_t$). Qualitatively similar predictions about the effect of control on Σ_{β_τ} are likely to be obtained for a wide range of prior estimates $\hat{\beta}_0$, but this must be checked in each nonlinear case by repeating the optimization for a range of $\hat{\beta}_0$ possibilities.

A final and rather more ambitious criterion would be the predicted decrease in expected value of perfect information EVPI (see above section). Recall that EVPI is calculated from $p_t(\beta)$ and from control choices u^* and u_β^{**} defined by the models considered. Thus, if we can predict how p_t will propagate, EVPI can be calculated at any time step τ. However, this means redoing one optimization (to determine u^* at time τ) for each control sequence examined. If this optimization is not too involved, then the optimal sequence can be sought as the one which *minimizes* EVPI at time τ.

None of these criteria gives much immediate insight about what policy sequence would be most informative, except when they are applied with very simple models and estimation schemes. However, in just trying to understand them one is forced to think a little more carefully about what uncertainty means, and about how we should measure it in management contexts where there are objectives other than just reducing uncertainty about unknown parameters.

Properties of informative policies

There has been some experience in numerically computing policies that optimize the information criteria outlined above (see, for example, references in Bard, 1974; Mehra, 1974). These studies indicate that the particular criterion used can make a big difference, but a few general (robust) properties of the solutions have been noticed. By a "solution" we mean an input or control sequence $u_1, \ldots, u_\tau$ that optimizes one of the criteria listed above, conditional on some reasonable prior estimate $\hat{\beta}_0$ if one is needed to predict the propagation of $p_t(\beta)$ over the τ "experimental periods."

The first noteworthy property of such solutions is that they are usually not unique. There will be several control sequences that give the same, or nearly the same, expected learning performance. These sequences are likely to represent a repeated cycle of inputs (like $u_1, \ldots, u_4$ = high, low, low, low; then $u_5, \ldots, u_8$ the same), and nonuniqueness arises from insensitivity to where in the cycle one starts at $t = 1$. A similar effect arises in classical experimental design, where it is usually best to replicate a small set of experimental combinations several times.

A second important property is that it is usually best to use only extreme controls, such as the lowest and highest feasible harvest rates. These extreme controls generate the greatest possible contrast over time in the system states achieved as well as the direct control responses observed. So, for example, if one wishes to estimate parameters like the logistic r and k, the best harvest sequence will first drive the stock down as rapidly as possible, so responses at low stock sizes (effects of r) are visible. Then it will be best to shut down harvesting so that the stock moves as rapidly as possible up to levels where the effects of k become visible. Obviously the best predicted sequence (high, low, low, low, ..., high, low, low) will depend on how hard it is feasible to harvest, on how long the recovery is predicted ($\hat{\beta}_0$!) to take during periods of low harvest, and on the length of experimental period (τ) available. Longer experimental periods will favor the use of longer, more extreme depletion and recovery cycles. When exploitation parameters (i.e., catchability q when effort is the control) are also uncertain, the best sequence may involve a repetition of short cycles (high, low, high, low, etc.) that hold the state nearly constant while providing contrast in direct control effects.

Finally, when the system model predicts strongly "nonlinear behavior," such as the catastrophic collapses and recoveries predicted for lake trout responding to lamprey control and harvesting (see Chapter 5), analyses by Mehra (1981) indicate that the best experimental policy will be one that keeps the system where it will exhibit the strong behavior as much of the time as possible. So if there is uncertainty about the location of a "cliff edge," the boundary of a domain of stability, or the outer rim of a management donut, the optimum experimental policy will be to drive the system back and forth across the state/control combinations that are thought to contain this location. This is obviously a very dangerous prescription from the management point of view, and highlights the basic conflict between learning versus maintenance of temporal stability and avoidance of risks.

A common management prescription that is definitely not optimal from the viewpoint of experimental design is to slowly increment the inputs over time so as to keep the system near equilibrium. Such policies result in strongly correlated inputs, and in state variables being correlated with inputs (i.e., stock size negatively correlated with harvesting effort), so the effects of each cannot be distinguished. These individual effects would not matter if we could assume the equilibria to be stable and repeatable over time (i.e., no parameter changes), and if τ were large enough to permit a leisurely search for the optimum equilibrium (which is a reduced function H of the original β). Unfortunately, neither of these conditions is met in practice.

A challenging area for future research is in the design of informative policies for problems where the appropriate model structure is highly uncertain, and where part of the design problem is to decide what set of output variables y_t to monitor when resources for monitoring are very limited. Most dynamic models for renewable resources contain functionally similar parameters for response under extreme states (r and k, for example), and so imply similar learning policies. But there is growing evidence that responses away from these extremes are not always "logistic-like" (dome-shaped), and may involve such things as multiple equilibria. If a design is adopted that assumes some logistic-like response when, in fact, the behavior at intermediate states is more complex, the extreme inputs associated with that design will either fail to detect the nonlinearities or drive the system into a state from which recovery (and therefore later learning) is impossible. Such arguments imply that robust design should involve less extreme input sequences than would be recommended on the basis of very simple models, but it is not yet clear how to strike a reasonable balance that avoids the confounding effects associated with incremental input changes.

Policies that permit natural variation

As we shall see in the next chapter, feedback policies to maximize average harvest may lead to reduced variation in the stock size after harvest (escapement) each year. Such policies will obviously reduce learning rates about the relationship between stock size and production, and Chapter 9 will argue that it is often worthwhile to deliberately reintroduce escapement variation through probing experiments. As a compromise between these extreme policy choices, one suggestion has been to use a constant harvest rate (constant proportion harvested) policy that will partially stabilize the escapement, but will also allow some informative variation through natural, random effects on the stock size before harvest each year.

Unfortunately, this compromise policy can lead to severe bias in estimates of production parameters (see Appendix 5A), in a direction that favors overexploitation. Intuitively, the basic problem is that each potentially informative natural variation has a double effect: it causes the stock size to change (as desired), but it also immediately favors parameter values or hypotheses for which the variation is considered more likely. In other words, it cannot immediately be determined whether the variation represents a large deviation from the correct model for average response, or a small deviation from a response model that will later be proved incorrect. This ambiguity is exaggerated in situations where the deviations are autocorrelated (tend to occur in runs or cycles). The bias disappears for large sample sizes, but over long time scales it is dangerous to assume that parameters are constant in the first place. Ultimately, "natural experiments" or "dithering" cannot be trusted as a source of informative variation.

Problems

7.1. As introduced above, the expected value of perfect information (EVPI) is calculated for a fixed set of alternative models and a fixed set of policy choices, by first finding the best choice if there is no learning (never learn which model is correct), then looking at how bad this choice is compared to the best choices for each of the alternative models. Identify the two basic reasons why EVPI may be small in some cases, and give graphical examples (e.g., alternative models of surplus production as a function of exploitation rate choice) of such cases.

7.2. Monte Carlo simulation studies have confirmed much sad experience with management experiments, demonstrating that it generally takes many years to determine which response hypothesis is correct. Management agencies generally cannot guarantee commitment of financial resources (manpower, equipment, etc.) over such long time scales, and the original plan of actions and patient evaluation may be forgotten as the personnel who conceived it are retired or move to new positions. Suggest practical steps for overcoming such institutional difficulties.

7.3. We have seen that it is possible for uncertainty to actually increase over time, if key "parameters" are changing slowly, but the system is managed so as to prevent these changes from becoming apparent. A particularly nasty and common example is with the "catchability coefficient" q (fraction of stock taken by one unit of harvesting effort), through which we interpret catch and effort data using assumptions like $H_t = qE_tN_t$ (catch H_t is proportional to effort E_t and stock N_t). Suppose effort is held constant over several years, and that catches likewise remain steady. What two hypotheses can equally well explain this temporal stability? Identify at least three strategies for resolving which hypothesis is correct, and identify which management actors (management agency, harvesters, future harvesters) would bear the direct "cost of learning" for each strategy. In view of who bears the costs, which strategy is likely to be the "course of least resistance" politically?

7.4. Develop a microcomputer simulation to demonstrate how harvest policies will affect learning rates for a stock–recruitment system, where recruits R_t are generated from previous spawners S_{t-1} by the stochastic Ricker model $R_{t+1} = S_te^{a-bS_t+w_t}$ with $a = b = 1$, w_t normally distributed with mean zero and standard deviation $\sigma_w = 0\text{-}0.5$. Consider two extreme cases for the initial stock size $S_0(S_0 = 0.1, S_0 = 1)$ so the stock may be initially overexploited or not yet touched. Estimate a and b over time by the recursive linear regression equations (7.12)–(7.15), with

$$\Sigma_{\hat\beta_0} = \begin{bmatrix} 10^6 & 0 \\ 0 & 10^6 \end{bmatrix}$$

$$\hat\beta_0 = \begin{bmatrix} a \\ b \end{bmatrix}_0 = \begin{bmatrix} 0 \\ 0 \end{bmatrix}$$

$$y_t = \ln \frac{R_t}{S_{t-1}}$$

$$Z_t = \begin{bmatrix} 1 \\ -S_{t-1} \end{bmatrix}$$

For the first four years of each simulation, use a constant exploitation rate of 0.5, so $S_t = 0.5 R_t$, $t = 1, \ldots, 4$. Then try several harvest strategies, ranging from continued constant exploitation (so S_t varies only due to random effects of w_t on R_t) to constant escapement (at $S_t = 0.5$ each year) to a "probing policy" where you attempt to have S_t range as widely as possible.

7.5.   Develop a program to recursively estimate the logistic parameters R_0 and R_1 for the George Reserve deer herd (problems 1.3, 6.4), assuming only process errors. How does your confidence in R_0 and R_1 improve over time, particularly over the long recent period where the population has been held relatively constant by harvesting? Then include a discount factor for older data [equation (7.22)]; as you vary λ from 0.8 (use 5 years' data) up to 0.95 (use 20 years' data), how do the parameter estimates vary over time? Can you see evidence that the "carrying capacity" (inversely proportional to R_1) has decreased over time?

7.6.   Estimate EVPI for the Fraser River sockeye salmon situation in Figure 1.1, while assuming that (1) the two models η_1 and η_2 are assigned equal prior probability; (2) the management objective is to maximize average annual yield; (3) the optimum policies u^*, u_1^{**}, and u_2^{**} are each a fixed escapement strategy. u_1^{**} is to use that fixed escapement that will maximize average catch if η_1 is known to be true, and u^* is to use the best escapement if there is no chance to learn which model is correct. Note that the best escapement for u^* is the one that will maximize the expected catch $0.5\,\bar{C}_1 + 0.5\,\bar{C}_2$, where $\bar{C}_1$ is the average catch if η_1 is true and $\bar{C}_2$ is the average catch if η_2 is true ($\bar{C}_1$ and $\bar{C}_2$ both depend on the escapement choice). Considering Figure 7.1, how long would it likely take to determine which model is correct if the u^* choice is actually adopted?

Chapter 8

Feedback Policy Design

In essence, our approach is that management of a natural resource, and also the sampling processes by which we learn about the resource, can be thought of as a game. The object of research is to gain insight into the structure of the game and determine the procedure that optimizes the outcome when the game is played.

Watt (1968)

The notion of a feedback policy was introduced in Chapter 2, then used rather loosely in later analyses, such as the section in Chapter 7 on the value of perfect information. Before proceeding with discussions about actively adaptive policy design in Chapter 9, we need to make the notion more precise and examine some of the problems associated with computation of optimum feedback policies in general.

The intuitive notion used so far is that a feedback policy is a *conditional program of actions*, where the management action at each decision point in time is chosen as a fixed function of the "system state" at that point. It makes obvious sense to develop such a program when the system state is not predictable in advance, if for no other reason than to force a strategic statement of how various management objectives should be weighed against one another in the event of (i.e., contingent upon) extreme changes that require painful trade-offs.

In the absence of a feedback policy, each unanticipated system change must be either ignored, or made the basis for an open-ended new evaluation of objectives and actions. Policies that simply ignore changes and stick to previously planned actions are known as "open-loop policies" (no feedback "loop" from system to actions). It makes little sense to even talk about open-loop policies in the usual resource management contexts, where at least some opportunities exist for monitoring and evaluation as management proceeds.

The other extreme, of wide reevaluation at each decision point, is much more interesting. Many management authorities, such as international fisheries commissions, actually proceed in this way; major decision points are associated with events, such as annual meetings, where there is a more or less thorough review of historical data and "models" for system response, and intensive debate about strategic objectives and the best immediate tactics to follow. The possibility of such reevaluation raises a serious question for anyone who proposes an analysis of alternative feedback policies: what factors should be treated as variables in the definition of system state?

For example, suppose we want to specify how annual harvests should be varied in response to changes in stock size due to unpredictable events, such as the appearance of strong or weak year classes. Should the catch be made a fixed function only of total stock size, or should recruitment estimates be treated as separate variables? Should the function have fixed parameters, based on some model(s) whose parameters have in turn been estimated from historical data, or should we assume that the parameter estimates will be updated each year as new information becomes available? (If so, the "parameters" are variables of the feedback policy function.) Particularly this latter question can have profound importance for adaptive policy design: if we admit that management actions may vary in an informative manner just due to revision of parameter estimates, we may well conclude that it is unnecessary to deliberately introduce further "experimental" variation. Policies that rely just on parameter revision to provide informative input contrast are called *passively adaptive policies*, while those that include deliberate probing for information are called *actively adaptive policies*.

In thinking about the questions raised above, it is useful to recognize very clearly that feedback policies are something more than just programs for responding to change. Rather, they are programs *designed with at least some anticipation of future responses to the actions taken*. That is, feedback policies are not strictly substitutes for accurate prediction, since we cannot even design them without some model(s) for the effects that they will have! A friend of mine who misunderstood this point once gave a definition of a feedback policy as any rule for adaptation to changing circumstances. The difficulty with this definition is that some responses to changing system state, all of which he referred to as "adaptation," can actually prevent learning about which responses would actually be best. For example, it is a feedback policy to allow the same fixed number of salmon to spawn in a river each year, in the sense that harvest is made a function of recruitment. But, if successful, this policy absolutely prevents learning about what spawning stock would result in the best recruitment. Feedback response to short-term changes should not be confused with longer-term learning and adaptation to changing system structure.

In this chapter we will first look at some alternative state measures that can be used as the basis for feedback policy design. Then we will examine practical problems of computing "optimum" feedback policies given a well defined model and management objective. Finally, we will look into the stability and passive adaptation properties of feedback policies that are designed without deliberately taking account of how actions affect learning rates.

Feedback and Information States

It is a truism to state that the best management decision to make at any point in time is some function of all the information available at that time. In terms of the shorthand introduced in Chapter 6, we say that the optimum vector of actions u_t^* is a function of the alternative models identified as plausible M, of the historical data available Y_t, and of prior information available about which model is correct, measured by probabilities $P_0(m_i)$. Of course, u_t^* is also a function of the management objective(s), which hopefully can be quantified (through measures such as total catch, employment, and profits to the harvesting industry), but for now let us concentrate on the dependence on M, Y, and P_0.

Sufficient information states

The first problem that we must face in feedback policy design is that the total "information state" $\{M, Y_t, P_0\}$ is growing in dimension over time. Obviously this happens with the data set Y_t (unless older data are discarded), and it is not unusual to see new models m_i proposed as scientists find fresh viewpoints about key variables or ways to deal with existing concepts. But P_0 can also change over time, as experience with and analysis of similar systems leads to greater confidence about common processes and parameters. We cannot, in principle, foresee all such changes, but we can at least try to find information state representations for Y_t that do not involve viewing the feedback policy as growing pathologically in complexity over time.

Much of the literature on feedback policy design begins by pretending this problem away. It is assumed at the outset that Y_t and P_0 are adequate to define the "correct" model m_i^* exactly, and that the system state x_t can be measured precisely. If m_i^* is of the form $x_{t+1} = f(x_t, u_t, w_t)$, then there is no reason to look back into the history of the system at all, since that history is fully summarized by the current state x_t. In other words, for any choice u_t, there is no information about the future of the system that is not contained in $f(x_t)$ and the probability distribution for random effects w_t. This very bold pretense has been justified in two ways. First, it makes the mathematical analysis more tractable, and hence the analyst can publish more papers. Second, and more defensible, is that the policy computed by assuming away uncertainties except w_t is often close to or identical to the actual optimum policy based on the full information state. This second argument should begin to sound suspiciously familiar; we have returned to the introductory remarks in Chapter 6, where I noted in passing that "certainty-equivalent" policies are optimum for a special class of problems, including linear-quadratic control systems. For a more precise and insightful discussion about certainty equivalence, see Bar-Shalom (1981). Unfortunately for resource analysts, a key requirement for certainty-equivalent policies to be optimal is that the learning rates about unknown states/parameters be independent of the control choices u_t.

But the previous paragraph makes a key point: the optimum feedback policy should be a function only of those variables that contain unique information about the future of the system. To clarify the point, let us think of the feedback policy as a "black box" that outputs a policy choice u_t whenever you give it a vector of inputs I_t (the input vector I_t is an arbitrary information state). Now, if the black box is to output the optimum u_t^*, then *either* it must be able to place the best possible odds on all future outcomes that might arise from applying each choice u_t, *or* it must be able to compute u_t^* without reference to these odds (certainty equivalence). It is the former case which is of interest to us here, and the question becomes: what is the minimal input (information) state I_t^* necessary for the black box to act as if it were reanalyzing the whole information state $\{M, Y, P_0\}$ anew at each decision point t, so as to place the best possible odds on all future outcomes? This question should sound quite familiar if you have read Chapter 6, where we discussed the notion of "sufficient statistics" that capture all of the information about $P_t(m_i)$ that is contained in the original data set Y. To the extent that "placing odds on future outcomes" means using the probabilities $P_t(m_i)$, the sufficient statistics for P_t are at least part of the minimal information state I_t^*.

When the system state x_t can be measured exactly (no observation model necessary), then placing odds on future outcomes (in the determination of u_t^*) involves knowing only x_t, $P_t(m_i)$, and the distribution of environmental effects w_t. We should be able to design a black box policy that has evaluations of the effects of w_t "built in," so the minimal information state I_t^* is $\{x_t, P_t(m_i)\}$. When P_t can be computed from sufficient statistics $s_1(Y_t)$, $s_2(Y_t)$, ..., then I_t^* reduces to $\{x_t, s_1, s_2, \ldots\}$. In this case the black box must *in effect* reconstruct P_t from s_1, s_2, ..., then look forward at x_{t+1}, x_{t+2}, etc., using the odds defined by P_t.

The situation is more complex when only indicator measurements y_t = $h(x_t)$ are available. Then there is uncertainty about the state x_t as well as model structure and/or parameter values, and looking ahead involves placing joint odds of the form $p_t(x_t) P_t(m_i)$ on both the starting point and the rules for state transition. The feedback black box must try to reconstruct $p_t(x_t)$ from available data Y_t, as well as $P_t(m_i)$; the black box designer must seek sufficient statistics for $p_t(x_t)$ if the information state I_t^* is to be kept constant in dimension. Notice that this "extra" problem of estimating $p_t(x_t)$ vanishes, at least symbolically, when the alternative models m_i are generated by varying unknown parameters β that can be viewed as part of the state variable set x_t; in this case all uncertainties are thought of as represented in $p_t(x_t)$, where x_t is called the "augmented state vector."

Approximate information states

Let us contrast the above view of a feedback policy as a black box against the behavior of an intuitive decision maker engaged in harvest regulation. To make a harvesting decision u_t (allowable effort, quota, etc.) in year t, the decision maker would typically begin by plotting historical records of harvest and abundance (indexed, perhaps, by catch per effort). Then he would examine the abundance trend for evidence of harvest effects, and thereby construct an intuitive "model" for the stock response x_{t+1} = $f(x_t, u_t)$. Recognizing the possibility of measurement errors in the abundance time series, he would probably decide on a best estimate $\hat{x}_t$ by visually smoothing the abundance curves, thereby making $\hat{x}_t$ a function of the historical points $y_1, \ldots, y_t$. Finally, using the best estimate $\hat{x}_t$ and the intuitive model, he would visually project abundance changes for a range of choices u_t, and select that choice u_t^* which he expected would give the best balance of objectives (immediate harvest versus maintenance of stock). I am deliberately being a bit optimistic here about intuitive decision makers, but at least the good ones do try to form response models, recognize measurement errors, and make alternative projections of some sort. So what then would the black box do that is so different? The answer is nothing, really, except to follow each of the steps precisely and repeatably: placing odds on models replaces intuitive model construction, state/parameter estimation replaces smoothing, and optimization replaces balancing of objectives. The key difference is that, provided the black box has a reasonably small input set I_t^* so it does not have to "reinspect" the entire data history at each step, we can repeatedly simulate how it would perform in response to various things that can go wrong with models, data, and objectives. That means we can "train" it, by adjusting its components and parameters, in order to do a better job. In such training exercises, we are almost bound to find ways to improve upon purely intuitive feedback behavior.

For practical game playing and formal optimization with feedback policies, the information state vector I_t^* required by the policy function (u_t^* as a function of I_t^*) should have a very low dimension. This means that if we throw together unknown parameters and state variables into a single uncertain vector x_t, the probability distribution $p_t(x_t)$ needed to place odds on future outcomes must be characterized by a few statistics in I_t^*. Three options suggest themselves immediately:

(1) $I_t^* = \hat{x}_t$, i.e., base the policy function only on the best estimates of states and parameters (assume certainty equivalence).

(2) $I_t^* = \{\hat{x}_t, \Sigma_{\hat{x}_t}\}$, where $\Sigma_{\hat{x}_t}$ is the covariance matrix of $\hat{x}_t$. This is known as the "wide-sense" information state (Bar-Shalom, 1981). The basic justification for option (2), as mentioned in previous chapters, is that $p_t(x_t)$ is expected to be approximately normal around the mode $\hat{x}_t$.

(3) $I_t^* = \{x_t, P_t(m_1), P_t(m_2), \ldots\}$, when the state x_t is measured exactly and model uncertainty is represented in terms of just a few discrete equation/parameter combinations m_i. An advantage of this third approach is that, if there are N hypotheses, only $N - 1$ values of P_t need be included (Wenk and Bar-Shalom, 1980); two-model analyses require only a single uncertainty measure $P_t(m_1)$.

The "wide-sense" information state $\{\hat{x}_t, \Sigma_{\hat{x}_t}\}$ has been used very successfully to gain understanding about a variety of adaptive control problems (see references in Bar-Shalom, 1981), and we shall return to it again in Chapter 9 along with the option of assuming only a few discrete alternative models. The wide-sense information state $\{\hat{x}_t, \Sigma_{\hat{x}_t}\}$ is essentially a description of the first and second moments (means, variances, covariances) of the historical data set Y, along with sensitivities of predictions to uncertain states. Recall from Chapters 6 and 7 that $\Sigma_{\hat{x}_T}^{-1}$ consists approximately of elements like

$$ s^{-2} \sum_{t=1}^{T} \left(\frac{\partial \hat{y}_t}{\partial \hat{x}_i} \right) \left(\frac{\partial \hat{y}_t}{\partial \hat{x}_j} \right) $$

which reduce to moment estimators $\Sigma\, x_i\, x_j$ for systems that are linear in parameters. The approximate behavior of $\hat{x}_t$ and $\Sigma_{\hat{x}_t}$ can be predicted *a priori* by filtering models such as the extended Kalman filter introduced in Chapter 7. It would, in principle, be possible to develop filters that predict the dynamics of higher moments of the data [and therefore of $p_t(\hat{x}_t)$], but such moments are unlikely to make much difference in calculations of the expected values of alternative policy choices.

Another possibility is to base the policy function on best estimates $\hat{x}_t$ plus a single measure $\sigma_{\hat{x}_t}$ of uncertainty about each (standard deviation, range limit, or whatever). The trouble with this approach is that it does not "tell" the function about correlations among parameters associated with lack of historical contrast in Y, and so the function cannot be constructed so as to account implicitly for (or, if you like, anticipate) how these correlations might be reduced through informative variations in u_t. By always "seeing" uncertainties as independent of one another, it will in effect miss opportunities to reduce uncertainty about one factor by taking actions that will reduce uncertainty about another.

Computation of Optimum Policies

Given an information state description I_t, we seek next to find a feedback policy function $u_t^* = u(I_t)$ that will prescribe the best possible u_t^* for any information state I_t that may arise as a result of management history and natural events. It may seem intuitively like an almost impossible task to find such a function, even for simple definitions of "best possible" in terms of management objectives like maximizing sustained yield. But it turns out that there is a method of optimization, known as dynamic programming, which produces exactly the desired function for small problems where I_t has only a few (≤ 5) variables. These results can then be used to obtain some idea of the form of the policy function for more complex problems, and once this form is defined we can seek best values of its quantitative parameters; this is called "optimization in policy space."

The principle of optimality and dynamic programming

There is an enormous literature on dynamic programming, and for detailed introductions I particularly recommend the books by Richard Bellman (1957, 1961) that set the field in motion, and the texts by Larson (1968), and Larson and Casti (1978, 1982). Here I shall give only a very brief overview for readers who want an intuitive idea about how it works and why it results in estimates of feedback policy functions. To make this overview more understandable, I will omit most of the mathematical notation that usually makes dynamic programming appear very formidable at first, and I will discuss only the simplest formulations for stochastic dynamic problems.

Consider what it means to make the "best possible" management decision at some point t in time. If we call the immediate "reward" (payoff, harvest, net economic return, etc.) from this decision v_t, then a reasonable objective function for management is to try and maximize the sum, say V_t, of these rewards from time t forward to some arbitrary "end time" T:

$$V_t = v_t + v_{t+1} + \cdots + v_T \tag{8.1}$$

Notice that this way of defining the total future value V_t makes it possible to write (8.1) as a recursive relationship:

$$V_t = v_t + V_{t+1} \tag{8.2}$$

Objective functions like this are called "separable," in the sense that we can break up the total value into an immediate component (v_t) that combines simply with (adds to, multiplies, etc.) a longer term component (V_{t+1}) that is of the same functional form as V_t. In dynamic programming we assume that the "best possible decision" is the one which maximizes the expected value of

$$V_t = \sum_{i=t}^{T} v_i = v_t + V_{t+1}$$

That is, the best decision is the one which gives the best combined value v + V, or, in our earlier terms, the best "balance" between short- and longer-term rewards when there is some trade-off between v_t and V_{t+1}.

As we would typically measure them, the immediate rewards v_t are a function of the system state x_t and the decision choice u_t:

$$v_t = v_t(x_t, u_t) \tag{8.3}$$

Furthermore, any choice u_t will in general affect the next state x_{t+1}, which will affect v_{t+1}, and so forth. In other words, V_{t+1} is a function of x_{t+1}:

$$V_{t+1} = V_{t+1}(x_{t+1}) \tag{8.4}$$

Now we can define the basic *principle of optimality* for dynamic programming. Suppose that someone hands you, or that you have somehow already computed, the *best possible long-term value* $V_{t+1}^*(x_{t+1})$ *for every state* x_{t+1} *that might arise.* (Be careful here: what you have been handed is not a single number, but rather a function of x_{t+1}; it may help to think of this function just as a graph of the best value V_{t+1}^* versus x_{t+1}.) Suppose further that you have a model that assigns odds $p(x_{t+1})$ to each possible next state x_{t+1} when given the current state x_t and any action choice u_t. Then the principle of optimality simply states that the best value that can be obtained *from time* t forward, $V_t^*(x_t)$, for any x_t, is the one that maximizes

$$v(x_t, u_t) + \sum_{x_{t+1}} p(x_{t+1}) V_{t+1}^*(x_{t+1}) \tag{8.5}$$

where the Σ is across all states x_{t+1} that are assigned nonzero probability given x_t and u_t. Operationally, the principle of optimality says that in order to find $V_t^*(x_t)$, you just search across action choices u_t while calculating the value of each choice using equation (8.5), and take V_t^* to be the highest value that you find in this search. By repeating this little optimization for every possible state x_t, you can build up the function $V_t^*(x_t)$. It is a tedious business to repeat all these little optimizations; dynamic programming is very much a technique of the computer age!

There are four remarkable features about the dynamic programming technique as outlined above. First, the calculation of $v_t + V_{t+1}$ depends only on having some consistent means for assigning odds $p_{t+1}(x_{t+1})$ to states that might arise given x_t and u_t. No other constraint whatsoever is placed on the dynamic model for x over time, or on the control set chosen for analysis. Thus, the dynamic model can be quite complex, provided it does not have many state variables x that link the system's behavior from t to $t + 1$. The controls may be tightly constrained to reflect practical limits on actions, and in fact this even simplifies the searches for optima.

Second, the technique may be applied repeatedly to move backward in time from T to the starting point t (so-called backward recursion), and thereby build up a picture of the best actions to use over all time steps. So if we start out with an *arbitrary* end value $V_T^*(x_T)$ for being in different states at the terminal time, we first look forward to this for each possible state x_{T-1} at the next to last time and calculate $V_{T-1}^*(x_{T-1})$. Then for each possible state x_{T-2} at time $T - 2$, we look forward to $V_{T-1}^*(x_{T-1})$ and thereby calculate $V_{T-2}^*(x_{T-2})$. Then we move back to $T - 3$, and so forth.

Third, at each time step we necessarily construct a feedback policy, because we find the best action u_t *for each state* x_t that might have arisen by time t, without ever saying how (or whether) that state will actually occur. The little optimizations for each x_t find the control that maximizes $v_t + \Sigma\, p V_{T+1}^*$, but by definition the V_{T+1}^* are the best that can be done in the long term. Thus, the feedback policy $u_t(x_t)$ resulting from doing these optimizations for all x_t's is optimal for the long term also. It is the best balancing policy (v_t versus V_{t+1}) in consideration of the best future balances for v_{t+1} versus V_{t+2}, etc., insofar as these future balances have been included in the calculation of V_{t+1}. Richard Bellman's genius in recognizing and promoting this remarkably simple logic will not soon be forgotten.

Fourth, the best policy u_t to use for each x_t quickly becomes *stationary* (independent of t), as we move backward from the end time T, provided the short-term payoffs $v_t(x_t, u_t)$ are calculated the same way each time up to constant factors, such as a discount rate ($v_{t+1} = \lambda v_t, \lambda < 1$ always results in a stationary policy), and that $p(x_{t+1})$ given x_t, u_t is constant over time (i.e., model is same for all times). This means that the requirement to assume some terminal time T, and value $V_t^*(x_T)$ in starting the calculations, has little or no effect on the final results, provided we move back sufficiently far from T (i.e., the horizon is far enough away). Unless a very high value is placed on some particular end state x_T^* and there is no discounting, the stationary optimum policy usually becomes evident within 2–10 backward steps.

The dynamic programming technique described above is known as "value iteration" since the value function $V_t^*(x_t)$ is built up by backward iterations in time. When it is known that a stationary policy will exist, another technique known as "policy iteration" can often be used to find the optimum stationary (feedback) policy more efficiently. For an example from fisheries, see Ludwig and Walters (1981).

An important special case: v_t linear in u_t

Some of the tedious, repeated optimizations described above can often be avoided by looking carefully at the structure of a problem. Of particular importance in resource management is the situation where the actions u_t can be defined so that the short-term rewards v_t are linearly related to them. For example, if x_t is the single variable "stock size" and u_t is defined as the stock to be left after harvesting (escapement), then the payoff measure v_t = total catch is given by

$$v_t = x_t - u_t$$

Suppose now we try to find the maximum of $V_t = v_t + V_{t+1}^*$ by standard calculus (differentiate with respect to u_t, set derivative to zero). The result is

$$\frac{\partial V_t}{\partial u_t} = -1 + \frac{\partial V_{t+1}^*}{\partial u_t} = 0 \tag{8.6}$$

which implies that the best u_t is at the point where $\partial V_{t+1}^* / \partial u_t = 1$. Notice that x_t does not appear anywhere here, since x_{t+1} will be a function of the escapement u_t $\{V_{t+1}^* = V_{t+1}^*[x_{t+1}(u_t)]\}$, except that x_t *constrains* $u_t (u_t \leq x_t)$. This means that the optimum policy will be to use a u_t as near to the point where $\partial V_{t+1}^* / \partial u_t = 1$ as permitted by the constraint $u_t \leq x_t$, i.e., to use $u_t = x_t$ for x_t below this point. This is just a "fixed escapement" policy: take no harvest (escapement = stock) unless stock size exceeds the point u^* where $\partial V^* / \partial u = 1$, and set escapement to this value u^* when stock size exceeds it. Notice that this argument does not depend at all on the particular model for x_{t+1} as a function of u_t, except implicitly in that $V^*[x(u)]$ must have a unique point where $\partial V^* / \partial u = 1$.

Similar results can often be obtained for more complex examples. The key determinants of whether or not a simple policy exists are (1) the short-term payoff function v_t, and (2) the dynamic model for x_{t+1} insofar as this model predicts state responses (x_{t+1} versus u_t) that are *not* dome-shaped. Models that predict multiple equilibria are likely to generate value functions V^* with several points such that $\partial V^* / \partial u = 1$, and each of these points must be checked for optimality.

State incompletely observed

The dynamic programming technique can be modified to handle situations where the feedback policy must be based on only part of the state variable set x_t. Such situations are especially important in adaptive policy design, where we must work with partial observations and parameter estimates. The key requirement to apply the modified technique is that it must be possible to assign a probability distribution $p_t(x_t|y_t)$ to the *unobserved state variables x_t given the observed variables y_t.*

The basic idea is to find a decision u_t^* that is optimal, in an expected value sense, with respect to the observed quantities y_t *only*. Then the joint optimum value function $V^*(x, y)$ is calculated from u^*. The little local optimizations seek to maximize

$$\hat{V}_t(y_t) = \sum_{x_t} p(x_t|y_t) \left[v_t(x_t, u_t, y_t) \right.$$

$$+ \sum_{x_{t+1}} p_t(x_{t+1}, y_{t+1}|x_t, y_t, u_t) \tag{8.7}$$

$$\left. \times V_{t+1}^*(x_{t+1}, y_{t+1}) \right]$$

Notice here that we must be able to assign odds to both x_{t+1} and y_{t+1} given x_t, y_t, and u_t [these odds are $p_t(x_{t+1}, y_{t+1}|x_t, y_t)$] and that otherwise $\hat{V}(y_t)$ is just a weighted average [by $p(x_t|y_t)$] of outcomes $v + V^*$ across the states x_t that might actually be present given y_t. When we have found the u_t^* that maximizes $\hat{V}_t$ for a given y_t, we then calculate

$$V_t^*(x_t, y_t) = v(x_t, u_t^*, y_t)$$

$$+ \sum_{x_{t+1}} p_t(x_{t+1}, y_{t+1}|x_t, y_t, u_t) \tag{8.8}$$

$$\times V_{t+1}^*(x_{t+1}, y_{t+1})$$

using this u_t^* for every possible value of x_t. Thus, in finding $V^*(x_t, y_t)$, we have to do a local optimization for each value of y_t, but only a summation for each associated value of x_t. Various computer programming tricks can be used to do this more complicated accounting almost as quickly as standard dynamic programming.

Equations (8.7)-(8.8) are the basic formulation used to develop feedback policies based on information states I_t^* as discussed in the previous section. The additional "variables" x_t are unobserved state variables and unknown parameters. Under the special condition that $p_t(x_t \mid I_t^*)$ is calculated by Bayes theorem [equation (7.6)], the double relationship (8.7)-(8.8) can be replaced by a simpler, standard value recursion for $V_t^*(I_t^*)$. For a more careful discussion of this, see Bar-Shalom (1981) and problem 8.8.

The curse of dimensionality

For all of its basic theoretical appeal, dynamic programming involves some really nasty practical problems. You may have noticed that I have repeatedly used statements like "for each x_t," and "for every y_t," rather loosely, as though the state variables could take only a discrete number of values. In principle, this is true if x is, say, the number of fish in a population, but it is hardly practical (or necessary) to do millions of little optimizations, one for each exact population size. For continuous variables x_t, u_t, and y_t the usual approach is to use "state increment dynamic programming" (Larson, 1968). We divide each state and action variable into a number of discrete levels, to give a grid of discrete state combinations. Then V^* is evaluated at these grid points. For discrete levels x_t such that some x_{t+1} has been calculated *not* to lie at a grid point, V^* values for these x_{t+1} cases are interpolated from V^* at nearby grid points. Unfortunately, this interpolation can require a lot of computing time. One way to avoid it is to reformulate the dynamic model as a "Markov decision process," which allows only discrete states and assigns probabilities to transitions between these states (Mendelssohn, 1980). Uncertain outcomes are represented by the "transition matrices" $p_{u_t}(x_{t+1} \mid x_t)$, where p_{ij} is the probability of going from discrete state j to discrete state i if control u_t is used.

The curse of dimensionality refers to what happens to the number of discrete states where V_t^* needs to be calculated, as the *number of state variables is increased*. So, if we need 10 levels of x_1, and 10 levels of x_2 for each x_1, then we need a grid with $10 \times 10 = 100$ discrete points. With six state variables each discretized at 10 levels, the grid will have $10^6 = 1$ *million points!* Remember, it is usually necessary to do a little local optimization at each grid point, for each backward time step. Brute force dynamic programming cannot be applied to problems with more than about eight state variables at 10 levels each, even using computers that approach theoretical limits of computation speed. This limitation has led to active research on methods such as "tunneling" in the state space, which means analyzing only a minimum number of states that are most likely to occur, and "differential dynamic programming" which avoids discretization and can handle some very large problems. For an especially clear review of the latter technique, I recommend Murray and Yakowitz (1979).

The curse of dimensionality has provided a strong motivation for research into model compression as discussed in Chapters 5 and 6. To be quite honest, I think many of us only started to recognize other benefits (credibility to decision makers, etc.) of model compression in policy analysis after we had developed some compressed models solely as a means to avoid technical problems in dynamic optimization. Even if various methods can be developed to deal with many-dimensional problems, we will still have to face the problem of finding compressed, understandable representations of the feedback policies resulting from these methods. That is, we will have to find ways to visualize $u^*(x_t)$ functions when there are many x_t variables, since it would be silly to expect any real decision maker or manager to blindly plug numbers into such a function and then follow its prescription.

Optimization in policy space

It was noted above that for relatively simple management objectives, such as maximum average yield, the optimum feedback policy is likely to have a correspondingly simple functional form. Even when the precise optimum function is complex, its qualitative prescriptions (fish, do not fish, etc.) will always be representable by a simple function with only a few unknown parameters.

Suppose we can find an arbitrary policy function $u_t = \hat{u}(x_t, \gamma)$ which has adjustable parameters γ. The simplest example would be when $u_t =$ escapement, $x_t =$ stock size before harvest, and $\hat{u}$ is the fixed escapement policy

$$u_t = \hat{u}(x_t, \gamma) = \begin{cases} x_t & \text{if } x_t \leq \gamma \\ \gamma & \text{if } x_t > \gamma \end{cases}$$

Here there is a single "policy space" parameter γ, which is the escapement sought in every year for which $x_t \geq \gamma$. A more complex example would be when x_t is an information state containing stock size x_t, an estimate of some unknown parameter $\hat{\beta}$, and its variance $\sigma^2_{\hat{\beta}_t}$. Then a simple adaptive probing policy would be to perturb the escapement u_t when $\sigma^2_{\hat{\beta}_t}$ is large:

$$\hat{u} = \begin{cases} \gamma_1 + \gamma_2 \hat{\beta}_t & \text{if } \sigma^2_{\hat{\beta}_t} \leq \gamma_3 \\ \gamma_1 + \gamma_2 \hat{\beta}_t + \gamma_4 & \text{if } \sigma^2_{\hat{\beta}_t} > \gamma_3 \end{cases}$$

$$u_t = \begin{cases} x_t & \text{if } x_t \leq \hat{u} \\ \hat{u} & \text{if } x_t > \hat{u} \end{cases}$$

This policy has four parameters: γ_1 and γ_2 relate the target escapement $\hat{u}$ to $\hat{\beta}_t$, and γ_3 sets a threshold $\sigma^2_{\hat{\beta}_t}$ above which a disturbance γ_4 is added to the target escapement.

Optimization in policy space means to seek the set of parameter values γ^* that are expected to give the best long-term performance for a well defined objective function $V = v_1(x_1, u_1) + v_2(x_2, u_2) + \cdots$, where v_t is the reward or payoff in time step t. It is usually not possible for stochastic systems to find an explicit relationship between V and the parameters γ; a brute force approach is to fix γ then to do repeated Monte Carlo simulations (dynamic model with actions at each step calculated using the function $\hat{u}$) to find an average $\hat{V}$ for this γ. Then γ is varied and the simulations to find $\hat{V}$ are repeated, until a best estimate γ^* is found. Various nonlinear programming algorithms can be used to vary γ efficiently.

Repeated simulations at each policy combination γ can be minimized by using the iterative procedure known as "stochastic approximation" (Robbins and Monro, 1951). The basic idea here is to do a few stochastic simulations around a starting value γ_1 to get a single (random) estimate g_{V_1} of the gradient of the value function at the point γ_1. {The ith element of g_{V_1} should be $\partial V/\partial \gamma_i$; this is approximated by $[V(\gamma_i + \Delta) - V(\gamma_i)]/\Delta$, where the same random sequence of inputs to the simulation is used at both points γ_i and $V_i + \Delta$.} Then a sequence of new policies γ_k is found by the simple rule

$$\gamma_{k+1} = \gamma_k + a_k g_{V_k}$$

where a_k is a positive constant that decreases as k increases, and can be chosen in various ways to make the sequence converge to γ^* more efficiently. This method and various extensions of it are discussed for a fisheries example in Ruppert et al. (1983). Gaivoronski and Ermoliev (1979a,b) have developed sophisticated software for applying it to problems in adaptive control; see also Ermoliev and Gaivoronski (1984). The general theory has been extensively developed by Soviet scientists, particularly Tsypkin (1971) and Ermoliev (1976); see the review in Poljak and Tsypkin (1980).

When the feedback function has only a few parameters γ, it can be very informative to systematically estimate $\hat{V}$ on a grid of γ combinations, and then plot the results using techniques like contour mapping. Indeed, this approach has been used a great deal in fisheries; for example "yield isopleth diagrams" are used to show how equilibrium harvest (a V measure) varies with exploitation rate and minimum size of fish harvested. For stochastic models, a very good idea of the average response surface can usually be obtained by doing only a few (5–10) Monte Carlo simulations at each grid point. A more elaborate variation on the theme is to plot a whole variety of performance or value measures as a function of the parameters γ, then arrange these plots as "nomograms" that show quickly how variation in the parameters will affect various trade-offs among the performance measures (for examples, see Peterman, 1975; Holling, 1978; Argue et al., 1983).

Properties of Feedback Policies

This section clears up a few common misunderstandings about what feedback policies do when applied in management, particularly to the temporal stability of the managed system, and about the robustness of feedback policies to errors in the formulation of models and objectives. Much intuitive discussion about feedback has unfortunately been borrowed from fields such as control system engineering, where a particular kind of control objective (regulation of system outputs to constant target levels) has been of predominant concern.

Stability of managed systems

Over a wide range of stock sizes, feedback policies for resource harvesting generally imply taking less when the stock is low and more when it is high. Such policies might thus be expected to reduce the range of variations in system state relative to the range that would be generated by "natural" bionomic processes of investment and stock response. Such policies as fixed escapement or fixed exploitation rate (harvest/stock) do, indeed, have this effect when the policy information state does not include uncertainty measures, such as variances of unknown parameters. However, it is a serious mistake to suppose that optimum feedback policies will, in general, stabilize the system state and/or output measures such as harvests. A few examples will serve to illustrate this point.

When harvesting a stock that has multiple age classes with the organisms becoming increasing valuable with age, two key policy variables are the annual harvest rate and the minimum age (or size) of animals harvested. (Usually it is physically impossible to prevent harvesting of several ages above this minimum.) However, often the minimum size cannot be closely regulated, and younger (smaller) animals than desired are inadvertently killed by the harvesting process, even when they are not retained as yield. In this case the best policy for maximizing total long-term value can be a periodic or "pulse fishing" policy, with high exploitation rates every few years and no harvesting in between (Walters, 1969). The idea is that it is better to wait for blocks of the young animals to grow to desirable ages, then take them at once, instead of steadily cutting away at potential production by killing some of them every year. Pulse harvesting is usually only economically practical where the resource can be divided into several spatial units (substocks, lakes, clam beaches, etc.), so that there is a pulse harvest in at least one unit every year. Nevertheless, the key point is that the best feedback policy for these situations is anything but state-stabilizing for each of the harvesting units involved. Pulse harvesting is not a new idea; there has long been clear-cut logging in forestry, where for various biological and economic reasons it often does not make sense to selectively take a steady harvest each year.

Harvesting policies designed to stabilize economic outputs will generally cause increased, rather than decreased, variation in stock sizes when the stock is subject to strong environmental effects. To see why this happens, just imagine how escapement levels for a salmon stock will vary if a constant catch is taken from the migrating recruits each year, while the abundance of these recruits is varying a lot from year to year. All of the variation in recruitment will be transmitted through the fishery into variation in spawning stock, which will in turn contribute to maintenance of variation in future recruitments. We will return to this basic trade-off between variations in output versus variations in stock size in the section below on the robustness of feedback policies to changes in management objectives.

Much of Chapter 9 will be concerned with demonstrating that optimum feedback policies will often deliberately induce variation in system states as a means of improving estimates of uncertain parameters or of testing for possible opportunities beyond the range of historical experience. There we will show, in fact, that the optimum policy is often much like a pulse harvesting one, in the sense that it involves periods of steady harvesting interspersed with strongly informative disturbances.

For at least some ecological systems, feedback policies that stabilize the system state in the short term can result in pathological growth of management costs in the long term. A widely discussed example of this is control of forest fires, which in many forest ecosystems can result in accumulation of understory vegetation that makes further fires more probable and more devastating when, inevitably, they do escape control. In fisheries, artificial "enhancement" programs are often justified as an interim means for making fish available to maintain high harvest rates after natural disasters or periods of overfishing; but the high harvest rates thus maintained are usually felt by the natural stocks as well, so they decline even further and the system becomes "locked into" artificial production.

Robustness to modeling errors

An important test of any feedback policy is its performance when used with other dynamic models than the one assumed in designing it. A robust feedback policy is one that will give near-optimum results for all, or at least a wide variety, of such alternative models. This is a rather vague definition; for a more precise analysis based on the notion of "nominal" (design) models versus "extended" (more realistic, elaborate) models, see Wierzbicki (1977). But the general idea is clear enough; it would be foolish to apply any policy whose performance is precisely dependent on the world behaving according to any particular model.

Feedback policies for harvesting are generally very robust to modeling errors in the biological production process, up to changes in the degree of polynomial equation needed to approximate equilibrium production rates as a function of stock size over the range of stock sizes generated by applying the policy. This rule of thumb is not so complicated as it first sounds. Most surplus production models, stock–recruitment equations, and dynamic pool models with constant growth curves and stock–recruitment relations use equations (logistic, Ricker, Beverton–Holt) for which the equilibrium production rate can be approximated by a quadratic polynomial. Models with two equilibria due to "predator pit" or cannibalism effects can be approximated by a cubic polynomial. Multispecies competition–predation interactions can generate high-order polynomial effects.

However, analyses such as that of Hilborn (1979) indicate that there is a strong asymmetry in the performance of "adaptive feedback policies" that involve sequentially fitting simple production models to the catch/effort time series, then modifying the feedback policy parameters at each time step in accordance with changes in the production model parameter estimates ("passive adaptation;" information state includes estimates $\hat{\beta}$, but not variance measures). Generally, such policies perform very well (i.e., are robust) if the stock is *initially unexploited*, even if stock behavior is affected by quite complex age–size structure effects not represented in the surplus production model assumed for the feedback policy. In contrast, they very often fail badly if the stock is already overexploited when data first become available for parameter estimation; the feedback policy often never "learns" that the stock is overexploited, and continues to hold it at unproductively low levels. Ludwig and Hilborn (1983) have shown that the only reliable way to prevent such failures is to initially ignore the feedback policy and not harvest the stock at all for a long period (10-20 years). The findings of Hilborn and Ludwig are particularly disturbing, because good data are usually not obtained during the early development of harvesting: the pathological situation is the most common one encountered in practice.

Robustness to management objectives

It is rather difficult to give a meaningful definition of the robustness of a feedback policy to changes in, or errors in the determination of, management objectives. But there is one very fundamental trade-off that has already been mentioned several times above, for which feedback policies are typically not robust at all.

This is the trade-off between the average reward rate (harvest, profit, etc.) versus its variability over time. Generally, policies that come close to maximizing the average will also result in high variability around this average. When the initial state is far from optimal for maximizing the average, such policies usually prescribe a "minimum time" action sequence for getting to the best state as quickly as possible. For overexploited stocks, this means taking no harvest at all (escapement = stock size) until the stock has recovered; for underexploited stocks, it means harvesting hard to drive the stock down to a more productive level as quickly as is feasible. For stocks with large variations in year-class strength, it means gearing up to crop the large year classes hard, then shutting down when only weak year classes are present.

There have been a few attempts to design suboptimal feedback policies that give some balance between objectives related to average yield versus variability (Allen, 1973; Walters, 1975). A remarkably simple policy that usually gives a good balance in stochastic simulations is to just maintain a constant exploitation rate. This means letting the harvests vary up and down with the stock size, and adjusting harvesting effort over time if the catchability coefficient is dependent on stock size (for example, when the search is nonrandom; see Chapter 4). A variation on this theme is to hold the exploitation rate constant except if stock size drops to, say, one half of the most productive level, but to reduce the exploitation rate in proportion to stock size below this level. For an example of performance analysis for this policy, see Walters (1975).

Let me inject a historical comment at this point. In the early 1970s, when a few of us began looking for tools like dynamic programming to gain some insights about how management should respond to variability and uncertainty, we intuitively expected the best policies to be quite complex functions of stock size. There were all sorts of discussions and heuristic prescriptions about what to do in the face of events like strong year classes and collapses due to environmental fluctuations (the Peruvian anchoveta and El Niño were a favorite test case for discussions). It was a pleasant surprise when the first results were published showing that a simple fixed escapement is likely to result in maximum average yield. But that result had been anticipated by earlier work. The really surprising discovery has been the one mentioned in the previous paragraph, that the best balance between mean and variance is likely to be well approximated by just holding the exploitation rate constant.

Performance of Passively Adaptive Policies

For the first time in this book, let us cast some doubt on the idea that actively adaptive, probing (experimental) policies need to be used in the ongoing management of a renewable resource. Here the doubt raised is not about the value of learning or the need to be adaptive, but rather only about the need to experiment deliberately with policy actions. What we will show is that strict, mechanical adherence to a passively adaptive (certainty equivalent) feedback policy can produce essentially as informative a sequence of actions as would deliberate experimentation, so even what is meant by an "experiment" becomes unclear. Then, as counter to this theoretical argument, we will simply rejoin that managers are highly unlikely to adhere mechanically to any procedure for calculating the next action to take, unless motivated to act experimentally; they instead act so as to filter out the informative variation in favor of more conservative, incremental policies.

Informative variation from passive adaptation

Let us look at the sequence of events that takes place when a passively adaptive policy is followed with care. The policy prescribes action as a function of the system state and current estimates of dynamic parameters, such as recruitment and mortality rates. It is assumed that during each period between actions, the historical data set Y is updated, new parameter estimates are calculated, and these new estimates are "plugged into" the policy function for the next action. For example, if a logistic surplus production model is assumed in the analysis of Y, and if the feedback policy is to try and maintain the harvesting effort at $E = r/2q$ (which would maximize the average long-term catch if the stock dynamics were deterministic with an intrinsic growth rate r and catchability q), then each year the new catch and effort data would be added to Y, parameters reestimated, and the resulting new estimates of r and q would be used to set the next year's effort.

The policy action calculated in this way can be informative for two quite different reasons. First, if earlier actions have been taken without analysis of data (a very common situation!), or were based on some other model and process of policy formulation, then the prescribed action may be well away from any taken so far. In this case we would tend to label the action an experiment with uncertain consequences, even though it was computed without giving deliberate thought to the value of information that might result from it. The most vivid example of this situation in my experience was with a major sockeye salmon stock in British Columbia. Spawning runs had been declining for more than a decade when we conducted a stock–recruitment analysis in about 1980. The decline was apparently due to a few natural accidents combined with a sequence of attempts to sustain catches without systematic examination of the incoming stock–recruitment data. Our analysis indicated that the "best" (for maximizing average yield) escapement would be at several times the 1980 level. Though there was direct historical evidence (from before the decline) favoring this conclusion, and no reason to suspect major recent changes in the spawning and rearing environments, our conclusion was hotly debated. It was finally decided to rebuild the spawning runs, but this was justified partly as an "experiment in rehabilitation."

Thus, we see that even the notion of an "experiment" is ambiguous. Should we call any policy change with uncertain consequences an experiment or probe? Or should we restrict the term only to changes that are planned and justified on the basis of calculations about the expected value of information associated with various alternatives? In Chapters 9 and 10, I will be careful to use the term "experiment" or "probe" only in this second, more restricted sense.

The second potential cause of informative variation from passive adaptation is that the prescribed policy actions are, in fact, *random variables*, since the parameter estimates used in calculating them are random variables. As shown, for example, in the simulations of logistic parameter estimation in Figures 7.2 and 7.3, parameter estimates can jump around violently, even when many years of data are available. This fluctuation is especially great when the state–action sequence has been relatively smooth (and uninformative), and when older data are discarded through procedures like discounting that presume some parameter change. Probing (fluctuation in actions) only when the historical data are uninformative is precisely the sort of policy that we might want to design into the feedback system, and to some degree we get it "for free" from a passive policy.

A way to find out about the robustness of parameter estimates derived from a given data set Y (i.e., the real amount of information in Y) is by what statisticians call "jack-knife techniques" (Tukey, 1977). The idea here is very simply to systematically reanalyze the data while leaving out various blocks of observations; if the whole data set is informative, parameter estimates will not change much unless very large blocks are omitted. When applied to the time series typically available for renewable resources, jack-knife techniques usually show that the estimates are not at all robust, and in fact usually depend very critically on one or a few outlying observations associated with short periods of disturbance. Let us turn this rather disturbing observation around and make it a prediction: when the inevitable disturbances do come, they are likely to result in big changes in parameter estimates and hence in the actions prescribed by passively adaptive policies.

There is a growing literature in control system theory about the "asymptotic behavior" of passively adaptive policies, as the quantity of data available becomes very large. Essentially, the idea is to discover whether policy changes due to random variations in parameter estimates will provide enough disturbance to guarantee finding correct parameter estimates (or at least the best policy) in the long term. Except for "pathological" cases, such as resources that are overexploited when data gathering begins (see above), it is usually found that passive policies are indeed "asymptotically optimal." However, I find this choice of terminology most unfortunate and even deceptive; asymptotic optimality of a policy says absolutely nothing about its performance on the relatively short time scales (30–60 years) that are of most interest in renewable resource management.

Why passive adaptation fails in practice

It is one thing to design an automatic control system that will blindly vary its actions as parameter estimates jump around; it is quite another to expect people to behave in this way when they have the time to inject judgment and intuition at each decision point. On several occasions I have been present when new data became available and were eagerly plugged into estimation algorithms, then watched the chuckles turn into groans when the new parameter estimates came out far from the old ones. Two things happen then. First, there is a frantic rechecking of the data and calculations, accompanied by fervent prayers that an arithmetic error will be found. Then when no errors are found, there is a rather remarkable reaction: the new estimates and/or the original model are simply *rejected*, and policy planning goes ahead on the basis of older information. In other words, actions are not changed drastically as required for passive adaptation to work well in theory. In the terms of statistical decision theory, people tend to act as though they had placed a tight prior probability distribution on the parameter estimates, even when there is absolutely no objective reason to do so.

People cannot be blamed for behaving more conservatively than a mechanical control system. Indeed, we would be fools not to question the accuracy of new data and validity of old models at each decision point. The difficulty is that such healthy skepticism is too easily made the basis of excuses for inaction, at times when bold changes are needed as a basis for more informed decision making in the long term.

Conservative, risk-averse decision making creates a particularly difficult situation for learning when environmental effects on stock size may be autocorrelated over time. In this case, each run of poor production years will lead to stock decline, which the manager cannot be sure is due to environment rather than overharvesting. His conservative response will be to cut back on harvests, rather than trust any hypothesis or model that predicts recovery due to a return to good environmental conditions. If stock recovery does occur following the conservative response, it is impossible to tell whether the recovery is due to harvest reduction or to a run of good environmental conditions. This confounding of environment versus management can persist over several "cycles" of decline and recovery, and lead to heated debate about the role of management. For an example, see the review by Skud (1975) of the famous "Thompson–Burkenroad debate" about causes of fluctuation in the abundance of Pacific halibut (Figure 2.2). In the end, such debates can only be resolved by an experimental decision to deliberately avoid reducing harvests; it is not clear that the risks of this decision are worth taking even in the management of major resources.

Problems

8.1. Consider the feedback policy "harvest $N - 1000$ animals whenever the population N exceeds 1000, and harvest none when N is less than 1000." This is a "fixed escapement" or "fixed base stock" policy, with a target escapement of 1000. Such remarkably simple rules are often optimum when the management objective is also simple, i.e., to maximize average yield. For a population that has decreasing percentage growth rate over the range $N = 0$ to $N = 2000$, draw a graph to show how the percentage harvest rate under the fixed escapement policy would compare to this growth rate; there is an equilibrium where the rates balance. Then plot the percentage harvest rate expected under two alternative policies, a fixed quota and a fixed percentage harvest rate. Will the fixed percentage harvest policy also act as a stabilizing feedback policy? Will it produce more or less variable yields over time than the fixed escapement policy?

8.2. Suppose someone tells you that the maximum long-term harvest V_{t+1} from a population, beginning next year ($t + 1$), can be approximated by $V_{t+1} = aS_t - bS_t^2$, where S_t is the population after harvest this year. Then if the population this year is N_t, the harvest will be $N_t - S_t$ and the total value from now forward will be $V_t = N_t - S_t + V_{t+1} = N_t - S_t + aS_t - bS_t^2$. For fixed $N_t = 1$ (and $a = 2$, $b = 1$), plot the value components $N_t - S_t$ and V_{t+1}, and their total V_t, as a function of S_t. This allows you to find the optimum S_t, S^*. Congratulations! You have just solved a simple dynamic programming problem. Now redo the plot for various values of N_t over the range 0–2, and find S^* for each case. Then plot S^* as a function of N_t, and you will have an optimum feedback policy: for any N_t that might arise, you have defined the best S to allow (and thus the best harvest in year t). The trick, of course, is to approximate V_{t+1}, and usually this requires an iterative computer procedure (backward recursion). Now that you understand how the "stage t iteration" is done, what additional relationship must you specify in order to move back another step and find V_{t-1}?

8.3. Suppose you try to manage an ungulate population according to a fixed escapement policy as in problem 8.1 above, after determining the optimum base stock from a short time series of data that shows density dependence in reproductive rates. Each year you estimate the stock size, then harvest accordingly, then wait to see the result next year when you make the next stock estimate. What will happen if your base stock is, in fact, too large for the range to sustain, so that the biomass and productivity of food plants decreases slowly over time? Now suppose range condition is also monitored, so you can relate reproductive rates to it as well as to population size. By plotting harvest rate as a function of both population size and range condition, can you suggest how the optimum feedback policy might look for this more complex situation?

8.4. A good way to understand dynamic programming is to try writing a computer program to do it for a deterministic, discrete model. The heart of this program should consist of three nested loops (1) an outside loop over time steps (stages), where you view time as moving backward as the loop counter increases; (2) an inner loop over states that might have arisen by each time step (generally, assume that all states are possible at each time step); and (3) an innermost loop over action choices, which allows you to test the value of each policy choice for each state. Within the innermost loop (state–action combination), you must calculate (a) the immediate payoff v_t in that time step as a function of state and action; (b) the state x_{t+1} at time $t + 1$ given the state at t and action choice (i.e., one dynamic model prediction); (c) the combined value $v_t + V_{t+1}(x_{t+1})$. To do step (c), you must have initially (before the first time step) set up a table of "terminal" $V(x)$ values (one for each discrete x), or set these all to 0. Also within the innermost loop, set up a test so that you can keep track of the largest combined value found, and the action associated with it. Then, after leaving the innermost loop, store this maximum in a separate array $VN(x)$ for use in step (c) in the next time stage. Outside the inner loop (after VN has been found for every x at time t), set $V(x) = VN(x)$ for every x before moving to the next time step. To test your program, let (1) $x = 0, 1, \ldots, 10$ be discrete population levels; (2) $h = 0, 1, \ldots, x$ be harvests; (3) $v_t = h_t$; (4) $V(x) = 0$ for all x at the first stage (end time); and (5) use the table below as your state–action dynamics model. In the table, 10 is the population carrying capacity, and escapements of 4 or 5 produce the largest sustainable net increase (i.e., leave 4 and have 8 from which 4 can be left again in the next stage).

Stock after harvest at stage t: $x_t - h_t$	Stock before harvest at stage $t+1$: x_{t+1}
0	0
1	2
2	4
3	6
4	8
5	9
6	10
7	10
8	10
9	10
10	10

8.5. Your computer program from exercise 8.4 can be easily modified to handle stochastic dynamics, where the stock before harvest at time $t + 1$ is uncertain. To calculate the expected value for each policy choice, modify step (c) within your innermost loop so that it calculates the combined values as an average across possible states at $t + 1$, with each possibility x_{t+1} weighted by its probability $p(x_{t+1} | x_t, h_t)$ of occurrence given x_t and the action choice h_t:

$$\text{combined value} = v_t + \Sigma\, p(x_{t+1} | x_t, h_t)\, V(x_{t+1})$$

(i.e., you need a loop over next states *within* your innermost loop from problem 8.4, to sum up the probabilities of next states times the values of those next states). Test your modified program using the following stochastic model (state transition probabilities)

Stock after harvest $x_t - h_t$	Probabilities of next state x_{t+1}										
	0	1	2	3	4	5	6	7	8	9	10
0	1.0										
1		0.2	0.5	0.3							
2		0.1	0.1	0.2	0.4	0.1	0.1				
3					0.1	0.2	0.3	0.2	0.1		
4						0.1	0.1	0.2	0.3	0.1	0.1
5						0.1	0.1	0.1	0.2	0.3	0.2
6							0.1	0.1	0.1	0.3	0.4
7							0.1	0.1	0.1	0.3	0.4
8							0.1	0.1	0.1	0.3	0.4
9							0.1	0.1	0.1	0.3	0.4
10							0.1	0.1	0.1	0.3	0.4

You should get essentially the same result as in problem 8.4.

8.6. The discrete models considered in problems 8.4 and 8.5 are somewhat unrealistic, and can be cumbersome to specify as tables of outcomes. The trouble with assuming that x_t is continuous is that your model [step (b) in problem 8.4] may specify an x_{t+1} that is between two of the "grid point" x values for which you have calculated $V(x_{t+1})$. To handle this difficulty, modify your programs from 8.4 to 8.5 so as to linearly interpolate V for any x_{t+1} from the values at surrounding grid points $V(x_-)$ and $v(x_+)$ where $x_- \leq x_{t+1} \leq x_+$, using the interpolation equation

$$V(x_{t+1}) \approx V(x_-) + \left(\frac{x_{t+1} - x_-}{x_+ - x_-} \right) [V(x_+) - V(x_-)]$$

Test this modification with the deterministic dynamic model $x_{t+1} = 10(x_t - h_t)/(4 + x_t - h_t)$. If you have done exercise 8.5, try also the stochastic model $x_{t+1} = w_t(x_t - h_t)/(4 + x_t - h_t)$ where the "carrying capacity" w_t is assumed to be stochastic and can take the values 7, 8, 9, or 10 with probabilities 0.1, 0.4, 0.4, and 0.1. Plot your estimates of the optimum discrete h_t after five backward steps, as a function of the discrete x_t values for which these h_t were estimated. Can you use this plot to interpolate optimum h_t values for x_t values not on the grid points? How does the "feedback policy function" h_t versus x_t change from stage to stage (does it become stationary, i.e., time independent)?

8.7. As noted in Chapter 2, resource harvesters are sometimes more concerned with avoiding very low harvests than with maximizing average or total long-term harvest. A simple way to represent both concerns is by assuming that the formal management objective should be to maximize the *product*, rather than the sum, of harvests over time. Then a single low (or zero) harvest can have devastating effects on the estimated total resource value ($V_t = h_t h_{t+1} \cdots h_T$ instead of $V_t = h_t + h_{t+1} + \cdots + h_T$). Modify your programs from problems 8.4 to 8.6 so as to test this assumption, by (1) changing the $v_t + V_{t+1}$ calculations [innermost loop, step (c)] to $v_t V_{t+1}$ where v_t is still just $v_t = h_t$, and (2) initially setting $V(x) = x$ [instead of $V(x) = 0$] for $x = 0$, ..., 10 before starting the loop over stages. Do you still obtain a stationary feedback policy (graph of optimum h_t versus x_t not changing with t) after several backward steps? How does it compare to the optimum policy when harvest values are additive? Do you still get similar policies for the stochastic and deterministic cases, or is the stochastic policy now more "cautious?"

8.8.   In the case of incomplete state observation, the dynamic program-
        ming recursion can often be simplified. Suppose the information state
        I_t^* is fully sufficient to characterize the prior distribution $p_t(x_{t+1})$ in the
        Bayes formula

$$p_{t+1}(x_{t+1}) = \frac{p(y_{t+1}|x_{t+1})\, p_t(x_t)}{\displaystyle\int_{x'_{t+1}} p(y_{t+1}|x'_{t+1})\, p_t(x'_{t+1})} \tag{8.8a}$$

The dynamic programming recursion can be written as

$$V_t(I_t^*) = \int_{x_t} p_t(x_t) \left[v_t + \int_{x_{t+1}} p(x_{t+1}|x_t) \right. \tag{8.8b}$$

$$\left. \times \int_{y_{t+1}} p(y_{t+1}|x_{t+1})\, V_{t+1}(I_{t+1}^*, x_{t+1}) \right]$$

where $V_t(I_t^*)$ is the expected value across possible states x_t and
$V_{t+1}(I_{t+1}^*, x_{t+1})$ is the conditional value given that x_{t+1} occurs (but is
not observed). By using (8.8b) to define $V_{t+1}(I_{t+1}^*)$ with (8.8a) substituted in for $p_{t+1}(x_{t+1})$, show that (8.8b) is identical to the simpler
unconditional recursion

$$V_t(I_t^*) = \int_{x_t} p_t(x_t) \left[v_t + \int_{x_{t+1}} p(x_{t+1}|x_t) \right. \tag{8.8c}$$

$$\left. \times \int_{y_{t+1}} p(y_{t+1}|x_{t+1})\, V_{t+1}(I_{t+1}^*) \right]$$

Hint: start with (8.8c), and show that it is equal to (8.8b) since the
denominator of (8.8a) cancels many of the probability terms.

Chapter 9

Actively Adaptive Policies

In the endless series of experimentation
in which they are caught, man and nature,
as a two-person game, constantly lead, mislead,
and surprise each other into revealing
their possibilities and their limitations.

LaPonce (1972)

Chapters 7 and 8 have provided some basic building blocks for looking at the question of how to design management policies for systems where the sequence of decisions made over time can have an important impact on uncertainty. When actions influence learning rates as well as obvious performance measures, such as harvests, we say that there is a "dual effect of control." In coining this phrase, the Soviet mathematician Fel'dbaum (1960, 1961) noted that there is often a basic conflict or trade-off between learning and short-term performance: actions that perturb the system state and output in an informative manner may require giving up immediate harvests, accepting the risk that the system will not recover after some perturbation, or simply living with temporal variation that is uncomfortable from a social and economic perspective.

Actively adaptive or "dual control" policies seek to establish some optimum, or at least reasonable, balance between learning and short-term performance. Notice here that I am being careful not to say "between learning and other performance measures," which would imply that learning (resolution of uncertainty) should be viewed as inherently valuable. In adaptive policy design, it is essential to begin by carefully rejecting the intuitive notion that learning is always valuable, and to instead view the resolution of uncertainty as an important step *only* insofar as it may help to improve long-term management performance in relation to objectives such as harvest and profits. When we fail to make this distinction, we invite the resource

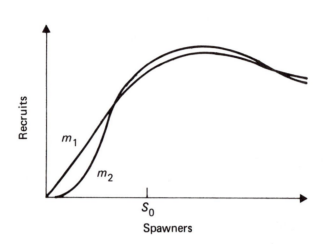

Figure 9.1. Two alternative models for the response of average recruitment to changes in parental stock. Both predict that S_0 is the best to allow in managing the system, but m_2 predicts an irreversible collapse if the stock ever gets very low.

manager to confuse personal curiosity with real needs in terms of his mandate as manager, and to advocate policies that would make him feel more comfortable at the expense of the clients that he serves. Consider, for example, the two alternative recruitment models shown in Figure 9.1. According to the model m_1, recruitment rate would be about proportional to parental stock at low stock sizes; according to m_2, there would be a depensatory collapse in recruitment at low stock sizes. However, according to both models the best parental stock would be up round the level marked S_0. Now, a manager might be curious or fretful about whether a response like m_2 would be obtained, but he certainly has no business *as a manager* advocating an experiment to resolve this uncertainty (except, perhaps, as a sacrificial "object lesson" for those who oppose regulations to keep the system at S_0).

Even when analysis proceeds by assuming that learning has value only insofar as it helps in achievement of simple management objectives, it turns out to be technically very difficult to calculate optimum adaptive policies by using techniques such as dynamic programming. The basic difficulty is dimensionality: in order to design a "closed-loop" feedback policy that anticipates the effects of learning as well as future system states, we must look forward in time to how uncertainty measures (such as parameter estimates and their covariances) propagate along with system state variables. Later in this chapter we will examine some very simple dynamic programming formulations, since they give some qualitative insights about the two basic questions of *when* to make probing experiments and *how drastic* to make each probe. However, as of this writing it is just not feasible to compute optimum adaptive policies for realistically complex models. We will examine one approximation scheme, known as the "wide-sense dual control algorithm," that has shown some promise for dealing with larger problems and gives further insights about the optimum balance between caution (avoidance of risks) and probing.

As a preamble and motivation to the discussion about formal optimization techniques for adaptive policy design, let us first try to define the general conditions under which it is worthwhile even considering an actively adaptive policy. These conditions are rather more restricted than one might expect, and in trying to pinpoint them we will see more clearly why it is dangerous to proceed with purely intuitive notions about the value of experimentation.

When Might Probing be Justified?

Let us define a probing action as any change in management tactics that is deliberately intended to produce an informative response in the system state or outputs, where, as in Chapter 6, we shall take "informative response" to be one that is assigned different likelihoods of occurrence by various models that are consistent with historical experience. We immediately encounter trouble with this definition, since it excludes policy changes associated with passive adaptation (see Chapter 8), that may well involve moving beyond the range of recent experience, but are prescribed on the basis of a single "best" model without explicit evaluation of their informativeness. So, as a first (negative) criterion, we might say that probing is not justified when an informative action would be prescribed in any case. Most actions taken in the "preadaptive phase" of management, when uncertainties are really gross, will obviously fall in this category.

In Chapter 6 we introduced the notion of expected value of perfect information (EVPI) as a measure of the importance of resolving uncertainty. EVPI was defined as the weighted average deviation between performance given the correct model and performance given the action that would be best in the absence of learning, with the average being across all models that might be correct and the weighting being the prior odds placed on each model. We noted that EVPI is different from zero only if (1) the models predict different performances, *and* (2) knowledge of the correct model would lead to a different action than would be best in the absence of learning. These criteria apply to the special case of probing, since it would not make sense to partially reduce uncertainty in situations where even having perfect knowledge would not make any difference. More succinctly, a necessary (but not sufficient) justification for probing is that EVPI be greater than zero. Unfortunately, the existence of a large EVPI tells us only that knowledge of the correct model m_i and action u_i^{**} would in expectation lead to a better result than using the no-learning action u^*; it says nothing about whether any particular action is best in the sense of providing that knowledge.

Pulling together arguments based on performance of passively adaptive policies and on EVPI, it appears that four basic conditions are necessary to justify a probing policy:

(1)   there must be uncertainty about the best action to take;
(2)   system performance must be sensitive to the action taken;
(3)   alternative actions must be differentially informative;
(4)   no single one of the alternative models for system response must be considered very probable.

The last of these conditions is obvious, and need not be elaborated. Let us examine each of the first three conditions in a little more detail, showing by example why it is important to be careful about each.

Uncertainty about the best action

The existence of large uncertainty about system response need not imply that the best action is also uncertain, as the example in Figure 9.2 shows. Here I have plotted three alternative hypotheses about how equilibrium yield will vary with harvesting effort, for a situation where the yield at a given harvesting effort is obviously very uncertain. However, all three models imply the same "optimum" effort. This situation can actually arise in practice, for instance, when the population is expected to exhibit logistic growth/production, and the r and q parameters are known in advance while k is very uncertain. Knowledge of r and q may have been obtained, for example, by studying the population for many years while it was severely depressed. Notice that EVPI = 0, since $u_i^{**} = u^*$ for all these models, and on that basis alone we would know that a policy experiment is not justified.

In situations like Figure 9.2, it is important to define precisely what actions and performance measures are of concern in policy design. To a government manager, Figure 9.2 presents the lucky opportunity to maximize average harvest simply by regulating the harvesting effort to a moderate level. To representatives of the harvesting industry, the uncertainty about yields under this policy would still represent a major source of concern about actions, such as investment in processing plant capacity and new vessels. It might be best for them to adopt actively adaptive policies in relation to these actions.

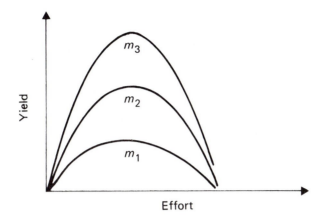

Figure 9.2. Three alternative models for yield as a function of harvesting effort. All three models predict the same optimum effort.

Sensitivity of system performance to actions

It has long been noted that for most stock–recruitment models, the average sustainable harvest (measured as recruitment minus spawning stock) is relatively insensitive to the spawning stock. For example, with typical parameter values for Pacific salmon, models like the Ricker and Beverton–Holt predict that the average yield will remain within 10% of its maximum if the spawning stock is kept within about 30% of the level that gives the maximum average yield. Indeed, there would not be many productive stocks still around if they were not forgiving of errors in policy choice.

Insensitivity of performance to action again implies that EVPI will be small. Each element in the EVPI average is $V(u_i^{**}|m_i) - V(u^*|m_i)$, and to say that performance is insensitive to action is the same as saying that V does not change much as u moves from u^* to u_i^{**}. Sometimes there can be a strong asymmetry in $V(u^{**}) - V(u^*)$ across the plausible alternative models, as shown for a yield/effort example in Figure 9.3. Here I have shown three hypotheses that imply only small differences in maximum yield, but at very different levels of effort. If m_1 is correct, operating at u_2^{**} or u_3^{**} would be disastrous. If m_2 is correct, operating at u_3^{**} would be disastrous. Yet operating at u_1^{**} gives a quite good performance even if m_2 or m_3 is correct. In this case, we would expect the best policy to be "cautious" and probe toward the level u_2^{**} only if detection and recovery would be rapid if m_1 turns out to be correct.

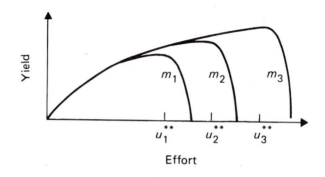

Figure 9.3. Three alternative models for yield as a function of harvesting effort. The models imply different optimum efforts u_1^{**} - u_3^{**}, but nearly the same yield no matter which policy is used. Thus, the obvious choice would be u_1^{**}.

Alternative actions differentially informative

Even when there is great uncertainty about system response to alternative actions, there are two possible reasons why no particular choices may stand out as candidates for the optimum probe:

(1) very slow learning due to high "environmental variation" or poor monitoring systems;
(2) similar learning rates under all choices.

If the first of these possibilities seems likely, the key policy recommendation should be to seek alternative monitoring systems that will give stronger signals of response: further analyses should include the costs of these alternatives as part of the costs of probing experiments.

Figure 9.4 shows a yield/effort example where most policy choices would be equally informative, in the sense that the three alternative hypotheses m_1 - m_3 predict quite different yield responses at most effort levels. Here the only partially uninformative choices are at effort levels where the curves cross. The best policy choice would most likely be at one of the three optima u_1^{**}, u_2^{**}, and u_3^{**}, since each is informative. An obvious bad choice (from the point of view of learning) would be a "conservative compromise" roughly halfway between u_1^{**} and u_2^{**}; this choice might be favored if learning effects are ignored, as a cautious way to achieve reasonable performance no matter which model proves correct.

The situations shown in Figures 9.3 and 9.4 are most likely to arise in systems that have never been heavily exploited. Thus, they represent the pattern of uncertainty expected in the preadaptive phase of management,

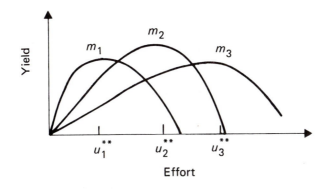

Figure 9.4. Three alternative models for yield as a function of harvesting effort, where most effort choices would be quite informative about which hypothesis is correct.

which we have already defined as the phase where practically all actions would be highly informative. Figure 9.4 is a good visual image to keep in mind when thinking about the preadaptive phase.

Probing for opportunity

The conditions and examples outlined above suggest that it is worthwhile to design deliberate policy experiments only when two rather special conditions are met: (1) the best policy choice based on analyses that ignore learning effects would be to continue operating well within the range of recent historical experience, and (2) at least one plausible alternative model predicts a higher performance by moving outside that range of experience, that is it represents a possible *opportunity for improvement*.

A prototypical yield/effort example with these properties is shown in Figure 9.5. Here I have assumed a very limited range of historical experience, centered around the effort level u_1^{**} that would be best if model m_1 is correct. Notice that if m_1 is taken as the most probable ("best") model based on experience, then the prescription would be to continue operating in the effort range where all models $m_1 - m_3$ predict similar outcomes. Now, compare Figure 9.5 with the examples in Figures 9.2 and 9.4. Notice that, as in 9.2, m_2 and m_3 in Figure 9.5 predict higher yields than m_1, *but the only way to find out about them is to probe* toward either u_3^{**} or u_2^{**}. In contrast, operating at the nominal effort level u_1^{**} in the Figure 9.4 example would give at least some information about which model is correct (lower yields than expected under m_1 if m_2 or m_3 is correct).

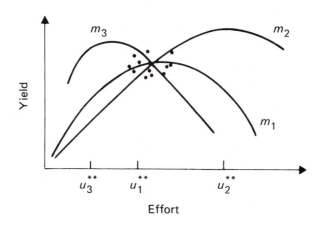

Figure 9.5. Three alternative models for yield as a function of harvesting effort, for a situation in which limited historical experience has been based on the assumption that m_1 is correct. Probing decisions (movement or effort toward u_3^{**} or u_2^{**}) might be justified in this case.

Figure 9.5 makes another point. Try to draw another two alternative models, say m_4 and m_5, that also go through the center of the historical data, but have lower maxima than m_1 (i.e., do not represent opportunities for improvement away from u_1^{**}). You will, of course, find this to be impossible. m_1 is a nominal model that justifies or rationalizes historical choices near u_1^{**}. The only plausible alternatives that you can construct besides the one m_1 that justifies (excuses) past actions are either indifferent to u (as in Figure 9.3), represent opportunities (Figure 9.5), or raise the possibility of dangers by acting differently (as in Figure 9.1).

Here is a key point: when the range of reliable historical experience has been narrow, there are always plausible alternative models besides the past-rationalizing choice (m_1) that predict opportunities for improvement. Thus, there is always a rational basis for probing policies when the range of historical experience is limited. If you buy the various arguments presented earlier (Chapters 4 and 7) about how parameters are likely to change over time, then you should agree to an extension of the point: long periods of "near-equilibrium" management imply a narrow range of *reliable* experience, hence plausible opportunities away from the current equilibrium. Thus, just to establish a sustained equilibrium (the basic goal of management according to some authors), is to set up conditions that will soon favor a probing experiment away from that equilibrium!

Legacies of Uncertainty and Dynamic Programming

The guidelines and examples presented above help to define situations where probing *might* be worthwhile, but it is necessary to look much more carefully at each case to decide whether probing is in fact optimum and how large the probes should be. In general, when we wish to talk about optimum sequences of decisions over time, it is natural to formulate the optimization problem in the framework of dynamic programming as introduced in Chapter 8. More particularly, we expect that the optimum decision will be specified by a feedback policy that can deal flexibly with various situations that might arise; if it is, indeed, optimum to probe, then probing actions should appear as behavioral features or "emergent properties" of the optimum feedback policy, without it being necessary to treat each probing change as a separate and explicit policy variable. This section develops a basic dynamic programming formulation and presents examples that do, indeed, show probing as an emergent feature of the optimum policy.

In its simplest form, dynamic programming is often thought of as a way to find an optimum sequence of decisions, u_t^*, u_{t+1}^*, u_{t+2}^*, ..., as though it were possible *today* (time t) to commit future decision makers to some "open-loop" prescription that we devise. A better way to think about stochastic dynamic programming is that it allows us to represent a *sequence of future decision analyses* conducted at each time step using the same rules for defining an optimum policy, but with different data (information states). To the extent that we can predict how actions taken today will affect the outcome of those future analyses, we may wish to take actions that will affect the degree of uncertainty present in them. More vividly, actions taken today contribute to the *legacy of uncertainty* faced in future analyses. If we can anticipate even roughly how much each future analysis will be improved by an action today that reduces the legacy, then we should include the improvement somehow as part of the value of taking the action.

The dynamic programming equation for actively adaptive management

Consider the decision analysis that might (can, should!) be made at any point in time during the ongoing management of a renewable resource. At this point, the analyst has a collection of past information Y_t from which he can assign probabilities $P_t(m_i)$ to alternative outcomes. Let us suppose that he has summarized this information in a vector of statistics of fixed dimension [say, y_t, $P_1(m_1)$, $P_2(m_2)$ or $\hat{x}_t$, $\hat{\beta}_t$, $\Sigma_{\hat{\beta}_t}$] that includes the best current estimate of the system state along with whatever measures are needed to reconstruct $P_t(m_i)$. Let us call this vector I_t, the decision information state. Without supposing that I_t is in any way an optimum or complete representation of the knowledge available for decision at time t, we may still use it to predict approximately how the analysis will proceed. The analyst will try to find a decision $\hat{u}_t$ that maximizes the expected total value $\hat{V}_t = v_t + v_{t+1} + v_{t+2} + \cdots$ of payoffs v_t, v_{t+1}, etc., into the future. If he bases the calculation of $\hat{V}_t$ only on I_t, without knowing the true state x_t (including true parameter values) and/or the true model m, then he should solve a dynamic programming problem with incomplete state information, as presented in Chapter 8. That is, he should try to maximize $\hat{V}(I_t)$ while looking forward in time to the conditional maximum values $V_{t+1}^*(x_{t+1}, I_{t+1})$ that might arise. If he solves the problem this way, he will "automatically" take into account his legacy of uncertainty by predicting I_{t+1}. The expectation he should try to maximize (by choice of u_t) is

$$\hat{V}_t(I_t) = \sum_{x_t} p(x_t \mid I_t) \times \left[v_t(x_t, u_t) \right. \tag{9.1}$$
$$\left. + \sum_{x_{t+1}, I_{t+1}} p(x_{t+1}, I_{t+1} \mid x_t, I_t, u_t)\, V_{t+1}^*(x_{t+1}, I_{t+1}) \right]$$

This equation is not as complicated as it looks. It says that $\hat{V}$ consists of a weighted sum of terms $v + \Sigma p V^*$, where each term admits one true state/parameter/model combination x_t and assigns it probability $p(x_t \mid I_t)$ (read "probability of the combination x_t given the information state I_t"). For this combination x_t, the total value is a short-term payoff $v_t(x_t, u_t)$, plus an expected long-term value $\Sigma p V_{t+1}^*$. Each element of this long-term value summation admits one possible next state x_{t+1}, I_{t+1} given the current state x_t, I_t, assigns it probability $p(x_{t+1}, I_{t+1} \mid x_t, I_t, u_t)$ (read "probability of x_{t+1}, I_{t+1} given x_t, I_t, and u_t"), and uses this probability to weight the future value $V_{t+1}^*(x_{t+1}, I_{t+1})$ that we assume has already been calculated as the best that can be done from time t_{t+1} forward if the state turns out to be x_{t+1}, I_{t+1}. Now, the analyst does *not* have to assume that the true state x_{t+1} becomes known at time $t+1$. Instead he can calculate $V_{t+1}^*(x_{t+1}, I_{t+1})$ in the same way that he can calculate $V_t^*(x_t, I_t)$ once he finds the $\hat{u}_t$ that maximizes $\hat{V}_t$ in equation (9.1). Using this $\hat{u}_t$, he then calculates

$$V_t^*(x_t, I_t) = v_t(x_t, \hat{u}_t) \tag{9.2}$$
$$+ \sum_{x_{t+1}, I_{t+1}} p(x_{t+1}, I_{t+1} \mid x_t, I_t, \hat{u}_t)\, V_{t+1}^*(x_{t+1}, I_{t+1})$$

This is just the summation element $v + \Sigma p V^*$ from (9.1) for x_t, evaluated at the decision $\hat{u}_t$ which is optimum given only I_t [i.e., maximizes (9.1)].

I have tried to simplify the notation in the above dynamic programming equations (9.1)-(9.2), by calling all unknowns x_t. So x_t includes unmeasured state variables and parameters. If there is a discrete number of alternative models m_i, then think of the index i^* of the correct model m^* as being one of the x_t variables. Thus, we see that equations (9.1)–(9.2) are quite general, since, in principle, they allow us to assign transition probabilities $p(x_{t+1}, I_{t+1} | x_t, I_t, u_t)$, even to such outcomes as change in the correct model from one time to the next.

Notice that to perform backward recursion with equations (9.1)–(9.2) is to build up a picture of a sequence of decision analyses, where each analysis involves the maximization of $\hat{V}_t(I_t)$ while looking forward to the possible outcomes already evaluated for similar analyses at $t+1$, $t+2$, etc. When the short-term payoff function v_t is of the same form at each step up to terms like a constant discount rate, we find, as noted in the previous chapter, that the optimum policy $\hat{u}_t$ becomes constant at each information state after a number of backward recursion steps have been performed. *This stationary function $\hat{u}(I_t)$ is a feedback policy,* with the remarkable property that it anticipates future information states I_{t+1}, I_{t+2}, etc., and future decision analyses based on them.

To gain a better intuitive feeling for what is going on in equations (9.1)–(9.2), it may help to think of them as representing a "branching tree" of possible outcomes associated with the decision u_t at time t (Figure 9.6). The first random branches away from each decision choice u_t represent alternative state/parameter/model conditions possible at the time of decision. Then the next set of branches represents states that might arise as of the next decision point due to random effects in the system itself. For each of these second-level branches, further random effects in the observation process may lead to various possible information states at the time of the next decision. Notice that the total future value V^*_{t+1} associated with the end of each three-way branch is conditional on the actual state achieved (second branch), which is in turn conditional on the first branch. This tree helps us to understand why we cannot simply use I_t as the dynamic programming state, as Walters and Hilborn (1976) incorrectly proposed. If we suppose that the branching tree ends on a $V^*_{t+1}(I_{t+1})$ that is not conditional on x_{t+1}, then we would admit in the next set of branches ($t+1$ to $t+2$) that various states x_{t+1}, x_{t+2} are possibly *independent* of the possibilities admitted in the tree for t to $t+1$. Then, to the extent that the odds on outcomes can be controlled by informative actions, we would be pretending some ability to control the actual parameters/model over time. The "optimum" policy would be to wait for a favorable parameter estimate in I_t, then try to narrow the odds on this estimate.

The feedback policy estimated by backward recursion of equations (9.1)–(9.2) is "globally optimal" (very best adaptive policy possible) only if three special conditions are met. First, the fixed-dimension information state I_t must contain all the information that the historical data Y_t has to offer about present and future states, i.e., about $p_t(x_t)$. In nonlinear systems, this condition cannot usually be met by any simple information state description

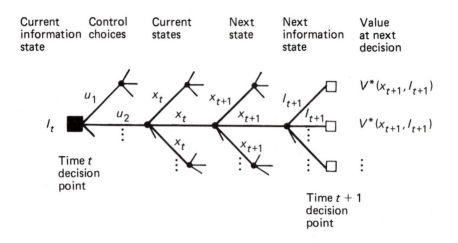

Figure 9.6. The dynamic programming equations (9.1)–(9.2) for adaptive decision making can be visualized as representing a branching tree of possible outcomes. The first branches represent decision choices, and the next three represent uncertain outcomes in (1) what the current state x_t actually is; (2) what next state x_{t+1} will arise given the x_t branch; and (3) what next information state I_{t+1} will arise given x_{t+1}, x_t, and u_t.

(see Chapter 7). Second, the decision-making process must consist of independent reevaluations of data and choices (maximization of $\hat{V}_t$) at each decision point, without it being possible for the decision maker at time t to impose "open-loop" choices u_{t+1}^*, u_{t+2}^*, ..., on his successors except through the effects he may have on the future states (x_{t+1}, I_{t+1}), (x_{t+2}, I_{t+2}), ... When this second condition does not hold, the optimal policy could conceivably involve commitment to a fixed "experimental" schedule of actions over several future time steps, followed afterward by a feedback policy based on the results from that schedule. Third, we must suppose that no new models (not admitted in x_t) and/or data-gathering systems (which influence the odds placed on I_{t+1} given x_{t+1}) will be developed at future time steps. This assumption of a closed, repeated decision process based only on options and outcomes recognized today is, of course, incorrect, but we cannot do much about it except to take care that the recognized range of possibilities is realistically broad.

   To apply the backward recursion for any particular example, we must precisely define (specify, model) how to make three basic calculations. First, we must specify how to calculate the probability $p(x_t \mid I_t)$ to place on each possible true state x_t, given any information state I_t that is considered possible. For example, if x_t contains an unknown parameter β and I_t includes its best estimate $\hat{\beta}_t$ and variance $\sigma^2_{\beta_t}$, then we might estimate $p(\beta|\hat{\beta}_t, \sigma^2_{\beta_t})$ as a normal distribution with mean $\hat{\beta}_t$ and variance $\sigma^2_{\beta_t}$. Second, we must specify how to calculate the payoff $v_t(x_t, u_t)$ in time step t as a function of the actual state and control choice. For example, if one element of x_t is N_t, the stock size, and if u_t is the harvesting effort, then we might take $v_t = \text{catch} = qu_tN_t$ as the simplest payoff measure. Third, we must specify how to calculate probabilities $p(x_{t+1}, I_{t+1}|x_t, I_t, u_t)$ to place on each possible next state x_{t+1} and information state I_{t+1}, given each possible combination of state, information, and action x_t, I_t, u_t at time t. As assumed in Figure 9.6, the joint distribution of x_{t+1}, I_{t+1} can usually be represented as the product of two independent distributions:

$$p(x_{t+1}, I_{t+1}|x_t, I_t, u_t) = p(x_{t+1}|x_t, u_t)p(I_{t+1}|x_{t+1}, I_t, u_t) \tag{9.3}$$

where the first calculation $p(x_{t+1}|x_t, u_t)$ places odds on x_{t+1} given x_t and u_t and the second $p(I_{t+1}|x_{t+1}, I_t, u_t)$ places odds on I_{t+1} given the action u_t, next state x_{t+1}, and last information state I_t. Notice that to specify $p(x_{t+1}|x_t, u_t)$ is to build a stochastic model for the system dynamics, independently of any measurements, while to specify $p(I_{t+1}|x_{t+1}, I_t, u_t)$ is to build a model for how statistics I_t about x will propagate over time.

Let us now outline three examples of these calculations for exploited populations. In the first two, it will be assumed that the stock size can be observed exactly prior to making the final action decision at each time step; with this simplification, I_t includes the stock size and x_t includes only information about which alternative model/parameter value is actually correct. In the third example, we will look at a surplus production situation with three unknown parameters and the stock size not measured directly. As we shall see, even this almost trivial biological representation of surplus production leads to a dynamic programming formulation that is beyond the power of existing computers to solve.

Example 1: Two state-transition hypotheses

Suppose we divide the range of preharvest stock sizes that a population might achieve in any year into 10 discrete levels or crude states: 1 = very low, 2 = low, 3 = medium, etc. Suppose further that the management decision is what discrete escapement level to leave behind after harvesting (1 = very low, 2 = low, etc.). Let us call the discrete population level N_t (N_t = 1, 2, 3, ..., 10) and the escapement choices u_t (u_t = 1, 2, 3, 4, or 5). Notice that $u_t \le N_t$. Let us measure the short-term payoff as $v_t = \lambda^t(N_t - u_t)$, i.e., harvest $N_t - u_t$ discounted by the factor λ^t, where $\lambda \le 1.0$. Suppose that there are two alternative models, m_1 and m_2, about how the next year's stock size N_{t+1} will depend on the control u_t; let us represent these models as two probability transition matrices, as shown in Table 9.1. The i,jth element of each table is the probability $p(N_{t+1}|u_t, k)$ of getting a stock size j next year given that escapement i was allowed this year, and if model k ($k = 1, 2$) is correct. Notice that the two models place exactly the same odds on different next states when escapement is low ($u = 1, 2, 3$), but differ in the odds placed on higher stock sizes next year when escapement is high. Thus, models m_1 and m_2 are a much condensed prototype of the situation in Figure 1.1; responses to low escapements are considered to be well known based on historical data, and there is uncertainty (represented by m_1 and m_2) about responses at high escapements.

Table 9.1. Two alternative models of population response, expressed as probabilities for next year's stock size given various escapement choices this year.

Model (m)	This year's escapement choice (u)	Next year's stock size (N)									
		1	2	3	4	5	6	7	8	9	10
m_1	1	0.5	0.5								
	2	0.3	0.3	0.5	0.2						
	3				0.2	0.6	0.2				
	4				0.1	0.4	0.2	0.2	0.1		
	5				0.8	0.1	0.04	0.02	0.02	0.01	0.01
m_2	1	0.5	0.5								
	2	0.3	0.3	0.5	0.2						
	3				0.2	0.6	0.2				
	4				0.1	0.1	0.2	0.4	0.2		
	5				0.01	0.05	0.8	0.05	0.05	0.03	0.01

Given an observed outcome N_{t+1} following a control choice u_t, Bayes' theorem can be used to calculate the new probability $P_{t+1}(m_1)$ that should be placed on model m_1 $[P_t(m_2) = 1 - P_t(m_1)]$. The formula is

$$P_{t+1}(m_1) = \frac{1}{\alpha_{t+1}} p(N_{t+1}|u_t, m_1) P_t(m_1) \qquad (9.4)$$

where $p(N_{t+1}|u_t, m_1)$ is the probability (likelihood) of N_{t+1} given u_t and m_1 from the first table in Table 9.1, and α_{t+1} is the total probability of getting N_{t+1}:

$$\alpha_{t+1} = p(N_{t+1}|u_t, m_1) P_t(m_1) + p(N_{t+1}|u_t, m_2)[1 - P_t(m_1)]$$

Let us take $P_t(m_1)$ as the single information statistic describing uncertainty about which model is correct; it is, in fact, a sufficient statistic for m_1, i.e., contains all the information about m_1 that is in the historical record (u_1, N_2), (u_2, N_3), ..., (u_{t-1}, N_t). It also includes a prior probability $P_0(m_1)$.

With the above definitions, the information state available for decision making at time t becomes $I_t = \{N_t, P_t(m_1)\}$. The only unknown x_t is k ($= 1$ or 2), i.e., which model k is correct. Let us assume that the correct model remains correct over time, i.e., $x_{t+1} = x_t$. Notice that the short-term reward $v_t = \lambda'(N_t - u_t)$ is independent of the correct model k. Under these conditions, the dynamic programming equation (9.1) reduces to just

$$\hat{V}_t(I_t) = N_t - u_t \tag{9.5}$$

$$+ \sum_{k=1}^{2} \sum_{N_{t+1}=1}^{10} p(N_{t+1}|u_t, k) V_{t+1}^*(k, I_{t+1})$$

where, for each element in the summation, I_{t+1} is the pair N_{t+1} for that element and $P_{t+1}(m_1)$; P_{t+1} is calculated by equation (9.4) using N_{t+1}, u_t, and $P_t(m_1)$ from I_t. The predicted $P_{t+1}(m_1)$ in I_{t+1} need not fall on any of the discrete grid points used to evaluate V_{t+1}^*. Thus, the numerical solution of (9.5) requires interpolation between V_{t+1}^* values in the $P_{t+1}(m_1)$ dimension. The reduced version of (9.2) is

$$V_t^*(k, I_t) = N_t - \hat{u}_t \tag{9.6}$$

$$+ \sum_{N_{t+1}=1}^{10} p(N_{t+1}|\hat{u}_t, k) V_{t+1}^*(k, I_{t+1})$$

where $\hat{u}_t$ is the escapement choice that maximizes $\hat{V}_t$ in equation (9.5).

Provided the information state variable $P_t(m_1)$ in I_t is represented on a fairly coarse grid (say five or ten levels), the recursive optimization equations (9.5) and (9.6) can be solved backward in time easily, even on a microcomputer. With a linear interpolation scheme for V_{t+1}^* in the P_{t+1} dimension and P_t discretized at five levels, an Apple II can do 15 backward steps [enough for the policy $\hat{u}(I_t)$ to become stationary even for $\lambda = 1$] in about 20 minutes. Note that at each step, the value function V_t^* must be estimated at $2 \times 10 \times 5 = 100$ points [two k values, ten N_t values, and five $P_t(m_1)$ levels]. This involves $10 \times 5 = 50$ maximizations of $\hat{V}$, one at each of the ten N_t values and five $P_t(m_1)$ levels.

An interesting feature of the problem is that the set of five u_t choices can be partitioned into four qualitative subsets, as shown in Figure 9.7. The actions $u = 1, 2$ are both uninformative as to which model is correct, and also not optimum unless the initial N_t is low (1 or 2). Action 3 is optimum if m_1 is known to be correct (and $N_t \geq 3$), but it is uninformative: $p(N_{t+1}|m_1, u = 3) = p(N_{t+1}|m_2, u = 3)$. Action 4 is optimum if m_2 is certain (and $N_t \geq 4$), and it is also informative about which model is correct. Action 5 is highly informative (see the transition probabilities in Table 9.1), but is not optimum if either model is known to be correct. The partitioning shown in Figure 9.7 appears to occur to some degree in all nontrivial adaptive control problems. When the optimum-given-m subset contains only one

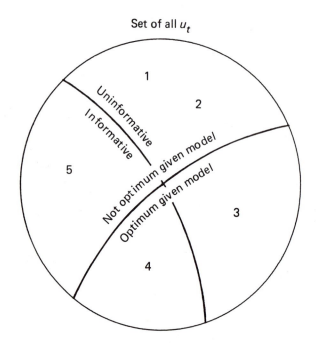

Figure 9.7. The decision choices $u_t = 1, 2, 3, 4, 5$ in a simple adaptive decision problem can be partitioned into four subsets. A key issue is whether decision 5 will ever be the optimum.

element, we get the degenerate situation of Figure 9.2. When the uninformative-optimum-given-m subset is empty (where $u = 3$ as in Figure 9.7) we get the degenerate situation in Figure 9.4; all optimum-given-m decisions are informative. Only one qualitative type of decision is missing from Figure 9.7, namely, a cautious intermediate experiment between the optima-given m_1 or m_2.

A critical question is: would it ever be optimum to choose the highly informative, "probing" decision $u = 5$ that is not optimum if either model is correct? That is, should we ever pu :h beyond $u = 4$, which is informative, might turn out to be optimum anyway, and would be the best "certainty-equivalent" choice if $P_t(m_1) < 0.5$? The answer is a resounding yes, for the information state $N_t = 5$, and $P_t(m_1)$ in the range 0.25–0.75. That is, when there is high uncertainty as to which model is correct and there exists an extreme decision that will resolve the uncertainty without risk of stock collapse, this extreme decision is part of the optimum stationary policy $\hat{u}(I_t)$.

While I have developed this example just to look at the question of whether extreme, but informative, decisions would ever be optimal, I sometimes wonder if such simple transition probability models (representing only a few aggregate stock levels and actions) might not be a more realistic way to deal with many practical situations than the more elaborate population equations that we usually assume. Populations are usually monitored only crudely (low, medium, high, etc.), and actions are often coarse (open season, close, etc.) as well. Moreover, it is easy to define rough transition probabilities from historical data and for extreme (and controversial) hypotheses about response; it is easy enough to show that the optimization results [feedback policy $\hat{u}(I_t)$] are not particularly sensitive to the probabilities assumed. For examples of much more detailed calculations using transition probability (Markov decision) models that do not involve learning effects, see Mendelssohn (1980). For further examples that do involve learning, see Walters (1981).

Example 2: Stock–recruitment model with an unknown parameter

Here we return to an example discussed in Chapter 7, namely the Ricker stock–recruitment model written as $N_{t+1} = u_t \exp [\alpha - \beta(u_t - \bar{u}) + w_t]$, where the mean historical productivity and escapement parameters α and $\bar{u}$ are well fixed by past data, u_t is the spawning stock, w_t is a normally distributed environmental effect, and β is an unknown parameter representing the sensitivity of productivity [$\ln (N_{t+1}/u_t)$] to spawning stock. Here we can generate an infinite number of models m_i by varying the parameter β, so we speak about uncertainty in terms of the probability distribution $p_t(\beta)$. Assuming that the prior probability distribution $p_0(\beta)$ given no data is normal, and that N_t and u_t are measured exactly at each time step (generation), then $p_t(\beta)$ is a normal distribution with mean $\hat{\beta}_t$ and variance $\sigma^2_{\hat{\beta}_t}$. The dynamics of $\hat{\beta}$ and $\sigma^2_{\hat{\beta}_t}$ are given recursively (for any step t to t_{+1}) by equation (7.17) ($\Sigma_{\hat{\beta}_t} = \sigma^2_{\hat{\beta}_t}$). As in the previous example, let us assume that the management payoff is measured simply by discounted catch $v_t = \lambda'(N_t - u_t)$. The example as outlined here has been analyzed in detail by Ludwig and Walters (1981), using a scheme devised by Ludwig for an efficient approximation of the dynamic programming value function V^*.

With N_t observed exactly, the only unknown x_t in this example is the parameter β. The information state I_t available for decision making at time t is assumed to be $I_t = \{N_t, \hat{\beta}_t, \sigma^2_{\hat{\beta}_t}\}$. The dynamic programming equation corresponding to equation (9.1) becomes

$$\hat{V}_t(I_t) = N_t - u_t + \lambda \int_\beta p_t(\beta) \int_{w_t} p'(w_t) V^*_{t+1}(\beta, I_{t+1}) \, dw_t \, d\beta \qquad (9.7)$$

where $p'(w_t)$ is the normal distribution for w_t (mean zero, variance σ_w^2). The computation of these integrals [which replace the sums over x_{t+1} in (9.1)] can be simplified since $p(\beta)$ and $p'(w_t)$ are independent normal distributions; their product gives a single normal distribution for the random variable $Z = \beta(u_t - \bar{u}) + w_t$, which is sufficient to calculate N_{t+1}, and hence $\hat{\beta}_{t+1}$ and $\sigma_{\beta_{t+1}}^2$ given u_t and I_t (i.e., to calculate I_{t+1}). The analogue of equation (9.2) is

$$V_t^*(\beta, I_t) = N_t - \hat{u}_t + \lambda \int_{w_t} p'(w_t) V_{t+1}^*(\beta, \hat{I}_{t+1}) \, dw_t \qquad (9.8)$$

where $\hat{u}_t$ is the escapement that maximizes (9.7) and $\hat{I}_{t+1}$ is the information state predicted given β, I_t, $\hat{u}_t$, and w_t [using the recruitment model and equations (7.17)].

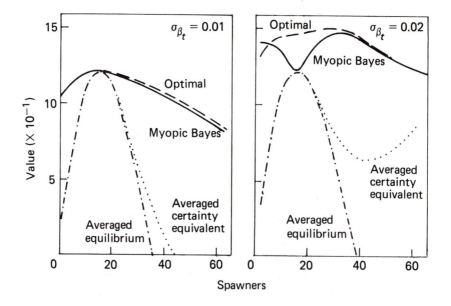

Figure 9.8. Various estimates by dynamic programming of the value $\hat{V}_t$ of allowing different numbers of spawners in a stock–recruitment example (for explanation see text). *Source:* Ludwig and Walters (1981).

Figure 9.8 shows examples from Ludwig and Walters (1981) of how $\hat{V}_t$ for this problem varies with u_t, for two different levels of uncertainty about β. Here we deliberately assumed that the system had been managed by a certainty-equivalent policy, so the historical $\bar{u}$ would coincide with any assessment of the optimum $\hat{u}$ made by ignoring learning. A discount factor $\lambda = 0.9$ was assumed, and $\hat{V}_t$ (marked "optimal" in the figure) is shown after enough iterations to assure that it had become independent of t (this always happens for $\lambda < 1$). For comparison, three alternative ways of estimating the optimum $\hat{u}_t$ are shown. Value functions estimated by ignoring all learning efforts are marked "averaged equilibrium" and "averaged certainty equivalent." They both predict that the best policy would be to allow around $u_t = \bar{u} = 15$ spawners. The value function marked "myopic Bayes" was computed by looking at learning effects (I_{t+1}) for only one step into the future, then assuming no further learning. Here it becomes apparent that the certainty-equivalent, historical choice $\bar{u}$ would be very bad; the increase in myopic Bayes value by moving away from $\bar{u}$ is due to the information value of taking just one informative decision. The optimal value $\hat{V}_t$ is much flatter for large $\sigma^2_{\beta_t}$, since in looking forward to future analyses that might involve informative probing it places a higher value even on staying at $\bar{u}$ for the first decision u_t. That is, it "recognizes" that some future decision maker will act as it predicts is best, which is to allow a much higher, probing escapement of around $u = 30$.

The most surprising discovery from this example was that the optimum adaptive policy apparently involves either assuming certainty equivalence (when $\sigma^2_{\beta_t}$ is low), or making a large and informative probe (when $\sigma^2_{\beta_t}$ is large). It never seems to be good to just "dither" the escapement level slightly in hopes of gaining information about β. One should either ignore the uncertainty, or take a substantial step to resolve it. K. Astrom (personal communication) has obtained similar results for a linear-quadratic regulator problem with a single unknown parameter: the optimal regulator acts to prevent output variation when the parameter variance is low, and induces strong variation when the parameter variance is large.

The optimal policy can be approximated as shown in Figure 9.9. For σ^2_{β} below some threshold γ_0, the best escapement is approximated by u_{ce} (ce, certainty equivalent), which depends only on $\hat{\beta}$ (and the known parameters $\bar{a}$, $\bar{u}$, and σ^2_w). For σ^2_{β} greater than γ_0, the best policy is approximated by $\gamma_1 u_{ce}$, i.e., a multiple of the certainty-equivalent escapement. The best value of γ_1 appears to be roughly 2.0. A. Gaivoronski (personal communication) has assumed a policy of this form and conducted optimizations in the policy space γ (see Chapter 8). For $u_{ce} \approx \bar{u}$, he also finds expected improvements in the total value function V by moving from only certainty-equivalent control ($\gamma_1 = 1$) to the approximate optimal policy, and finds the best value of γ_1 to be around 2.0.

Example 3: Surplus production model

Let us assume that the biomass dynamics B_t of a stock can be approximated by the stochastic logistic model

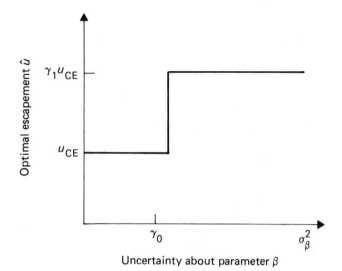

Figure 9.9. The optimum escapement policy for a stock–recruitment system with unknown parameter β (for response of productivity to stock size) can be approximated by the simple feedback policy shown here. When uncertainty is low, the best policy is to allow the certainty-equivalent escapement u_{ce}. When uncertainty is higher than γ_0, the best escapement jumps to $\gamma_1 u_{ce}$, where $\gamma_1 \gg 1.0$.

$$\frac{dB}{dt} = \beta_1 B - \beta_2 B^2 - \beta_3 Bu_t + dw_t \qquad (9.9)$$

where $\beta_1 = r$, $\beta_2 = r/k$, $\beta_3 = q$, and u_t is harvesting effort. Assume further that u_t is held constant over each year (t to $t + 1$), and that an abundance index $y_t = \beta_3 B_t$ (catch/effort) is measured at the start of each year. Then Schnute (1976) notes that (9.9) can be integrated to give (see Chapter 4):

$$\ln \frac{B_{t+1}}{B_t} = \beta_1 - \beta_2 \bar{B} - \beta_3 u_t + w_t \qquad (9.10)$$

where $\bar{B} = \int_t^{t+1} B_t \, dt$ is the mean biomass over t to $t + 1$. Assuming further that there is no observation error (y_t is exactly equal to $\beta_3 B_t$) results in a linear regression model for estimating $\beta_1 - \beta_3$:

$$\ln \frac{y_{t+1}}{y_t} = \beta_1 - \frac{\beta_2}{\beta_3} \bar{y} - \beta_3 u_t + w_t \qquad (9.11)$$

where a simple approximation of $\bar{y}$ would be the mean of the end points $(y_{t+1} + y_t)/2$. Using this regression, uncertainty about $\boldsymbol{\beta}$ is represented by the covariance matrix $\Sigma_{\boldsymbol{\beta}}$, which has six distinctive elements: $\sigma_{\beta_1}^2$, $\sigma_{\beta_2}^2$, $\sigma_{\beta_3}^2$, $\sigma_{\beta_1\beta_2}$, $\sigma_{\beta_1\beta_3}$, and $\sigma_{\beta_2\beta_3}$ (think of the $\sigma_{\beta_i\beta_j}$ as correlations among the parameter estimates). The regression estimates $\hat{\beta}$ and Σ_β can be updated from t to $t +$ 1 using the standard recursive linear regression equations (7.12)-(7.15), given u_t and y_{t+1}. Thus, it would seem that equations (9.10), (9.11), and (7.12)-(7.15) provide all the machinery needed for a dynamic programming formulation.

Now suppose we wish to design an adaptive feedback policy that specifies the effort u_t to apply during year t as a function of the estimate y_t of relative abundance at the start of the year, along with information about future production based on the stock history Y_t as summarized in the statistics for β. The "wide-sense" information state would then be:

$$I_t = \{y_t, \hat{\beta}, \Sigma_\beta\} \tag{9.12}$$

$$= \{y_t, \hat{\beta}_1, \hat{\beta}_2, \hat{\beta}_3, \sigma^2_{\hat{\beta}_1}, \sigma^2_{\hat{\beta}_2}, \sigma^2_{\hat{\beta}_3}, \sigma_{\beta_1\beta_2}, \sigma_{\beta_1\beta_3}, \sigma_{\beta_2\beta_3}\}$$

that is, 10 variables are needed to characterize the current state estimate ($\hat{B}_t$ $= y_t/\beta_3$) and uncertainty about future production. If we discretize all these variables for dynamic programming at only five levels each, we would end up having to maximize $\hat{V}(I_t)$ of equation (9.1) at $5^{10} = 9\,765\,625$ grid points. The value function V^*_{t+1} has four additional dimensions, since it includes the unknowns $x_{t+1} = (B_{t+1}, \beta_1, \beta_2, \beta_3)'$. This 14-dimensional function would be difficult to store in today's biggest computers even at only five levels per variable, and would take many hours to calculate for each time step. After making such a massive investment in computer time, we would have a feedback function $\hat{u}(I_t)$ defined over 10 variables, and it is not clear how we would go about trying to visualize (or otherwise understand qualitatively) the prescriptions of this policy.

If we are willing to assume that the historical data are adequate to determine one of the three parameters exactly (for example, if we assume the equilibrium combination $0 = \beta_1 - \beta_2/\beta_3\,\bar{y} - \beta_3\,\bar{u}$ is known, we can solve for one parameter given the others), then the required information state can be reduced to

$$I_t = \{y_t, \hat{\beta}'_1, \hat{\beta}'_2, \hat{\sigma}^2_{\hat{\beta}_1}, \hat{\sigma}^2_{\hat{\beta}_2}, \sigma_{\beta_1\beta_2}\}$$

where β'_1 and β'_2 are the two parameters that are assumed to remain unknown. This 6/9-variable problem would be barely within the feasible computation limits for dynamic programming on large computers, and there would still be a serious problem in visualizing and qualitatively interpreting the feedback policy function $\hat{u}(I_t)$.

Why is it that such an apparently simple dynamic model like logistic surplus production leads to gross difficulties in computation and interpretation of an optimal adaptive policy? For an intuitive answer to this question, look back again at what we seek to accomplish with an optimal adaptive policy: the anticipation of future learning and its value to management.

Simply to *measure* learning rates and the moving state of uncertainty, we need to add at least six extra variables (Σ_β) to the obvious state description $(\hat{B}_t, \hat{\beta}_1, \hat{\beta}_2, \hat{\beta}_3)$. In general, if there are n uncertain parameters, the adaptive optimization needs to look at roughly $(n^2 + n)/2$ additional variables beyond the state and parameter estimates needed to define a certainty-equivalent policy. To avoid this explosive growth in the information state dimension, we must either assume away most uncertainties or be content with some approximate calculation that does not require to build up the full feedback policy $\hat{u}(I_t)$ for all possible future information states. The best approximate scheme currently available is the wide-sense dual control algorithm, which we will introduce in the next section.

Wide-Sense Dual Control

Tse et al. (1973) introduced an algorithm that shows considerable promise for adaptive policy design in systems that have more than one unknown parameter. Their "wide-sense dual control" algorithm has been applied to a variety of examples, including economic models (Bar-Shalom, 1981; Pekelman and Tse, 1980) and fisheries stock–recruitment analyses (Smith and Walters, 1981). Reviews of the algorithm can also be found in Bar-Shalom and Tse (1976a,b). Basically, their approach is to begin with the current wide-sense information state $(\hat{x}_t, \Sigma_{x_t})$ for a problem, where x_t includes all unknown states and parameters. Without pretending to define $\hat{u}(I_t)$ for all possible information states, they then seek an efficient way to approximate the best *current* control $\hat{u}_t$ while taking immediate and future learning effects into account.

Given $\hat{x}_t$ and Σ_{x_t}, the algorithm proceeds by first calculating a nominal or certainty-equivalent trajectory $x_{t+1}^{(o)}, \ldots, x_T^{(o)}, \Sigma_{x_{t+1}}^{(o)}, \ldots, \Sigma_{xT}^{(o)}$ using the state and observation dynamic models

$$x_{t+1} = f(x_t, u_t, w_t) \qquad y_t = h(x_t, v_t) \tag{9.13}$$

with the stochastic effects w_t, v_t set to zero and with a nominal (open-loop) control sequence $u_{t+1}^{(o)}, \ldots, u_T^{(o)}$. If possible, the nominal control sequence should be chosen so as to be optimum for the nominal deterministic dynamic $x_t^{(o)}$. It is important to keep in mind that x_t is an "extended-state" vector including all unknown parameters. For example, with the Ricker model $N_{t+1} = u_t \exp(\beta_1 - \beta_2 u_t + w_t)$, the extended state would be $x_t = (N_t, \beta_1, \beta_2)'$, for which the state equations $f(x, u, w)$ are

$$x_{1,t+1} = u_t \exp(x_{2t} - x_{3t} u_t + w_t)$$

$$x_{2,t+1} = x_{2t}$$

$$x_{3,t+1} = x_{3t}$$

The nominal dynamics of the covariance matrix Σ_{x_t} can be approximated by various filtering models, such as the extended Kalman filter (Chapter 7). The key objective is to get at least some assessment of how uncertainty will propagate along the nominal or "most likely" future path of states $x_t^{(o)}$ and controls $u_t^{(o)}$.

It is assumed that the management objective is to maximize a value function V_t of the form

$$V_t = \sum_{t}^{T-1} v_t + \psi(x_T) \tag{9.14}$$

where in the simplest case the stepwise payoffs v_t are of the form

$$v_t = l(x_t, t) + \varphi(u_t, t) \tag{9.15}$$

(i.e., a term l related to the state and a term φ related to the control). $\psi(x_T)$ is a "terminal value" associated with ending up in different possible states x_T at the last time step. For example, taking annual catch as the payoff measure for the Ricker model results in $l(x_t, t) = x_{1,t}$ (i.e., N_t) and $\varphi(u_t, t) = -u_t$ (i.e., minus the escapement); a simple terminal value would be $\psi(x_T) = x_T$, i.e., just the last period stock size. Let us denote the total value for the nominal trajectory $x_t^{(o)}, \ldots, x_T^{(o)}, u_{t+1}^{(o)}, \ldots, u_T^{(o)}$ by $V_t^{(o)}$. Usually $V_t^{(o)}$ would be a most pessimistic assessment of the total future value of the system, since it ignores all potential gains due to improvement of state/parameter assessments and cautious anticipation of environmental effects.

As an aside, let me remind the reader that many resource assessments stop at this point, with the assessment of a most likely future $x_t^{(o)}$, $u_t^{(o)}$, and $V_t^{(o)}$. Indeed, decision makers often demand that *only such assessments be presented, without confusing and worrisome "hedging"* or qualifications about various uncertainties. As we shall see, the wide-sense algorithm at least allows the analyst to present a compressed and understandable representation of the effects of major uncertainties.

Effects of uncertainty are included in the wide-sense algorithm by carrying out a perturbation analysis (Taylor series expansion, to second order) of the value function in the neighborhood of the nominal trajectory, for "small" disturbances (in future states and controls) away from it due to stochastic effects and learning in future time steps ($t + 1$, $t + 2$, etc.). The perturbation analysis involves some rather complex algebra that need not be repeated here; the clearest available summary is in Bar-Shalom and Tse (1976b). Its basic result is that the dynamic programming value function from time t forward can be approximated by $V_t^{(cl)}$ (cl is for "closed-loop"), which consists of five terms:

$$V_t^{(cl)} = v_t + V_{t+1}^{(o)} + V_{t+1}^{(c)} + V_{t+1}^{(p)} + \gamma_{(t+1)}^{(o)} \tag{9.16}$$

Here $V_{t+1}^{(o)}$ is the value achieved along the normal trajectory $x_{t+1}^{(o)}, \ldots, x_T^{(o)}$ following the application of any choice of control u_t. The other terms represent stochastic effects. $V_{t+1}^{(c)}$ is a "caution term" representing the effects of uncontrollable future uncertainties (w, v) on the long-term value. $V_{t+1}^{(p)}$ is a "probing term" representing the future value of the reduction in uncertainty due to u_t (i.e., the effect of u_t on future covariance matrices Σ_{t+1}, Σ_{t+2}, etc.). $\gamma_{t+1}^{(o)}$ is a term representing the nonoptimality of $u_{t+1}^{(o)}$, ..., $u_{T-1}^{(o)}$ with respect to the deterministic dynamics; it is equal to 0 if $u^{(o)}$ is optimal for the deterministic problem (f with $w_t = 0$).

Formulae for computing the caution, probing, and nonoptimality value terms $V_{t+1}^{(c)}$, $V_{t+1}^{(p)}$, and $\gamma_{t+1}^{(o)}$ are presented in Bar-Shalom and Tse (1976b). Basically, each of these consists of a sum of components over future time steps, where each component involves a "weighting" placed on uncertainties predicted at that time step along the nominal trajectory. The weightings are calculated from sensitivities (partial derivatives) of the value function v_t and dynamic model f_t to the future states x_t and controls u_t. Of particular interest for us here is the probing term $V_{t+1}^{(p)}$. It is calculated as

$$V_{t+1}^{(p)} = - \sum_{j=t+1}^{T-1} \text{tr} \left[A_{0j} \Sigma_j^{(o)} \right] \tag{9.17}$$

(Here tr means "trace" of the matrix $A_{0j} \Sigma_j^{(o)}$, the sum of diagonal elements of the matrix.) The weighting matrices A_{0j} are measures of the value of information associated with choosing a control u_t that will reduce Σ_{t+1}, Σ_{t+2}, ..., Σ_T. $V_{t+1}^{(p)}$ is negative, since the matrices A_{0j} are positive definite: thus, it is measured as a "loss" that can be reduced by actions that reduce $\Sigma_j^{(o)}$.

Variation in value components with action choice

A simple use of the wide-sense algorithm is to compute the value components "deterministic" $V_{t+1}^{(d)} = v_t + V_{t+1}^{(o)} + \gamma_{t+1}^{(o)}$, "caution" $V_{t+1}^{(c)}$, and "probing" $V_{t+1}^{(p)}$ for a range of action choices u_t. This gives a graphic picture of performance for the "baseline model" ($V^{(d)}$) and the values $V^{(c)}$ and $V^{(p)}$ associated with choices u_t that are different from the nominal choice $u_t^{(d)}$ that would maximize $V^{(d)}$. Figure 9.10 shows two purely hypothetical examples of how we expect the value components to vary for a harvesting problem where u_t is a measure of exploitation rate and performance is measured in terms of long-term harvest. First, we expect a dome-shaped relationship between nominal harvest value $V^{(d)}$ and harvest rate; the peak of this dome defines the certainty-equivalent optimum rate $u^{(o)}$. Next, we expect the caution component $V^{(c)}$ to be the largest when the harvest rate u is smallest, and to decrease as u_t increases. Finally, we expect the probing value $V^{(p)}$ to be lowest if u_t is chosen to be near the historical average $\bar{u}$, and to increase for informative choices away from $\bar{u}$. Notice that the certainty-equivalent and average historical choices $u^{(o)}$ and $\bar{u}$ need not coincide (case B); when they do not, we get one of the degenerate cases mentioned earlier, where it is

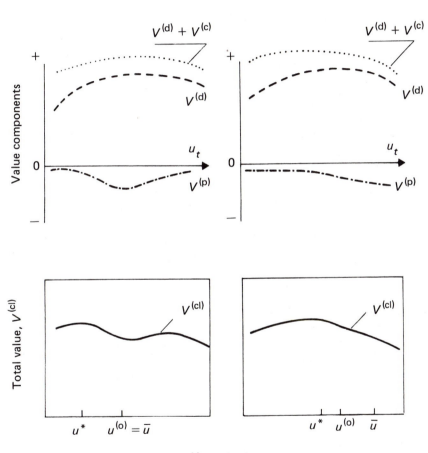

Case A: $\bar{u} = u^{(o)}$ Case B: $\bar{u} \gg u^{(o)}$

Harvest rate, u_t

Figure 9.10. Examples of how the long-term value of harvests can be broken down into components as functions of harvest rate. The total value is $V^{(cl)}$, which is equal to a deterministic value prediction $V^{(d)}$ plus a caution effect $V^{(c)}$, less a probing value effect $V^{(p)}$. In case A, higher probing values away from the nominal $u^{(o)}$ imply that the optimum u^* is far below $u^{(o)}$. In case B, even using $u^{(o)}$ is informative since it is far from the historical average $\bar{u}$.

unlikely to be worthwhile making a deliberate experiment since the certainty-equivalent action $u^{(o)}$ is itself informative. The probing value $V^{(p)}$ is usually not zero even at $u_t = \bar{u}$, since operation at this point may continue to induce informative changes in the system state and improvements in estimates of mean response, even when the state is not changing.

As I have drawn the value components for case A in Figure 9.10, the value of information is large enough to make $u^{(o)}$ a locally worst policy (compare Figure 9.8), and there are two informative local optima. The best of these is at the low harvest rate marked u^*, where the probing and caution components combine to produce a total expected value $V^{(cl)}$ that is higher than at $u^{(o)}$. In case B, the probing value is nearly the same for harvest rates $u^{(o)}$ and below, and the caution component $V^{(c)}$ "pushes" the closed-loop optimum u^* to somewhat below $u^{(o)}$. Obviously many other situations involving $u^* - u^{(o)}$ are possible, depending on the magnitude of the $V^{(p)}$ and $V^{(c)}$ components relative to $V^{(d)}$ and on how much they vary with u_t. To the extent that the wide-sense approximations can be trusted for action choices that deviate strongly from the nominal $u^{(o)}$, pictures like Figure 9.10 can be used to visualize all of the qualitative conditions and arguments presented earlier about when probing may be worthwhile. To clarify the possibilities, I have deliberately drawn the hypothetical values in Figure 9.10 so as to exaggerate the probing effect $V^{(p)}$ compared with that found in resource examples analyzed to date.

Figure 9.11 shows wide-sense value components estimated for the Ricker model with one unknown parameter, as discussed earlier $\{R = S \exp [a' - b(S - \check{s}) + w_t], b$ unknown, a' and $\check{s}$ known, R and S observed exactly$\}$. For this case the calculations can be done easily on a microcomputer. Here I assumed $a' = 1$, $\hat{b}_0 = 1$, $\sigma_{b_0}^2 = 0.5$, $\lambda = 1$, $t = 0$, $T = 10$, and $\sigma_w^2 = 0.1$. I took $\check{s}$ so that it would coincide with $u^{(o)}$, the most favorable condition for probing. As expected from earlier discussions, the optimum action u^* involves a strong probing increase in escapement level. The value of information is relatively small ($V^{(p)}$ not a large loss), as predicted from EVPI arguments in Chapter 7 (see Table 7.1). However, notice that $V^{(d)}$ and $V^{(c)}$ are almost independent of the initial choice u_0, since the nominal feedback policy $u^{(o)}$ (fixed escapement, $u^{(o)} = \min \{R_t, \bar{u}\}$) gives good catches for most years along the nominal trajectory except when u_0 is very small (then a recovery period is required). Thus, even small changes in $V^{(p)}$ can shift the maximum of the total value $V^{(cl)}$ substantially away from its peak considering only $V^{(d)} + V^{(c)}$. This is essentially the same conclusion reached by dynamic programming in Ludwig and Walters (1981), shown by their $\sigma_{\beta_t} = 0.02$ case in Figure 9.8; note that the wide-sense value estimate of Figure 9.11 is roughly halfway between the myopic Bayes and optimal estimates in Figure 9.8 (allowing for differences in the spawning stock scales).

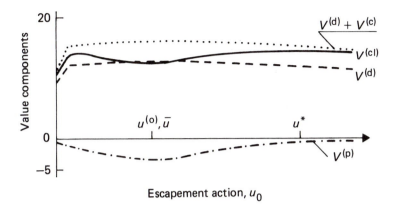

Figure 9.11. Effect of spawning stock choice on components of long-term harvest value, estimated by the wide-sense dual control algorithm. The total value estimate $V^{(cl)}$ is equal to a deterministic estimate $V^{(c)}$ accounting for uncontrollable environmental effects, plus a value $V^{(p)}$ of probing away from the historical average escapement $\bar{u}$.

Caution and probing in stock–recruitment

A detailed analysis of the Ricker stock–recruitment model with two unknown parameters [a and b in $R_t = S_{t-1} \exp(a - bS_{t-1} + w_t)$, $u_t \equiv S_t$] has been carried out using the wide-sense algorithm by Smith (1979) and Smith and Walters (1981). For this case the extended state description is $x'_t = (R_t, a, b)$. Assuming R_t and u_t (spawners) can be observed exactly, the covariance matrix of a and b, Σ_t, can be estimated (and future values predicted along a nominal trajectory $R_t^{(o)}$, $\Sigma_t^{(o)}$) by recursive linear regression [equations (7.12)–(7.15)]. The three different elements of $\Sigma_t(\sigma_a^2, \sigma_{ab},$ and $\sigma_b^2)$ are equivalent to (can be computed from) three more intuitively meaningful statistics:

t the number of $\{R_t, u_{t-1}\}$ observation pairs available in the historical record Y;

σ_u^2 the variance of $u_0, \ldots, u_{t-1}$, a measure of how much informative variation in u_t has already occurred;

$\bar{u}_t$ the mean historical level of spawning stock.

For typical values of $\hat{a}$ and $\hat{b}$, it is particularly interesting to ask how the closed-loop optimal control u_t^* varies with σ_u^2 and $\bar{u}_t$.

By conducting little optimizations as in Figure 9.11 across a grid of initial combinations of the uncertainty measures σ_u^2 and $\bar{u}$, Smith was able to determine how the optimum probing deviation $u_t^* - u_t^{(d)}$ depends on these measures. Figure 9.12 shows a perspective plot of the probing deviation as a function of σ_u^2 and $\bar{u}$. This picture shows that it is not worth probing (deviation negligible) when σ_u^2 is large, i.e., there has already been much informative variation. When σ_u^2 is small, it is best to probe, and the direction (higher or lower escapement) is determined by the historical average $\bar{u}$. If the stock has generally been low (relative to the estimated "equilibrium" stock $\hat{a}/\hat{b}$), so that $\bar{u}$ is small, the best probe is upward to higher escapements. When $\bar{u}$ has been large, it is best to probe down with lower escapements. These prescriptions make very good intuitive sense for stock–recruitment management in general.

By doing a large number of Monte Carlo simulations using randomly chosen true values of a, b (from a normal distribution with means $\hat{a}_0$, $\hat{b}_0$, variance Σ_0) and random disturbances w_t, it is possible to construct probability distributions of total value V_t associated with using a policy, such as in Figure 9.12, and for alternatives, such as the certainty-equivalent regime (use $u_t^{(o)}$ every year). As shown in Figure 9.13, such simulations show that the expected improvement from using the dual control (closed-loop) policy comes largely from reducing the odds of very *bad* outcomes (V_t much lower than the potential). For most trials, the certainty-equivalent policy (passively adaptive) does about as well as the dual control policy. But occasionally, certainty-equivalent actions result in the spawning stock being "locked in" at a low level so good parameter estimates are not obtained. The dual control policy never gets caught in this "trap."

Figure 9.12 implies a very interesting management pattern when parameters are expected to vary over time and this is reflected in the parameter estimation procedure (older data discarded, data discounting, etc.). The basic effect of assuming parameter variation is to make σ_u^2 *decrease* over time during any period when u_t is held relatively steady. Without saying anything about exactly how rapidly this decrease (increase in uncertainty) will occur, we can nevertheless see what its qualitative effect will be: *the optimal policy will involve alternating periods of certainty equivalent versus probing management*, as shown in Figure 9.14. A few probing decisions will increase σ_u^2, thus making it optimal not to probe. Then σ_u^2 will decrease over the following period of certainty-equivalent actions, until it again becomes optimum to probe. Notice that if the first probe is upward ($\bar{u}$ initially low) the next is likely to be downward since the first will have caused $\bar{u}$ to increase; this results in the "chattering" probing pattern during the probing periods shown in Figure 9.14. Of course, the basic productivity of the stock will set strong limits on how much probing is necessary or tolerable during the probing periods.

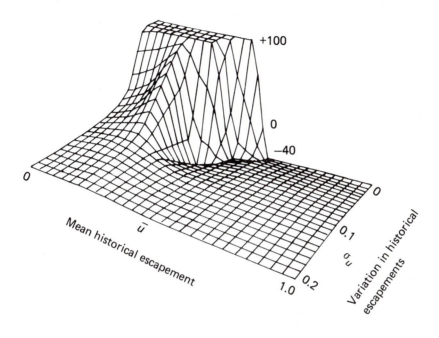

Figure 9.12. Optimum escapement level measured as a percentage deviation from the best point estimate based on historical data, for a stock-recruitment system. When the mean historical escapement $\bar{u}$ has been low and steady (variability σ_u low), it is best to probe upward with higher escapements. When $\bar{u}$ has been high and steady, it is best to reduce escapements, in this example to 40% of the best point estimate. Redrawn from Smith and Walters (1981).

In terms of the "management donut" idea discussed in Chapter 7, Figure 9.12 implies staying within the donut hole (domain of increasing uncertainty) while σ_u^2 is large, then probing outside of it when uncertainty increases sufficiently (σ_u^2 decreases sufficiently). It is impossible to give any precise and general prescription about how often it is optimum to move outside the donut hole, since this frequency will depend on (1) how rapidly parameters are assumed to change, and (2) learning rates following informative decisions, as affected by noise levels (process and observation variances). It is unlikely that the optimum policy will involve continual probing (u_t^* far from $u_t^{(o)}$ in every time step); this would imply such rapid and frequent parameter changes as to make it best to seek some alternative model that explicitly represents why and how the original "parameters" vary. Likewise, it is probably unwise to assume that parameters are constant over long enough time scales to make probing periods not worthwhile at all. Based on various simulation experiments with time-varying parameters in

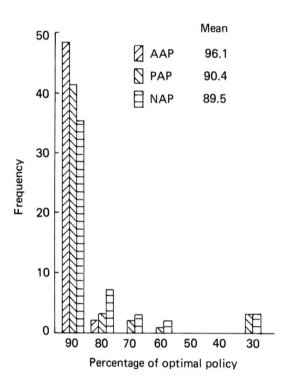

Figure 9.13. Over 50 simulation trials, an actively adaptive policy like in Figure 9.12 was always able to achieve at least 80% of the maximum yield obtainable if the stock–recruitment model parameters were known in advance. Passively adaptive policies (PAP) without probing, and nonadaptive policies (NAP) without any learning usually also perform well *on average*, but occasionally do very poorly. *Source:* Smith and Walters (1981).

stock–recruitment systems, I would generally recommend a probing period about once every 10–20 generations of the organisms, whether or not there is obvious evidence of parameter change.

Caution and probing in a surplus production model

As noted in the above section on dynamic programming, the logistic surplus production model leads to an adaptive policy design problem that is too large to solve directly. The wide-sense algorithm avoids the dimensionality problem by trying to find only the next effort u_t to make, given a fixed information state. The wide-sense extended state becomes $x'_t =$

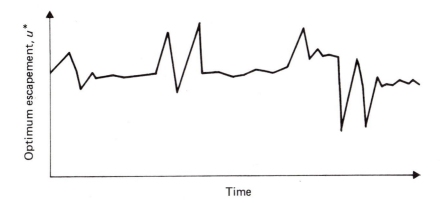

Figure 9.14. The adaptive policy in Figure 9.12 can involve alternating periods of stable (certainty-equivalent) escapements and strong probing episodes, if recruitment parameters are admitted to change over time so that old data are discarded (σ_u decreasing during periods of stable escapement).

(B_t, r_t, k_t, q_t). This state has covariance Σ_t, whose critical elements are σ_r^2, σ_k^2, σ_q^2, σ_{rk}, σ_{rq}, and σ_{kq}, if it is assumed that measurement errors are negligible (if not, Σ_t includes $\sigma_{B_t}^2$, σ_{Br}, σ_{Bk}, and σ_{Bq} as well). Again, Σ_t can be estimated (and predicted) by recursive linear regression if the measurement errors are negligible. There arises an "opportunity" for learning not found in the stock–recruitment examples, since learning rates depend on the states B_t as well as on the effort actions u_t: a single, large value of u_t may be used to generate a whole sequence of informative values of B_t. The wide-sense algorithm accounts for such future learning due to current actions, since the nominal trajectory $x_{t+1}^{(o)}$, $x_{t+2}^{(o)}$, ..., $u_{t+1}^{(o)}$, $u_{t+2}^{(o)}$, ..., and nominal covariance predictions $\Sigma_{t+1}^{(o)}$, $\Sigma_{t+2}^{(o)}$, ..., both depend on the choice u_t.

As in the stock–recruitment case, the covariance matrix Σ_t can be represented in terms of a set of more intuitively meaningful statistics:

t	the sample size;
$\bar{y}_t$	the mean index of past stock size;
$\bar{u}_t$	the mean past effort level;
$\sigma_{y_t}^2$	the variance in past stock size (biased upward if measurement errors are present, but not assumed in the estimation);
σ_u^2	the variance in past effort levels;
r_{yu}	the historical correlation coefficient between past stock size and past efforts (usually large and negative).

Of particular interest in this example is the correlation statistic r_{yu}, which reflects the impossibility of controlling the stock size and effort independently, as would be desirable in classic experimental design (we must live with the fact that increasing u will drive y downward). More generally, the whole state space defined by the six statistics (or Σ_t) is not reachable through controls; increases in σ_u^2 imply increases in $\sigma_{y_t}^2$, increased $\bar{u}$ implies lower $\bar{y}$, and so forth. Since the whole space of combinations cannot be reached in practice, there is no need to seek multidimensional analogues of Figure 9.12; instead, we can deal just with representative trajectories (historical patterns).

Smith (1979) has calculated trajectories of effort and relative abundance implied by the wide-sense algorithm and by passive adaptation (choose u_t to maximize $V^{(d)}$ each year), for several prototypical situations in terms of the initial information state Σ_0. Figures 9.15 and 9.16 show results from two of his case simulations. In case A (Figure 9.15) he began with 10 years of data (open circles) on an "underexploited" stock, as might be available during the early development of a resource. The wide-sense algorithm then gives the trajectory marked AAP (for actively adaptive policy), and passive adaptation gives the effort-abundance combinations marked PAP. The AAP initially *shuts down* harvesting for three years, then moves to high effort levels when catch per effort does not respond. The stock is pushed down rapidly, and within a few years the best equilibrium [marked by an asterisk in Figure 9.15(a)] is approached. In contrast, the PAP spends several years continuing to harvest at low rates, then increases efforts more smoothly, eventually to end up near the best equilibrium. As shown in the plot of cumulative catches [Figure 9.15(b)], the AAP gives up early catches in favor of higher ones (closer to optimum marked OP) later. The sharp disturbances in effort by AAP result in some reduction in the abundance-effort correlation r_{yu}, and much improved parameter estimation.

Figure 9.16 shows a case where the simulated stock was initially overexploited, and was being held near a low equilibrium by high effort levels. In this case the AAP immediately lowered effort levels, "watched" relative abundance increase for five years, then increased efforts again and moved quickly to the best equilibrium. In contrast, the PAP spent several years continuing to overexploit, then reduced effort drastically (as had the AAP earlier), and finally moved up smoothly to near the optimum. As in the other case, the AAP gave up early catches in favor of gaining information about response, then took higher catches, later to end up with a better cumulative result than the PAP.

Unfortunately, Smith found that both the active and passive policies fail with discouraging regularity when the stock is initially overexploited, especially when the data are, in fact, generated by a more complex population model. As noted by Hilborn (1979), the failures are basically due to

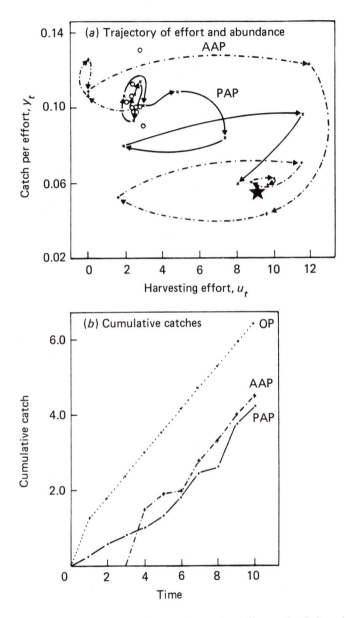

Figure 9.15. Successive combinations of harvesting effort and relative abundance (catch per effort), and cumulative yields, for a simulated comparison of actively adaptive (AAP) and passively adaptive (PAP) policies for management of a logistic surplus production system. In this case the stock was initially underexploited, $\bar{y} \approx$ 0.1, $\bar{u} \approx 2$. Note how the AAP gives up early yields in order to improve catches later. The cumulative yields marked OP would be best if the parameters were known to the decision maker. *Source:* Smith (1979).

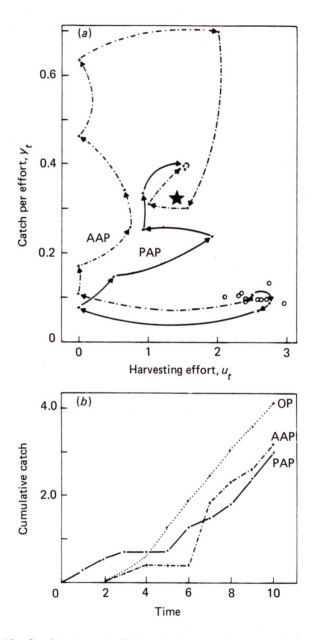

Figure 9.16. Combinations of effort and relative stock size generated by actively adaptive (AAP) and passively adaptive (PAP) policies, and cumulative yields compared with the theoretical maximum (OP) achievable if the parameters were known. In this case the stock was initially overexploited ($\bar{y} = 0.1$, $\bar{u} = 2.5$); see Figure 9.15. *Source:* Smith (1979).

biases in parameter estimation, which are most extreme when the stock size has been held low for some time. As noted earlier, Ludwig and Hilborn (1983) have found that the only consistent way to avoid these biases is initially to stop harvesting for a long period. The wide-sense algorithm prescribes exactly this, but often does not wait long enough for a response. The likelihood of this error, of course, increases for less productive (low r) stocks, since they take much longer to give recovery "signals" against the backdrop of random variation in both production and observation processes.

Difficulties with the wide-sense algorithm

Although the wide-sense algorithm has strong intuitive appeal as a way of estimating the benefits of caution and probing in the face of uncertainty, its predictions must be interpreted with care. One weakness in the algorithm is quite obvious. Its predictions are strictly valid only for small disturbances from the nominal "best guess" trajectory $x_t^{(o)}$, $u_t^{(o)}$. In the presence of large stochastic disturbances whose effects cannot be fully controlled, and/or when parameter uncertainty is very high, the nominal trajectory may be almost meaningless. In particular, the wide-sense algorithm may underestimate how much learning will occur just due to passive adaptation (see the final section, Chapter 8) and will therefore prescribe probing when it would not really produce any net benefit.

Another potentially serious weakness is with the *a priori* prediction of future uncertainties as measured by $\Sigma_{t+1}^{(o)}$, $\Sigma_{t+2}^{(o)}$, ... For computational efficiency, it is usual to predict Σ using simple filtering models such as the extended Kalman filter. These predictions can be very bad when the dynamic equations for the extended state x_t and observation process are highly nonlinear. In the two cases above, I assumed away some of the worst problems in estimating Σ, by pretending that stock sizes and actions can be measured exactly. The basic consequence of rejecting this assumption is to make state "reconstructions" part of the estimation problem, which results in slower learning of unknown parameters. Simple filtering models may either under- or over-estimate this reduction in learning rates. Poor estimates of learning rates have two conflicting effects. When learning rates are underestimated, so is the immediate effect of a probing decision. But, on the other hand, probing appears more important because future learning rates are predicted to be slow. When learning rates are overestimated (for example, when measurement errors are ignored), then probing a decision appears immediately more valuable than it should. But against this is balanced the prediction that future learning rates will be higher. Thus, the net effect of error in learning rate predictions is difficult to determine without trying some numerical calculations with different filtering equations and estimates of process and measurement error variances.

Conflicting Objectives

Let me close this chapter with a word of warning about actively adaptive policies. In the various examples discussed above, I have taken the maximization of harvest as a reasonable optimization objective. This happens to be about the only payoff measure (in renewable resource management) for which probing decisions are often optimal. When the management objective reflects strong risk aversion, the wide-sense "caution" term dominates, and learning is seldom seen as worth the risks that it entails. When discount rates for future payoffs are high, there is likewise little point in giving up immediate yields in favor of possibly detecting better opportunities for future yields.

Indeed, any objective function V or reward measure v_t that places a premium on maintaining the current state of affairs (reducing variability in catches, incomes, etc.) will generally lead to cautious actions unless there is gross uncertainty about whether the state of affairs can be sustained at all. Now, suppose you are dealing with a managed system for which historical actions have been highly cautious. This will have led to low variance in past controls (σ_u^2 small) and possibly system states as well. Thus, you are likely to encounter precisely the conditions most favorable for adaptive probing, *if* the management objective is really to maximize long-term yield. If you do design such a new policy, you will have to justify it either by arguing that (1) there has been misunderstanding in the past about what actions would, in fact, maximize the long-term yield objective, or (2) whatever management objectives led to the past policies should now be discarded in favor of the objective you assumed in concluding that there should be active adaptation. Either way, you are inviting a lively debate: you must either question past methods, or past motives. The very conditions under which active probing is most "needed" are the conditions under which it is going to be the most difficult to sell to the various interest groups already involved with management.

It is hardly worthwhile to engage in complicated optimization exercises to find precisely the right probing decision for a particular management objective, when there is great uncertainty about what that objective should be in the first place. Therefore, I suggest proceeding in a stepwise fashion. First, use the criteria outlined early in this chapter to decide whether deliberate probing could be worthwhile in principle. Second, use relatively simple procedures like the wide-sense algorithm to explore whether information values and learning rates are likely to be large enough to be concerned about, and to get a rough idea of, the best direction and magnitude of disturbance under simple management objectives. Third, develop simple adaptive policy options based on rough calculations, and discuss these options with the management authority and associated interest groups; here it is especially helpful to have vivid visual demonstrations, such as microcomputer simulations, of how the options might work in practice. Fourth, if the discussions lead to focused debate on management objectives, then try to obtain consensus about the importance of improving long-term performance. If this consensus favors the use of an actively adaptive policy, then finally proceed with more precise calculations about the best disturbance regime to implement initially.

Problems

9.1. The three situations shown in Figures 9.1 to 9.3 are common in practice. For each case, assume that the curve marked m_1 is considered the most likely, and describe precisely why it is not worthwhile trying to design a deliberately experimental policy to determine whether one of the other curves might be more accurate.

9.2. Every experimental policy is a gamble, since it involves an "investment" (reduction in immediate catch, capital investment, or risky stock reduction) now that has no guaranteed rewards except reduced uncertainty for future decision makers. The value of this reduction can be measured only in a probabilistic sense. In most human affairs (business decision-making, personal decisions), we accept such gambles as routine and unavoidable "facts of life." Discuss factors that have led to more conservative attitudes, and often unreasonable demands for certainty and risk avoidance, in relation to public decisions involving renewable resources. Does it matter that many actors are involved in such decisions? Has the scientific community been realistic in its arguments about the value of doing research and monitoring before making risky decisions?

9.3. The wide-sense dual control algorithm provides a means to supplement "best" (most likely, certainty-equivalent) estimates of future resource value with an additive estimate of the value of information associated with alternative policy choices. This estimate of information value is obtained by predicting how uncertainty (measured by variances of unknown parameters) will propagate over time in relation to the state (stock) changes induced by each policy choice. As we saw in Chapter 7, learning rates are usually predicted to be quite slow except after dramatic policy disturbances. How is this observation reflected in the shape of the information value (probing value) curve $V^{(p)}$ in Figures 9.10 and 9.11? Can you explain why the nominal (most likely) value curves $V^{(d)}$ are so flat in these figures? (Hint: $V^{(d)}$

estimates the total future value while assuming future decisions are made optimally with respect to the most likely model, no matter what is done at the current decision point.)

9.4.   Develop a dynamic programming algorithm to estimate the harvest policy for maximizing long-term yields from populations where uncertainty about future production can be captured in terms of two alternative models, as in Figure 1.1. This is a three state-variable problem, with two continuous variables (stock size and probability placed on first model). The value recursion equations are given by (9.1) and (9.2), where stock size at $t + 1$ is defined by your two models and p_{t+1} is calculated from Bayes' theorem [equation (9.4)]. Use your program to find optimum policies for a variety of two-model cases as in Figures 9.1-9.5, involving both surplus production and stock–recruitment dynamics.

9.5.   Write a computer program to estimate how alternative choices of γ_0 and γ_1 in the Figure 9.9 policy should influence long-term yield from a stock that follows a Ricker stock–recruitment relationship. Set up a grid of γ_0 and γ_1 choices. At each choice, do 100 Monte Carlo simulations of 40 years each, taking total yield for the 40 × 100 years as a performance measure for the choice. For each simulation (1) choose the Ricker a and b parameters from normal distributions with means of 1.0 and standard deviations 0.3; (2) let the initial stock size be $a/2b$; (3) harvest at a rate of $a/2$ for the first five years, then use the Figure 9.9 policy afterward; (4) assume environmental effects w_t are normally distributed with $\sigma_w^2 = 0.2$ (as in problem 7.4); (5) estimate a and b over time within each simulation using a recursive linear regression, and approximate the certainty-equivalent optimum escapement for each year as $(0.5 - 0.07\hat{a})\,\hat{a}/\hat{b}$ (note that according to Figure 9.9, you use this escapement whenever $\sigma_b^2 < \gamma_0$).

9.6.   Develop a computer program to implement the wide-sense dual control algorithm, as presented in Bar-Shalom and Tse (1976b), for the logistic surplus production model $N_{t+1} = R_1 S_t - R_2 S_t^2 + w_t$ (see Chapter 5), where S_t is the stock size after harvest in year t and w_t is normally distributed with mean zero and variance 0.1. Assume that N_{t+1} can be measured exactly, and that S_t can be chosen at will each year (except $0 \leq S_t \leq N_t$) and also measured exactly. Assume that the annual value v_t is simply $N_t - S_t$ (the harvest) and that the planning period is $t = 1, \ldots, 30$ with $V_{31} = N_{31}$. Note that your extended state vector is $x_t = (N_t, R_1, R_2)'$, and $u_t = S_t$, so your state equations are

$$x_{1,t+1} = x_{2t} u_t - x_{3t} u_t^2 + w_t$$

$$x_{2,t+1} = x_{2t}$$

$$x_{3,t+1} = x_{3t}$$

Assume that the covariance matrix $\Sigma_t^{(o)}$ can be predicted by an extended Kalman filter [equations (7.18)–(7.21)], with $y_t = N_t$,

$$V_v = 0 \quad \text{and} \quad V_w = \begin{bmatrix} 0.1 & 0 & 0 \\ 0 & 0 & 0 \\ 0 & 0 & 0 \end{bmatrix}$$

Assume initial state estimates $x_0 = (0.5, 1.5, 0.5)'$ and the nominal optimal policy $u_t^{(o)} = \min(0.5, x_{1t})$, $t = 1, \ldots, 30$. Assume that the initial covariance matrix for x is

$$\Sigma_0^{(o)} = h \begin{bmatrix} 0 & 0 & 0 \\ 0 & 1 & 0.95 \\ 0 & 0.95 & 1 \end{bmatrix}$$

and obtain results for various h (parameter variance) values in the range 0.01–0.5. For the initial time step ($t = 1$), plot the wide-sense value components $V^{(o)}$, $V^{(p)}$, and $V^{(c)}$ as a function of u. For larger values of h, you should find that $V^{(cl)}$ is maximized at a u value larger than $u^{(o)}$, implying that the population should be increased to test its carrying capacity.

Chapter 10

Adaptive Policies for Replicated Systems

We inherit the earth, but within the limits of the soil
and the plant succession we also rebuild the earth—
without understanding of the increasingly coarse
and powerful tools which science has placed at
our disposal.

A. Leopold (1933)

An implicit assumption in Chapters 4-9, and, indeed, in most of the theoretical literature on resource management, is that every unit of analysis (stock, spatial area) is somehow unique in terms of how we value its performance and learn about its responses over time. But management authorities are usually responsible for whole collections of ecological units that are structurally similar (same species, etc.). In the statistician's terms, we might call each of these units a "replicate;" we do not expect that replicates will be quantitatively identical (same parameters), but we do hope for similarity in response to factors like harvesting and also the possibility that some quantitative differences can be ascribed to measurable "covariates" like size of area. Often the harvesters may move about more or less freely among such replicates, testing and choosing so as to achieve much more consistent performance over time than they could obtain by staying with any single replicate. So replicate management units provide both opportunities for controlled experimentation with policy options, and also opportunities to distribute the immediate costs and risks of these experiments so as to make them less objectionable to resource users who are not much concerned with the long-term value of better information. This chapter is intended to provide only a very broad introduction to some of these exciting opportunities, and to some of the frustrating technical difficulties that are encountered in attempts to design optimum adaptive plans.

The notion that various managed units may somehow be informative about one another is certainly not unfamiliar to practicing resource managers. Most management agencies have established experimental management units (lakes, rivers, game management areas, etc.) where research and monitoring investments are concentrated and new management tools are evaluated. In North America the most extensive experimental networks are related to agriculture and its subsidiary resources, such as farm pond fisheries. Experimental forest units are also common. Most fisheries and wildlife agencies can afford to maintain only a few experimental units for long-term studies, and tend to spread their experimental investments across a variety of short-term studies aimed at evaluation of specific tools, such as lake fertilization, special harvesting regulations, and provision of improved breeding habitats (exhaustive descriptions of such tools can be found in most applied ecology texts). Unfortunately, many of these studies are initiated without any clear idea about, or means to estimate, the "universe" of replicates to which the experimental results might be extrapolated, without careful consideration of sampling requirements and the duration of study needed for unambiguous results, and without any systematic procedures to ensure that favorable results are used more broadly as they arise. Thus in the end, short-term results tend to end up buried in student theses and agency reports; very few have a discernible impact on management practices.

There has also been much agonizing about how to allocate monitoring resources across many replicates. Some agencies have opted for relatively inaccurate sampling of all replicates, with more detailed monitoring of only the largest and most economically important units. Others have elected to concentrate on better sampling of some subset of "index" replicates that are thought to be representative or are convenient in terms of cost factors like physical accessibility. Index monitoring programs are sometimes accompanied by research studies aimed at finding inexpensive measurements that are highly correlated with the detailed index results, but can be gathered from all replicates. In light of recent discoveries about severe bias in population parameter estimates from inaccurate survey data (Walters and Ludwig, 1981; Uhler, 1979), the option of sampling all replicates may turn out in some cases to have been the worst possible choice. At this extreme we face the risk that none of the parameter estimates can be trusted. With a small number of index systems, we face an alternative risk: parameter values from the index replicates may not be representative of, or may vary so widely as to be uninformative about, other replicates.

A point that has not been critically examined by many agencies is that replicate units are "connected in value," in terms of how the agencies evaluate their own performance and in terms of how resource users view the opportunities available to them. It is common practice to present time series of aggregate harvest statistics, then point to the stability of these totals as evidence of success in management for sustained yield. But often the totals hide a pathological pattern of deterioration, in which harvesters concentrate first on easily accessible replicates (and deplete these) then move on to new areas as access or harvesting technology develops over time. When it is feasible to regulate the distribution of harvesting effort among replicates, by measures ranging from direct closures to area licenses to cost subsidies for operating in unattractive locations, the management agency can induce informative patterns of harvest variation within some replicates without greatly changing overall performance levels. But harvesters can seldom be treated as pieces in a board game, to be moved about freely at the manager's discretion; the design of effective programs for redistributing effort requires at least some understanding of how harvesters perceive the choices available to them in terms of resource abundance, competition with other harvesters, and differential costs related to factors like distance from home or port.

Obviously it would be possible to write whole books about each issue raised in the above paragraphs. In this chapter, I will attempt to clarify what the issues are, by providing (1) an overview and classification of the types of linkages (among replicates) that should be of concern in adaptive policy design; (2) a qualitative analysis of how dual effects of control can propagate among replicates, and thereby imply different optimal adaptive policies than would be designed by treating every replicate as unique; (3) a review of factors that should be considered in decisions about the allocation of monitoring resources (funds, manpower); and (4) a summary of key ingredients needed for evaluating alternative management plans that involve experimentation on some or all replicates.

Linkages among Replicates

We have noted that replicate units can be linked or connected from a management viewpoint in two obvious ways: *information* from each replicate may help in management of others, and *value* from each replicate may be traded off against (or accumulated across) value from other replicates in measurement of management performance. This section tries to classify more precisely the types of linkages that should be of concern in policy development, and reviews some mathematical tools that can be used to study linkage patterns and their effects on parameter estimation and feedback policy performance.

Similarity in parameters

In the design of scientific experiments, replication is necessary to ensure that observed responses are not simply due to "local" factors that the experimenter cannot anticipate or control on each of his experimental units. Also, we take it as an obvious matter of good planning to make the replicates as similar as possible, by regulating the experimental environment, choosing units that appear physically similar, and controlling treatment conditions. This freedom to carefully plan and select units for study is seldom available to the resource manager; he usually must confront a potentially heterogeneous collection of units that have not been treated similarly in the past and will not be subject to a controlled environment in the future. Also, selection of units on the basis of similarity and convenience is an invitation to results that are not representative at all of the universe of units for which he is responsible as a manager. Where the scientist would try to physically eliminate various factors that might be important in determining responses, the manager must instead seek ways to live with their effects.

The temporal responses (abundances over time, catches, etc.) of replicates may vary widely due to differences in (1) exploitation histories, (2) time-varying environmental factors, and (3) values of functional parameters, such as intrinsic rate of increase, carrying capacity, and vulnerability to harvesting. Think of each replicate as having a "production function" with possibly unique parameter values, and a temporal response pattern that depends on this function and on how the system is disturbed over time. In previous chapters, our emphasis has been on how to estimate (reconstruct, retrieve, infer) the production function for any one unit, given a set of data Y about its temporal behavior. Let us suppose now that estimations have been carried out for a series of replicate units, and we treat the resulting set of parameter estimates as "data" for further analysis. This first step in living with variation can have a dramatic result: we often find that some functional parameter estimates vary much less among replicates than would intuitively be expected from differences in temporal performance. In such cases, the replicates are "linked" through similarity in production functions.

When the replicates represent different stocks or subpopulations of a single species (or communities with the same species composition), my experience suggests the following pattern. First, parameters that determine maximum rates of population growth tend to be similar among replicates. Examples of these parameters are the logistic model r, the Ricker model a, calves or fawns per female in ungulate populations at low densities, and natural mortality rates in animals of intermediate age. Second, parameters that measure "natural" population sizes in the absence of harvesting tend to be extremely variable among replicates. Examples are the logistic model k, the Ricker model b, and the slopes of graphs relating ungulate reproductive and mortality rates to population density. Third, parameters that measure vulnerability to harvesting effort offer no consistent patterns; estimates are similar from replicate to replicate in some cases and vary enormously in others. Vulnerability is usually measured by a catchability coefficient q.

Variation among maximum rates of population growth tends to increase as the universe of replicates is extended to larger geographic scales. So, for example, we find very similar Ricker a values for sockeye salmon stocks from the same watershed, yet great variation when stocks from southern British Columbia are compared with stocks from southeast Alaska. Such latitudinal variations in maximum productivity are common in species of recreational and commercial importance, and most often are associated with changes in juvenile survival rates rather than fecundity. Similarity in some components (such as fecundity) of a population rate process does not in any way imply that the net rate (sum or product of all components) will be similar across replicates.

Substantial variation in "carrying capacity" parameters is to be expected, simply because replicate natural stocks generally occupy geographical areas (or critical habitats like spawning areas) of varying size. Also, trophic conditions (habitat productivity, food supplies, abundance of competing species) often vary greatly across replicates, and exert their effects mainly when population sizes are relatively large.

Variation in vulnerability to harvesting effort is expected when (1) harvesting technologies and experience of harvesters differ across replicates; (2) there are differences in such conditions for search as the fraction of area accessible, visibility of animals within the searched area, and patterns of clumping or schooling; and (3) animals from different replicates behave differently in response to harvesting (for example, have become more wary in units where harvesting has historically been intense). The first of these sources of variation can be partly eliminated by calibrating the harvesting effort data in terms of some standardized gear type or searching unit. The other two sources vary greatly in importance from species to species.

It is useful to think of variations in parameter estimates among replicates as consisting of three "variance components:" (1) deviations of estimates from the true values for the replicates; (2) deviations among true values due to unknown natural factors; and (3) deviations among true values that can be "eliminated" or "factored out" by blocking the replicates into subsets based on measurable factors (covariates), such as latitude, size of area, and type of harvesting technology. More precisely, suppose $\hat{\beta}_i$ is the estimate of some dynamic parameter for replicate i, based only on analysis of the available data from that replicate. We can write $\hat{\beta}_i$ as

$$\hat{\beta}_i = \beta + (\beta_g - \beta) + (\beta_i - \beta_g) + (\hat{\beta}_i - \beta_i) \qquad (10.1)$$

where β is the true mean value of the parameter across all replicates in the management universe, β_g is the true mean of the parameter for the more homogeneous subset (stratum) of units to which replicate i belongs, and β_i is the true parameter value in replicate i.

Equation (10.1) implies that we may see large variations among the estimates $\hat{\beta}_i$ even when the subsets are quite homogeneous (small $\beta_i - \beta_g$) and there is little variation due to factors like latitude (small $\beta_g - \beta$), just due to errors $\hat{\beta}_i - \beta_i$ in the estimations for some replicates. As noted in earlier chapters, estimation procedures usually result in deviations $\hat{\beta}_i - \beta_i$ that are approximately normally distributed with mean zero and variance of $\sigma_{\hat{\beta}_i}^2$. High values of $\sigma_{\hat{\beta}_i}^2$ are expected for replicates that have been poorly monitored and/or not subjected to informative variations in management policy.

Let us examine for a moment the simplest possible situation, when the replicates are not blocked into any subsets and it is reasonable to assume that the actual parameter values β_i were "drawn by nature" from a single normal distribution with mean β and variance σ_β^2. Then, assuming further that the $\hat{\beta}_i - \beta_i$ are normally distributed with mean zero and *known* variances $\sigma_{\hat{\beta}_i}^2$, Bayes' theorem can be used to find the most probable value $\tilde{\beta}_i$ for each replicate β_i; after algebraic simplification, the result is:

$$\tilde{\beta}_i = W_i \hat{\beta}_i + (1 - W_i) \tilde{\beta} \tag{10.2}$$

where W_i is a weight associated with the independent estimate $\hat{\beta}_i$:

$$W_i = \frac{\sigma_\beta^2}{\sigma_\beta^2 + \sigma_{\hat{\beta}_i}^2} \tag{10.3}$$

and $\tilde{\beta}$ is a weighted estimate of the mean β around which nature's "sample" β_i were drawn:

$$\tilde{\beta} = \frac{\sum_i W_i \hat{\beta}_i}{\sum_i W_i} \tag{10.4}$$

Equation (10.2) is a striking result, for it says that the best estimate of β_i for each replicate is something other than the independent estimate $\hat{\beta}_i$, if we are willing to believe that the β_i actually represent samples from a distribution of natural possibilities. If we believe *a priori* or have some evidence that the actual variance σ_β^2 among the β_i is small, then W_i [equation (10.3)] is small and we should assume that β_i is close to the average $\tilde{\beta}$ across replicates. But notice that we should place less weight on $\tilde{\beta}$ if $\sigma_{\hat{\beta}_i}^2$ is also relatively small, that is, if we are sure about $\hat{\beta}_i$ being close to β_i.

There is, in principle, nothing wrong with viewing the replicate parameter values β_i as having been drawn from some natural distribution. The key issue raised by equations (10.2)-(10.3) is the magnitude of the "natural" variance σ_β^2. As we shall see in a later section, σ_β^2 is critical not only in the estimation of β_i, but also in the determination of optimum experiments and monitoring allocation among replicates. Note that σ_β^2 is not measured directly by the deviations $(\hat{\beta}_i - \tilde{\beta})$, since these contain components $\hat{\beta}_i - \beta_i$ as well as $\beta_i - \beta$. Also, standard analysis of variance procedures cannot be used to estimate σ_β^2, since the replicates generally have inhomogeneous variance. Using Monte Carlo experiments to generate data sets with known variance characteristics, I have found that an approximately unbiased estimate of σ_β^2 can be found by a simple iterative procedure. Given a data set $(\hat{\beta}_i, \sigma_{\hat{\beta}_i}^2, i = 1, \ldots, R)$, use a trial estimate of σ_β^2 to calculate W_i and $\tilde{\beta}$ [equations (10.3)-(10.4)], and obtain a new estimate as

$$\sigma_\beta^2 = \frac{\displaystyle\sum_{i=1}^{R} W_i^2 (\hat{\beta}_i - \tilde{\beta})^2}{\displaystyle\sum_{i=1}^{R} W_i} \qquad (10.5)$$

Then use this new estimate as the trial estimate for a second iteration [W_i, $\tilde{\beta}$, equation (10.5)], and repeat until the successive estimates stop changing or approach zero. Getting a zero estimate of σ_β^2 implies that the $\hat{\beta}_i$ vary less than would be expected just on the basis of $\sigma_{\hat{\beta}_i}^2$; this can happen by chance alone, but it can also be due to underestimation of the $\sigma_{\hat{\beta}_i}^2$—a common problem when $\hat{\beta}_i$ is obtained by some nonlinear estimation procedure that produces only an approximate estimate of $\sigma_{\hat{\beta}_i}^2$. Intuitively, equation (10.5) apportions the deviations $(\hat{\beta}_i - \tilde{\beta})^2$ according to the ratio $\sigma_\beta^2/(\sigma_\beta^2 + \sigma_{\hat{\beta}_i}^2)$, while using this ratio to weight each variation; deviations with smaller $\sigma_{\hat{\beta}_i}^2$ are given greater weight in the overall estimate.

The theory of linear statistical models (experimental design, analysis of variance and covariance, variance component models) can be used to develop estimation procedures analogous to equations (10.1)-(10.5) for more complex situations where the replicates can be blocked into more homogeneous subsets, or part of the variation among $\hat{\beta}_i$ ascribed to covariates. However, such procedures should be used with care since they involve a basic trade-off noted in the final section of Chapter 6 (Figure 6.4); as the replicates are blocked into smaller sets, or more of the variation is ascribed to covariates, the accuracy of estimation for each "structural parameter" (block mean, slope of response to covariate, etc.) may improve at first, but will eventually deteriorate if too many parameters are included in the analysis. In my experience the best balance usually involves only a few (two or three) blocks and/or covariates. Let me put this in more vivid terms: just because you can think of many factors that might cause variation in a parameter across replicates, do not suppose that the inclusion of all these factors in the analysis will automatically make for a better prediction or estimate of the value for each replicate; the effects of most factors will be badly confounded (equally well explained by many different response models), or reasonably ascribed to other factors that you have not yet thought to include.

Correlation in response to unmodeled variables

When the time series of deviations from model predictions are compared for several replicates, it often turns out that the deviations are correlated. This can mean either that the replicates are all responding to some common, large-scale factors, such as climatic change, or that changes in management policy have been applied to all replicates simultaneously and have produced effects (deviations) not accounted for by the model(s). If these two possible sources of error can be separated from one another, model parameter estimation for each replicate may be improved by sorting out (estimating, eliminating) components of deviation about which all the replicates give some information.

One might expect intuitively that the use of data from several replicates to jointly estimate common environmental effects would always result in improved parameter estimation for each replicate. But two things work against this improvement. First, estimating common effects is a special case of estimating more parameters (the effects themselves), which is not always a good idea (see previous section). Second, common environmental effects are not logically separable from effects of policy actions, if all replicates have been subject to the same management regime. This second point may be especially important for adaptive policy design, since it hints that in order to reap some benefits from joint estimation, we should consider deliberately creating contrast in management actions among replicates.

There are two possible assumptions with which to begin analysis of correlation in response to unmodeled factors. The simplest is that all replicates are subject at each time to a common effect, measured as a component of the process error around model predictions. We will develop an example of this approach below. A more difficult approach is to assume that there is a possibly unique correlation in deviations for every pair of replicates, so some replicates may have very similar deviations while others behave independently. Details about one nonlinear estimation procedure for this second assumption can be found in Gallant (1975); the procedure involves the simultaneous estimation of model parameters for all replicates, while weighting the data by covariances estimated from prediction errors.

Let us develop a very simple example, in which stock–recruitment data are available for several salmon stocks that spawn in a single watershed and are therefore likely to share various factors ranging from river flow conditions during spawning to variations in the marine ecosystem where all the fish grow to maturity. Suppose we are willing to believe that the behavior of each stock i can be approximated by a Ricker model with parameters a_i and b_i. Treating the stocks as independent, we would estimate these parameters from the linear regression

[Handwritten annotation at top: $a_i + b_i$ are different for each replicate i. Why expect joint estimation to reduce their variance? $[a_i, b_i]$ - like random effects — not estimated in common across reps]

$$\ln \frac{R_{it}}{S_{i,t-1}} = a_i - b_i S_{i,t-1} + w_{it} \quad (t = 1, \ldots, T) \tag{10.6}$$

where R_{it} is recruitment to stock i for generation t, $S_{i,t-1}$ is spawning stock at the end of generation $t - 1$, and w_{it} is a normally distributed process error that represents the combined effect of all variable survival factors. Now suppose we assume that w_{it} consists of two components:

$$w_{it} = \bar{w}_t + w'_{it} \tag{10.7}$$

where $\bar{w}_t$ is a common effect shared by all replicates during generation t, and w'_{it} is an independent effect due to unique conditions encountered by replicate stock i. Equations (10.6)-(10.7) define a general linear model with $2R + T$ parameters (Ra_i's, Rb_i's, and $T\bar{w}_t$'s), for which unique maximum likelihood estimates do not exist unless we add some further condition such as $\Sigma \bar{w}_t = 0$. Adding this condition just means that we measure the time mean of $\bar{w}_t$ as part of the a_i, from which it is, in principle, indistinguishable anyway. For simplicity of presentation, we shall assume in the following analysis that the w'_{it} have the same variance in all replicates, and that the S_{it} are measured without error. Notice that we are not assuming that the $\bar{w}_t$ values follow any particular pattern or probability distribution over time; we are simply measuring them around a mean of zero, and throwing trends into the estimates of a_i.

Using the theory of general linear models (Graybill, 1961), we find that the maximum likelihood estimates of b and $\bar{w}_t$ are given by the matrix equation

$$\begin{bmatrix} \hat{b} \\ \cdots \\ \hat{w} \end{bmatrix} = \begin{bmatrix} A & \vdots & B \\ \cdots & & \cdots \\ B' & \vdots & RI \end{bmatrix}^{-1} \begin{bmatrix} d \\ \cdots \\ W \end{bmatrix} \tag{10.8}$$

where (1) A is a diagonal matrix having elements $A_{ii} = \Sigma_{t=1}^{T} (S_{it} - \bar{S}_i)^2$; (2) $B_{it} = (S_{it} - \bar{S}_i)$; (3) RI is the $T \times T$ identity matrix multiplied by R (the number of replicates); (4) $d_i = \Sigma_{t=1}^{T} (y_{it} - \bar{y}_i)(S_{it} - \bar{S}_i)$; and (5) $W_t = \Sigma_{i=1}^{R} (y_{it} - \bar{y}_i)$. In the above, $\bar{S}_i$ is the arithmetic mean of S_{it} over time for replicate i, $y_{it} = \ln(R_{it}/S_{i,t-1})$, and $\bar{y}_i$ is the mean of y_{it} over time for replicate i. Formulae for inversion of partitioned matrices can be used to solve equation (10.8) efficiently on a computer. Then the estimates of a_i are found by

$$\hat{a}_i = \bar{y}_i - \hat{b}_i \bar{S}_i \tag{10.9}$$

as in standard linear regression. Though equation (10.8) looks quite messy, it ends up being like the standard regression formula

$$\hat{b}_i = \frac{\sum_t (y_{it} - \bar{y}_i)(S_{it} - \bar{S}_i)}{\sum_t (S_{it} - \bar{S}_i)^2}$$

except that correction terms are included for covariances $(S_{it} - \bar{S}_i)(S_{jt} - \bar{S}_j)$ among the spawning stocks and for deviations $(y_{it} - \bar{y}_i)$ that are attributable to the common environmental effects. Without these correction terms, some variations due to simultaneous changes in spawning stock size would incorrectly be interpreted as common environmental responses. The key parameters for setting optimum spawning stock levels for the replicates are the b_i (see Chapters 8 and 9), and the variances of these parameters are given by the ii element of Σ_{ℓ}, where

$$\Sigma_{\ell} = s^2 \left(A - \frac{1}{R} BB' \right)^{-1} \tag{10.10}$$

where A and B are matrices as defined for equation (10.8) above,

$$(BB')_{ij} = \sum_{t=1}^{T} (S_{it} - \bar{S}_i)(S_{jt} - \bar{S}_j)$$

are covariances among spawning stocks, and s^2 is the common estimate across all replicates of the variance of w'_{ij} (process effects not shared among replicates); it is calculated as

$$s^2 = \frac{\sum_{i=1}^{R} \sum_{t=1}^{T} (y_{it} - \hat{a}_i - \hat{b}_i S_{it} - \bar{w}_t)^2}{RT - 2R - T + 1} \tag{10.11}$$

Equation (10.10) implies that the estimates of the b_i obtained by analyzing all replicates simultaneously are not necessarily more accurate than would be obtained by analyzing the replicates independently. $\sigma_{\ell_i}^2$ for the joint analysis depends on (1) how large the common ($\bar{w}_t$) effects are compared to the independent errors w'_{it}; (2) how accurately the $\bar{w}_t$ are estimated, which depends partly on how many replicates are included; (3) how much informative variation in the S_{it} there has been within each replicate, and (4) how much covariation there has been in the S_{it} among replicates. Independent analysis of replicates will generally give larger values of s^2 than the joint analysis (i.e., $\bar{w}_t$ effects not removed from the residuals), and the variance of the independent estimates is given by $s^2 A^{-1}$ [equation (10.10)] without the covariance terms $[(1/R) BB']$. There are two opposing effects of joint analysis: (1) s^2 may get smaller, but (2) the elements of $[A - (1/R) BB']^{-1}$ will generally be larger than comparable elements of A^{-1}. Notice that $[A - (1/R) BB']$ approaches A only as the number of replicates R is increased, and the correlation among spawning stocks (measured by BB') is reduced. This is the same pattern of opposing effects as is shown graphically in Figure 6.4.

Figure 10.1 summarizes qualitatively how key factors interact to determine the relative performance of joint versus independent estimation procedures. First, note that when all replicates are subject to relatively large shared effects ($\bar{w}_t$ large compared with w_{it}'), stock sizes are bound to be correlated unless there is very tight control over each replicate; thus, there is a region of situations that is not feasible to produce in practice. Then there is a domain of situations within which joint estimation will result in better parameter estimates; this domain is larger when more replicates are available and when shared effects are larger. Finally, there is a domain of situations involving small shared effects and/or high correlation in stock sizes, within which it is best to just do independent estimations. The domain appropriate for any particular case study can only be determined by trying both joint and independent estimation procedures, then comparing the resulting parameter variances and covariances.

If it is found that the joint estimation procedure gives poorer results, yet there do appear to be substantial shared effects that cannot be accounted for by correlated changes in stock sizes, then a next step should be to examine how future performance of the joint estimation scheme might be affected by policy options that deliberately reduce correlation in stock sizes. Using reasonable estimates of dynamic model parameters and future $\bar{w}_t$ values, generate a variety of simulated future data sets for several policy options that are considered technically/economically feasible. Append the simulated data to the original data set, and run through the joint estimation procedure on these extended "synthetic" data sets. If such Monte Carlo trials consistently show that some options would result in significantly better parameter estimates for most replicates, then it is worth looking further at whether the increased management costs of implementing one of those options might be more than balanced by the expected value of having the better parameter estimates.

The model defined by equations (10.6)–(10.7) is in one sense a worst-case assumption about shared effects, since it assumes no temporal structure in the $\bar{w}_t$. In some cases, it may be reasonable to construct models with fewer parameters (less deterioration in joint estimation), by assuming that the joint effects are functionally related to particular covariates (such as river flow for salmon, winter snowfall for ungulates) that have been measured for all times and replicates.

Ecological coupling among replicates

In the previous section we discussed the possibility of learning more about each replicate by using information from other replicates to sort out some unmodeled effects that all replicates might display. Obviously the replicates can be coupled more intimately than as echoers of a common environment; they may directly influence one another through ecological processes, such as dispersal between areas, competition for mobile resources,

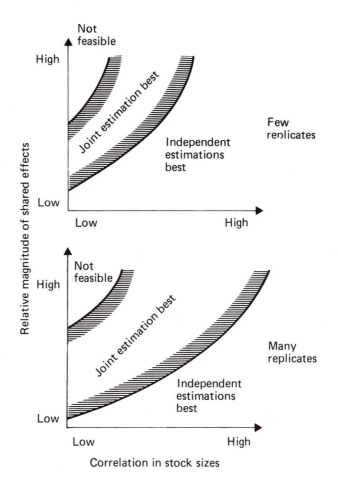

Figure 10.1. In replicated systems, there is no clear answer as to whether joint estimations of process error components shared by all replicates will result in improved parameter estimation. If the replicates have been highly correlated in their time dynamics, shared effects cannot be separated from the effects of changing state.

and support of predators that end up dispersing among the areas. While the effects of some processes (such as loss to mobile predators) may be viewed as shared "environmental effects" and treated as in the previous section, others imply a more careful look at the model structure used for each replicate. We should, for example, account for dispersal among replicates through accounting terms that make each emigration rate show up as immigration rates to other areas.

Ecological isolation of replicates is likely to be strongest for freshwater and island ecosystems. But even in these situations, resource managers have had to deal with some big surprises as organisms have shown remarkable powers of dispersal. A few of these surprises have been pleasant, as, for example, when pink salmon reinvaded and rapidly increased in the upper Fraser River of British Columbia more than 30 years after a disaster had virtually destroyed the stocks. But most of the dispersal surprises have involved exotic species with undesirable properties or even catastrophic impacts on the invaded ecosystems. Major examples here include sea lamprey invasion and rapid dispersal through the Laurentian Great Lakes, and the spread of brook trout through high mountain watersheds of the western United States. Considering the variety of human activities that can deliberately or accidently cause transport of organisms among replicates, resource managers should never assume that the replicates are fully isolated or protected from structural change through the introduction of new species.

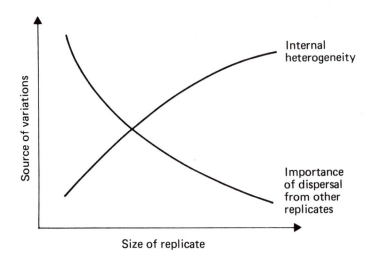

Figure 10.2. For terrestrial systems, an increase in the size of each replicate management unit means reduced importance of immigration–emigration rates as determinants of changes in stock size, but increased complexity in observed responses and in monitoring systems needed to measure average response accurately.

For terrestrial ecosystems, the importance of ecological linkages is very much dependent on the size of replicates as defined arbitrarily for administrative or experimental purposes. Generally, we expect the basic trade-off shown in Figure 10.2. As the size of each replicate is increased, processes such as dispersal are expected to become less important (probability of emigration decreases, number of immigrants smaller relative to births within the unit) as determinants of changes in stock size. But larger units are expected to be internally more heterogeneous, and to consist of a mosaic of subpopulations that have different productivities and exposure to harvesting. The larger the area, the more difficult it is to design monitoring programs that will representatively sample the mosaic to give accurate estimates of overall states and rates. Also, the internal structure is expected to evolve over time (overall parameters not constant), as the harvesting process depletes more accessible and less productive parts of the mosaic. When a choice of replicate size is available (i.e., when historical data can be variously aggregated or when new management plans are being developed), it is probably best to start with smaller units that can be aggregated later as sampling strata; however, experimental policy tests should not be conducted on single replicates so small that responses might be dominated or masked by immigration rates from untreated units (in other words: try to monitor by smaller units, but apply management changes to larger aggregations of these units).

Consider for a moment the problem of estimating dispersal rates among replicates. The usual approach here would be through tagging studies, which have a number of well known difficulties, such as tag loss, differential mortality of tagged organisms, and sampling for tag recoveries. Suppose we try to avoid these difficulties by using parameter estimation methods to infer dispersal rates directly from abundance changes among replicates. Here we would begin by formulating a dynamic model for abundance changes in each replicate, with dispersal rate terms included. For example, we might try the logistic model

$$N_{i,t+1} = R_{1i} N_{it} - R_{2i} N_{it}^2 - H_{it} \tag{10.12}$$
$$- \left(\sum_j m_{ij} \right) N_{it} + \sum_{j \neq i} m_{ji} N_{jt}$$

where N_{it} is population size in replicate i, R_{1i} and R_{2i} are internal production parameters for the replicate, H_{it} is the harvest from i in year t, and m_{ij} are dispersal rates from area i to area j. Notice first that R_{1i} and Σm_{ij} cannot be separated by regression of $N_{i,t+1}$ on N_{it} alone; that is, emigration from i cannot be distinguished from natural mortality in i (a component of R_1; see Chapter 5). We must use all of the stock sizes N_{jt} in predicting each $N_{i,t+1}$, and the regression coefficient for N_{jt} is interpreted as m_{ji}. A basic condition for this approach to work is that the N_{jt} should not be too highly correlated with N_{it}; otherwise we cannot logically distinguish between the effects of $(R_{1i} - \Sigma m_{ij})$ and each of the m_j. But if the dispersal rates m_{ij} are, in fact, much greater than zero, they are almost sure to cause N_{it} and N_{jt} to be correlated unless replicates i and j have been subjected to very different harvest policies or environmental effects.

So we come to basically the same conclusions as in the previous section on shared environmental effects: an elaborate estimation procedure is unlikely to help unless the replicates have been deliberately managed so as to produce contrasts (low correlations) in abundances over time. This conclusion is even stronger for more complex hypotheses/models of interaction among replicates than is presented in equation (10.12). For more complex interactions, such as competition for shared resources, we end up replacing terms like $m_{ji} N_{jt}$ first by products $a_{ij} N_{it} N_{jt}$, and then by messier nonlinear functions; if N_i and N_j are correlated, most functions of them (like $N_i N_j$) will also be. But such difficulties should not be construed as an argument in favor of using methods such as tagging experiments instead of elaborate estimation; it can be cheaper in the end to manage so as to deliberately produce contrasts in abundance, especially considering that the estimation of other parameters (besides interaction rates) is dependent on having such contrasts in the first place.

Coupling in aggregate value

A management agency responsible for many replicate systems has several choices about how to value the performance of each replicate. At one extreme, the agency may choose to treat each replicate as a unique and irreplaceable resource, to be managed in a highly conservative (risk-averse) manner. In this case the agency will act roughly as though it measured overall performance across replicates as the *product* of values (yields, net recoveries, or whatever) obtained from the units; failure of any one unit will be seen as a failure of the whole management system. Precisely this sort of performance evaluation can be forced on management agencies by the political environment within which they operate; it happens when harvesters and/or public interest groups treat each replicate as a battleground for bringing pressure on the agency. It is, of course, quite natural for people to notice and bring attention to failures, even when such failures are very localized and rare. However, really strong pressure to avoid local changes (management experiments, etc.) is mainly a problem in situations where each replicate is large enough to support a "dependent community" of economic interests (local fishermen, resort owners, hunters with limited mobility) for which use of or movement to other replicates means considerable risk and hardship.

At the other extreme, the agency may act as though its performance is judged as the *sum* or mean of values across replicates. In this case there is obvious flexibility and opportunity to trade-off performance among units in order to achieve informative variability. Such an ideal situation for adaptive management is likely to exist only where the replicates are quite small and numerous, and do not individually support dependent communities of economic interest. For example, it is hard to drum up much opposition to an experimental management scheme involving a few small trout lakes in the Sierra Nevada mountains of California, simply because there are hundreds of lakes where recreational fishermen may go if the experiment involves reduced harvests.

Most management agencies in North America operate somewhere between these extremes, with some replicates where change would be strongly opposed and others where adaptive management schemes might even be welcomed by public interest groups. Unfortunately, monitoring efforts are usually concentrated on the larger replicates that support dependent communities. This means that good pretreatment (baseline, historical "control") data are seldom available for the replicates where change is politically and economically most feasible. I have seen three reactions to this state of affairs: (1) increased investment in baseline monitoring programs intended to provide a more solid foundation for experimentation, without trying to anticipate what actions might eventually be tried; (2) refusal to support adaptive programs, simply on the grounds that it would take too long to gather baseline information; and (3) initiation of experimental programs without concern for the possibility that response might be due to historical factors rather than the experimental treatments (i.e., without adequate baseline or control data). It is easy to find fault with each of these reactions, but the key point is to recognize that they are often motivated not by a concern with long-term understanding of the resource as a whole (across all replicates), but rather by a desire to avoid confrontation over disturbances of the most popular replicates.

For replicates whose values can be treated as roughly additive, it is tempting to seek policies that generate variable harvesting within each replicate while maintaining near constancy in total harvesting effort and/or yield across replicates. Consider, for example, a situation where there are two replicates, with stock sizes N_{1t} and N_{2t}. Suppose the harvest in each area in any year can be expressed as $q_i E_{it} N_{it}$, where q_i is a catchability coefficient for unit i, and E_{it} is effort in that unit. If we try to maintain a constant total effort E^*, then $E_{1t} + E_{2t} = E^*$, and total harvest is given by

$$C_t = q_1 E_{1t} N_{1t} + q_2 E_{2t} N_{2t} \qquad (10.13)$$

Notice that if we also require harvest to be constant at some level C^*, equation (10.13) has a unique solution for E_{it}; we are forced to choose

$$E_{1t} = \frac{C^* - q_2 E^* N_{2t}}{q_1 N_{1t} - q_2 N_{2t}} \quad \text{and} \quad E_{2t} = E^* - E_{1t} \qquad (10.14)$$

and to depend on changes in N_{1t} and N_{2t} as determinants of permissible variation in effort levels. Using simulations with various models for the dynamics of N_{1t} and N_{2t}, it is easy to show that constraining both total catch and total effort generally leads to highly correlated stock sizes over time, which would preclude the use of joint estimation procedures as introduced in the last two sections. The situation is not much improved when only one of the factors (total catch or effort) is held constant, since the stocks tend to be consistently pushed in opposite directions (negative correlation) if total effort (and catch) is moderate, or driven down together (positive correlation) when total effort is high.

Correlations among stock can be reduced when several or many replicates are managed together, but only by resorting to concentration of effort on one or a few replicates each year. Since concentration of effort leads to exploitation or even interference competition among harvesters (crowding effects in the replicates where effort is concentrated), they are unlikely to cooperate in the experiment unless some inducement is provided. One inducement is to offer direct compensation for extra operating costs and/or lost harvests. Another is to rotate the effort concentration over time so that each replicate has a long "fallow period" to become especially attractive (high abundance, larger fish, more trophy animals) for the time when it is to receive the concentration. We noted earlier (Chapter 8) that such "pulse harvesting" regimes can even produce the highest average yield per year from each replicate, when it is impossible to regulate the size/age distribution of animals harvested.

Sharing of monitoring and control resources

Biological monitoring and enforcement of regulations are typically the two largest budget items for resource management agencies. Occasionally some activities are combined, by making enforcement personnel responsible for various sampling chores, but most often there is strong (or even vicious) competition for budget shares between groups (divisions, departments, etc.) responsible for the functions. The easiest function to justify to the public is enforcement, so it is often an uphill battle to maintain even the most basic monitoring programs. As noted in the previous section, there is a tendency for all agency activities to be concentrated on the most accessible and publicly visible replicates.

So far as I am aware it has never been possible to demonstrate quantitatively how reallocation of agency resources among replicates will affect the performance of each replicate. We can build statistical arguments about how changes in monitoring programs will affect the accuracy of estimation for parameters needed to set key regulations, but such arguments are complex (hard to present in debates about funding) and require risky assumptions. It is even more difficult to build a convincing case about the effects of incremental changes in enforcement effort among replicates, since the response of harvesters to enforcement activity is very poorly understood. There are generally no data, for example, about how the occasional presence of a "game warden" would affect the behavior of hunters who have previously enjoyed unchallenged access to an area or stock.

Thus, in the end, management resources are allocated to activities like monitoring and enforcement largely on the basis of historical precedent, public pressure involving very naive arguments ("more wardens, less poachers"), and the competence of activity leaders at bureaucratic infighting for budget shares. The basic outcome of this allocation process is that the share of "attention" devoted to each replicate need bear no measurable relationship to the importance of that replicate either as a source of immediate yields or as a potential site for future development or experimentation. Moreover, we should expect this state of affairs to persist even if clever experiments are conducted to demonstrate empirically how important each activity really is, since there are strong personal motives for some management actors to ignore or deplore such experiments in favor of "business as usual."

Realistic planning for adaptive management of replicated systems should either (1) include strong efforts to demonstrate vividly and simply why a change in agency resource allocation would be worthwhile, or (2) treat existing allocation patterns as severe constraints on where to conduct experiments and on the type of regulatory actions that are practical to enforce in effecting experimental changes. Where this second attitude is adopted, it will often be best to pass up the best plans from a scientific viewpoint (nice experimental designs, choice of representative replicates, etc.), in favor of using existing opportunities, such as replicates that already have good monitoring programs. Let me put this another way: the selection of replicates to include in an adaptive plan may involve a number of difficult compromises, in terms of replicate attributes like accessibility, ease of enforcement, and likelihood that responses will be representative (informative about how replicates not included in the initial plan would respond if results from the plan are later applied more broadly). It is doubtful that formal optimization procedures (multiattribute decision analysis, etc.) will be of much help in finding the "best" compromise, but it is important to begin by laying out the available choices in some systematic way (like lists of replicates that would be favored on the basis of different criteria) in order to promote informed judgment.

Dual Effects among Replicates

Chapter 9 emphasized that management actions often have "dual effects of control," by influencing not only the immediate performance of the managed system, but also the rates of learning about parameters that are important to long-term performance. In single systems that are managed over time, the effects generally conflict: the most informative action is generally not the best for short-term performance. This conflict leads to a very difficult optimization problem for which there is not yet much computational experience. In this section we shall look at what happens when actions are taken across several replicates that may be more or less informative about one another; there is essentially no computational experience about this much larger optimization problem, and the discussion will be restricted to a qualitative review about its formulation and likely results.

Let us review for a moment how dual effects of control are represented for single replicates in the wide-sense algorithm (Chapter 9), which has promise for computation of optimum policies for larger systems. The decision (control, action) u_t at time t is seen as having three basic effects: (1) along with x_t, it determines a short-term value component v_t of the total resource value V_t from time t forward; (2) it affects the next state x_{t+1} which will in turn influence v_{t+1} and V_{t+1}; and (3) directly and through its effects on x_{t+1}, it influences parameter estimation performance and hence the level of uncertainty faced by the decision maker at time $t + 1$ as measured by the parameter covariance matrix Σ_{t+1}. We may represent these effects diagrammatically, as

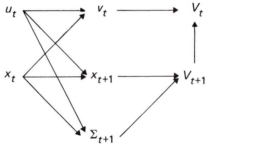

$$(10.15)$$

where an arrow $a \rightarrow b$ symbolizes "a influences b" (or "the calculation of b depends in part on a"). Recall from Chapter 8 that a crucial connection in this diagram is $V_{t+1} \rightarrow V_t$ (since $V_t = v_t + V_{t+1}$), which points out that the total value V_t of any decision choice u_t is a combination of a short-term payoff v_t and future returns V_{t+1}; the diagram shows that u_t affects both components (v_t directly, V_{t+1} through effects on x_{t+1} and Σ_{t+1}).

Now, suppose we have several replicates each with decision choices u_{it}, state variables x_{it}, and parameter uncertainties measured by Σ_{it}. We wish to define an optimal choice $u_{1t}^*, \ldots, u_{Rt}^*$, while recognizing at least four basic interactions among the replicates:

(1)   the contribution of each to some aggregate performance measure $_A V_t$ $= f(v_{1t}, \ldots, v_{Rt}) + _A V_{t+1}$;

(2)   possible constraints on $u_{1t}, \ldots, u_{Rt}$ associated with factors such as the total harvesting effort available to apply across replicates;

(3)   biological interactions and/or shared environmental effects that link the dynamics of x (make x_{it+1} depend on x_{jt} as well as on x_{it}, and on $\bar{w}_t$ shared random effects); and

(4)   effects of joint estimation, where Σ_{it+1} is possibly influenced by all state-control choices and outcomes $(u_{jt}, x_{jt}, x_{j,t+1})$. In terms of the diagrammatic scheme (10.15), we have the "within-replicate" influences shown there plus a collection of "cross-replicate" influences:

$$u_{it} \longrightarrow u_{jt} \text{ (control constraints)}$$

$$x_{it} \longrightarrow x_{j,t+1} \text{ (biological interactions such as dispersal)}$$

$$\left.\begin{array}{c} u_{it} \\ x_{it} \\ x_{it+1} \end{array}\right\} \Sigma_{jt+1} \text{ (joint estimation effects)}$$

Obviously, such influences can lead to a very complex web of calculations for predicting the effects of each control choice $u_{1t}, \ldots, u_{Rt}$, which can make the search for an optimum choice very difficult. Below we will be content to examine three special cases where the interactions are simple enough to permit some qualitative arguments and conclusions. We will concentrate on arguments about adaptive learning as represented by propagation of the Σ_{it} over time, and say a little about the interesting effects that control constraints and biological interactions can have even in the absence of uncertainty.

Value independence, no joint estimation

Suppose the replicates are assumed to contribute independently to the aggregate resource value, so the value obtained in year t is simply the sum $v_{1t} + v_{2t} + \cdots + v_{Rt}$. Suppose further that (1) the variation of actual parameter values among replicates is thought to be very large, so joint estimation equations like (10.2) would place very little weight on the common estimator $\tilde{\beta}$, and (2) shared environmental effects $\bar{w}_t$ appear to be very small, so there is not likely to be an improvement in parameter estimation by considering these effects.

In this extreme case it is best to manage the replicates independently so as to maximize each of their long-term performances V_{it}, unless there are strong biological interactions or control constraints. When dispersal rates among replicates are very high, they should be managed as a single unit with the best total harvest allocated among replicates according to convenience or access cost criteria. The effect of intermediate dispersal rates on optimum policies has not been thoroughly analyzed, and we will not consider it in this review. Control constraints (enforcement resources, minimum tolerable harvests, etc.) will generally favor a concentration of management activity (probing policy changes, etc.) on those replicates having the highest expected performances V_{it}. Since, by assumption, the replicates are not informative about one another, there is no point in shifting management resources and experimental activities toward less-valued replicates so as to minimize risks and direct costs to harvesters.

Value independence, joint estimation

In this case we assume again that replicates contribute additively to aggregate value, but assume that replicates may have similar parameter values and/or large shared effects. This means assuming small values for σ_β^2 (variance among replicates) in weighted estimation models like equations (10.2)-(10.3), and/or large shared effects $\bar{w}_t$ in models like equations (10.6)-(10.7), and/or significant connection parameters m_{ij} in models like equation (10.12). In all of these situations, there can be substantial effects of decisions and responses for each replicate i on the uncertainty Σ_{jt+1} of parameter estimates for other replicates. Then, if we also estimate significant values of information for reducing Σ_t in some or all replicates [using equations like (7.2) and (9.17)], it will be worth considering the development of a coordinated probing policy that involves (1) deliberate actions to reduce Σ_{it} on *some* representative subset of replicates if σ_β^2 is small, and (2) variation of actions across all replicates in a way that will reduce covariances in temporal behavior.

Recall from Chapter 9 that the long-term value from each replicate in isolation can be approximated by the wide-sense components

$$V_{it} = v_{it} + V_{i,t+1}^{(o)} + V_{i,t+1}^{(c)} + V_{i,t+1}^{(p)} + \gamma_{t+i}^{(o)} \tag{10.16}$$

When values are additive across replicates, we can combine the nominal and caution terms $V_i^{(o)}$, $\gamma_i^{(o)}$, and $V_i^{(c)}$ to obtain an aggregate base (nominal) value $V_{t+1}^{(b)}$ representing all long-term effects besides learning. Then

$$_A V_t = \sum_i v_{it} + {}_A V_{t+1}^{(b)} + \sum_i V_{i,t+1}^{(p)} \tag{10.17}$$

where $_A V_{t+1}^{(b)}$ is simply the sum of wide-sense value components from independent analyses of the replicates, unless there are strong direct (biological) interactions. The $V_{i,t+1}^{(p)}$ terms are more difficult to assess, since each depends on all the decision choices $u_{it}, \ldots, u_{Rt}$ (through the effect of these choices on all the $\Sigma_{i,t+1}$).

Two situations are possible, depending on whether it is important to avoid statistical correlation among replicates in states and controls over time. When shared effects or direct interactions are not so important, probing decisions on individual replicates simply reinforce one another by reducing the variance of weighted estimators [for example, variance of $\tilde{\beta}_i$ in equation (10.2) is reduced by any action that reduces $\sigma_{\tilde{\beta}}^2$, which in turn depends on the variances of all the $\hat{\beta}_i$ from which $\tilde{\beta}$ is estimated]. When shared effects are important, it may be necessary or worthwhile to avoid simultaneously probing in the same direction (for example, increasing escapements in several stock-recruitment replicates simultaneously) since the probing effects will be confounded with the shared $\bar{w}_t$ effects.

To visualize the difference between these situations, consider a stock-recruitment system with two replicates, where application of the wide-sense algorithm separately to each replicate would give a pattern of value components as in Figure 9.11 (the optimum escapement for each replicate would be assessed at u_i^*, much greater than the historical average $\bar{u}_i$). We can plot the combined value $_A V_t$ [equation (10.17)] as a function of the two escapement choices u_{1t} and u_{2t}, to give a contour map of value for various decision combinations. Figure 10.3 shows roughly how these contours are likely to look for the two situations. In the absence of shared effects (case A), increasing escapement away from $\bar{u}_i$ will increase the total value for both replicates and the optimum combination will be somewhere near (u_1^*, u_2^*) as determined by separate analyses. When shared effects are present (case B), (u_1^*, u_2^*) will be a bad choice and it will be best to hold one of the escapements near $\bar{u}_i$ while moving the other to u_j^* or beyond. The replicate held near its historical average then acts as a backdrop or "control" against which to measure the shared effects. If the systems are similar in nominal value ($V_i^{(o)}$), the best replicate to probe will be the one with the largest Σ_{it}. In successive time steps, the best replicate to probe may then change, so as to give an alternating pattern over time until probing is not worthwhile for either replicate.

It is interesting to speculate about the best probing policy when there are strong shared effects across many replicates. If the responses of all replicates are highly uncertain, the best policy is likely to involve probing on most of them while holding a small subset constant as controls for shared effects. But suppose that the responses of a few replicates have been thoroughly tested (Σ_i small), and they are found to have almost identical parameter values ($\sigma_{\tilde{\beta}}^2$ small). Suppose further there is no reason to believe that these replicates are somehow not representative of the system as a whole. Should we then proceed with probing experiments on the other replicates as well, thus sacrificing yields during the experimental periods, or instead take it on faith that they are all similar and can be managed accordingly? At this point a strict application of equations (10.2) and (10.3) would

Case A: No shared effects considered

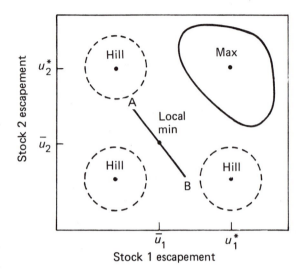

Case B: Strong shared effects admitted

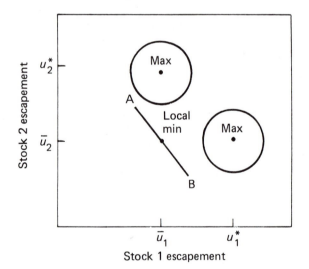

Figure 10.3. Qualitative patterns of combined long-term value when two stocks are managed as replicates. In case A, the value is maximized by pushing escapement for both stocks toward the probing optima u_1^* and u_2^* (away from the historical average escapements $\bar{u}_1$ and $\bar{u}_2$). In case B, shared environmental effects could be confused with the effects of simultaneous probes toward higher escapement, so it is best to increase only one stock while holding the other near $\bar{u}$ as a "control" for the shared effects. Line A–B shows escapement combinations that are likely to produce total catch (both replicates) near the historical average without substantially affecting future recruitments to either stock.

tell us not to probe (W_i near zero and $\tilde{\beta}$ well determined, so $\sigma_{\hat{\beta}_i}^2$ very small). Here I suspect most people would begin to seriously (and correctly) question the basic assumption underlying equation (10.2), namely that the replicate parameter values were "drawn by nature" from a simple distribution. In other words, we would act as though σ_β^2 were larger than indicated by the few replicates tested to date, on the off-chance of discovering some replicates with atypical or special opportunities for increased yield.

It appears that the single most critical factor in probing policy design for many replicates is the pattern of variation in parameter values among replicates, measured most simply by σ_β^2 [equation (10.3)]. To the extent that replicates can be classified into more homogeneous subsets by using structural covariates, such as area size, the best plan may involve a corresponding stratification of replicates into test and control groups for each subset. But the classification process requires that at least some replicates be tested sufficiently (Σ_{it} small) to make covariate effects visible against the background of variation caused by estimation errors within replicates. Thus, in a "new" resource where the behavior of all replicates is highly uncertain, it will be best initially to assume large σ_β^2 and choose replicates at random for probing tests. Then, if there is simultaneous investment in the measurement of covariates that are likely to "explain" some parameter variation, the random test pattern will give way over time to a more focused (structured, fragmented) probing design where homogeneous subsets are managed as learning units. To a distant observer, the overall management program will then exhibit a mixture of passive adaptation or evolution (responses to the lucky discovery of patterns attributable to covariates) and active probing to uncover local opportunities.

Value dependence

The inferences about probing policy design presented in the previous section depend critically on having additivity of value across replicates, so that the payoff v_{it} from any replicate need not be immediately (in year t) balanced against or compensated by payoffs from other replicates. However, most management authorities dare not operate with the assumption of additivity, and must, instead, try to maintain at least some stability in the total annual performance across replicates. This means that probing experiments that reduce any v_{it} are penalized (by forcing changes in some other v_{jt} payoffs), unless the compensating changes v_{jt} also lead to informative responses. Unfortunately, the best probing changes are usually highly directional, toward untested opportunities for increased production (as discussed in Chapter 9).

One obvious way to represent value dependence is to assume that payoffs v_{it} are additive, but that the sum is constrained to lie within a narrow range around its historical average. This constraint then defines a set of feasible policy choices. An extreme example is shown in Figure 10.3, where the line A–B defines escapement policy combinations that are expected to produce exactly the historical average yield in year t from two stock–recruitment replicates, if recruitments are at their historical average. In this example, the length of the line A–B is set by the range of escapement levels that are likely to produce recruitments at time $t + 1$ large enough to allow total catch at $t + 1$ also to be in the tolerable range. We see in Figure 10.3 that probing toward either of the feasible extremes (A, B) will produce some increase in expected value, but will not be toward the best combinations estimated by assuming value independence. There is one consolation: if one or both stocks have higher than average recruitment in year t, the A–B line moves up and right; thus, the best adaptive choice may be feasible in some good years.

The situation is not quite so grim when there are many replicates, unless variability in parameter values (σ_β^2) among them is so high that tests on a few will provide no insights about how the others can be better managed. There are two extreme strategies for meeting overall harvest constraints while introducing as much variation as possible: (1) a *diffuse* strategy of disturbing all replicates a little; or (2) a *focused* strategy of probing hard on a few replicates while making up losses by small changes in the rest. I suspect that a focused strategy will generally be best, since there is high risk in any diffuse strategy that the small disturbances will produce no useful information at all (see the shape of probing value component in Figures 9.10–9.11). Notice that a focused strategy does not imply that only a few replicates are ever disturbed, since different units can be tested over time; however, sequential probing patterns will require some careful planning to minimize the chance that various cumulative effects from unsuccessful tests (for example, increase in the number of stocks recovering from low levels) will suddenly make it impossible to meet overall targets or constraints. With luck, favorable results from the first few probing tests (increased stock sizes, etc.) will result in increased flexibility to disturb other replicates later. Indeed, there would be no point in undertaking the first focused tests unless there were a good chance of higher payoffs from these replicates later.

It is a pity that adaptive policy design cannot guarantee increasing flexibility to meet harvest constraints over time, due to improved performance in replicates for which uncertainty is reduced early in the experimental program. But improvement in expected performance, a basic requirement for probing in the first place, does not imply improvement in every instance. Instead, we expect to make matters worse in at least a few replicates, and the balance can go this way for any finite sequence of examples.

All we can honestly claim is that spreading the risk across many replicates will greatly reduce the odds of an overall loss; this spreading of risk is retarded by constraints that prevent variation in short-term payoffs for more than a few replicates at a time.

Allocation of Monitoring Resources

It is obviously silly to talk about various adaptive probing strategies unless there is substantial commitment to monitor dynamic responses over time, and even the most rudimentary feedback policies require some monitoring of state changes in the replicates to which they are applied. However, in situations where harvesters have some flexibility to choose the replicates in which they will exert effort, it can be difficult to justify even that first step of rudimentary feedback regulation; the harvesters may generate a kind of feedback dynamics on their own, by shifting away from replicates with low stock sizes (if low stock size implies low success rate) and toward replicates where abundance is high. In the following discussion it will be taken as given that the harvesting system is large and efficient enough to severely deplete many replicates if there is no active intervention by the management authority.

An initial step in the allocation of monitoring resources is to ensure that either (1) enough information is gathered from every replicate to allow detection of and response to extreme state changes, or (2) state changes are sufficiently correlated among replicates so that a feedback policy can be applied to all of them based on changes measured in a representative subset. Then the basic issue is whether further investments should be directed to providing very accurate assessments on a few replicates, or a bit more baseline information on all of them.

We might expect some guidance about monitoring allocation from statistical sampling theory, where a standard problem is the allocation of sampling effort among strata, clusters, and so forth (see Cochran, 1963). A typical result from this theory is that sample size should be proportional to overall importance (stratum size) and expected variance among observations. Unfortunately, such results are generally based on the assumption that there is a single overall quantity or parameter that is of interest, rather than the performance of individual sampled units. In the notation of equation (10.2), sampling allocation formulas usually specify how to best estimate $\tilde{\beta}$, but the resource manager is interested in this quantity only insofar as it helps him estimate the parameters β_i that characterize replicate performances. It is hardly wise to allocate resources so as to best estimate a quantity that may not be correct for any of the individual replicates.

Monitoring and parameter estimation

It is important to recognize that monitoring and dynamic parameter estimation are not the same thing. Monitoring is certainly required for estimation, but precise monitoring is not sufficient to guarantee precise estimation. Consider the trivial example of a population that has been constant over time, and where a substantial investment has been made to measure birth–death rates and the population size very accurately. If we try to fit even a logistic model to these data, we will find that only one linear combination of the r and k values can be estimated; the data contain no information about density-dependent effects (variation in rates with population size) that are critical in determining whether the population is at its best level. Indeed, the dual control calculations presented in Chapter 9, where probing experiments were found to be worthwhile, were all based on the simplifying assumption that the system state (catches, escapement, relative stock size) is measured exactly over time, with *no* monitoring errors.

As noted in Chapter 6, it is extremely difficult to find even unbiased parameter estimates when both process and measurement errors are present. When we try to establish functional relationships of responses (rates, state changes) to state variables, it is necessary to be very careful about statistical "errors-in-variables" effects (bias) due to measurement errors in the state variables. Even simple stock–recruitment analyses give downright deceptive results when spawning stocks are measured with the level of counting/sampling error that is practical to achieve with inexpensive field surveys.

Biases due to state measurement errors can be reduced not only by better monitoring, but also by increasing the range of states tested through more violent probing experiments [see Walters and Ludwig (1981) for examples of bias calculations]. Since state variation is important anyway for parameter estimation, it is quite possible that increasing variation will turn out to be less costly than improved monitoring in many contexts. The simplest way to check this possibility is by Monte Carlo methods, where parameter estimation performance is simulated for various combinations of state variation (probing) and investment in monitoring. However, the trade-off between probing and measurement will involve a conflict of interest in most North American situations, where the costs of increased state variation would ordinarily be borne by harvesters (through variations in harvests) while monitoring costs would be borne by management agencies that are partially funded from sources other than harvesting. Increased harvest variation can be a way of avoiding higher public expenditures for management monitoring, but I would hate to try and sell this rather esoteric point in any context of political debate.

Since increased state variation can potentially compensate for the effects of measurement errors, it is not obvious how to allocate monitoring activity in coordination with adaptive plans, as discussed in the last section. Intuition would suggest larger allocations to the replicates that are subject to probing tests, but this means that poorer control (pretreatment) data will be gathered on other replicates that might be subject to test later in time. Again, a key planning variable is the prior estimate of variance in actual parameter values among replicates (σ_β^2). If this estimate is small (replicates expected to be similar), the best strategy is probably to focus both monitoring and probing on a few replicates. If σ_β^2 is large, so every replicate will have to be treated as unique in the long term anyway, then the best monitoring policy may be diffuse: give each replicate equal attention, or allocate in proportion to expected long-term value if the replicates appear to differ greatly in potential.

A very dangerous monitoring strategy is to allocate measurement effort in response to large state changes that are due to natural or unmanaged human factors, and come to the management agency's attention only after they are well under way. It is a very common "firefighting" behavior for agencies to set up monitoring schemes after big changes, simply as a way to demonstrate visible "concern for the problem" or in the mistaken belief that the cause of the change can be inferred from measurements after the fact. A case can be made for the notion of exploiting "natural experiments" as a source of informative variation in system state (i.e., study transitional responses after the unplanned disturbance). But, unfortunately, natural disturbances large enough to be helpful for parameter estimation are also likely to be associated with changes in system structure (new species, new harvesting tactics, loss of substocks, etc.) and new parameter values that are not representative of normal replicates.

Adaptive allocation of monitoring resources

The previous paragraph points out that management agencies often engage in a kind of adaptive allocation of monitoring effort, by shifting attention to crisis situations as they arise. Such strategies are intriguing, even though they are unlikely to result in clean, consistent time series that are convenient for statistical analysis. Parameter estimation procedures and Bayesian probability assessments can be formulated so as to account for differences in measurement precision over time [see Bard (1974) for details], though the assessment of changes in measurement error variances then becomes a more important component of the monitoring program.

Suppose it is planned to use a joint parameter estimation scheme, such as equation (10.2), and that we can make some estimate of how the variance of $\tilde{\beta}_i$ for any replicate will be affected by allocating an increasing share of the total monitoring effort to it. Then, if we ask how to minimize some simple measure of overall information performance, such as the sum of variances of $\tilde{\beta}_i$ across replicates, formal optimization procedures are likely to tell us that an extreme strategy is best: either diffuse the allocation across all replicates in rough proportion to the $\sigma_{\beta_i}^2$, or else focus all of it on a subset of replicates with the highest initial uncertainty. The latter outcome is more likely if σ_{β}^2 is small and focused monitoring can result in quick reduction in uncertainty about $\tilde{\beta}$. Thus, it is quite possible for the optimum allocation to flip suddenly in time, from diffuse to focused (or vice versa) as the $\sigma_{\beta_i}^2$ are reduced and more information becomes available about σ_{β}^2.

In the face of very large uncertainty, diffuse monitoring will provide little information about the β_i, which in turn means a poor estimate of the actual variation among replicates (σ_{β}^2). Thus, I would recommend to gamble initially that σ_{β}^2 is small, and focus the monitoring (and experimental probing) on a few replicates while gathering only very crude information for feedback control on the other replicates. The initial test replicates should be chosen so as to be as different as possible with respect to structural covariates, in order to provide both a conservative (high) estimate of σ_{β}^2 and a basis for later classifying replicates into more homogeneous subsets. Then, as good parameter estimates begin to emerge from the test set, two evaluations should be initiated: (1) exploratory analysis of the detailed data already available, to see if there are simpler (less expensive) measurements that are highly correlated with system state, and (2) calculation of expected results from diffuse versus focused monitoring, to see if [in conjunction with the results of evaluation (1)] a shift in attention would be worthwhile. At some point in time, evaluation (2) is likely to call for a major, adaptive reallocation of monitoring activity that permits better feedback management of all replicates simultaneously.

It is possible that precise monitoring of state responses in even one replicate would require most of the resources available to the management agency. In this case it would be unwise to use a focused monitoring strategy, since there is no way to assure that the replicate chosen for the study is representative. Instead, the agency should seriously consider either (1) diffuse monitoring coupled with hard probing experiments on many replicates simultaneously, or (2) placing its attention initially on the development of innovative monitoring tools and techniques, or (3) shifting the burden of monitoring more to the harvesters, such as by requiring them to do sample searching in patterns that will allow better mapping of relative resource densities.

Design of Experimental Plans

It is not yet possible to compute optimal adaptive policies for replicated systems, even when each replicate can be assumed to follow some very simple dynamic model. Still, it should be possible to develop management plans that improve substantially on the extreme alternatives of either treating every replicate as unique, or allocating management activity solely in response to political pressure and to the detection of local crises. In this closing section, I will simply identify five basic ingredients that will be needed in the design of experimental plans that involve deliberate attempts to probe for information.

Systematic evaluation of available data

The starting point for experimental design should be a systematic synthesis of historical data, and from this the estimation of model parameters and measures of uncertainty (or place odds on a variety of alternative models) for each replicate. If reasonably accurate parameter estimates are available for some replicates, a further component of the historical analysis should be a search for patterns in the parameter estimates that might be attributable to structural covariates and provide a basis for stratifying the replicate set into more homogeneous response groups.

Historical databases can be surprisingly difficult to assemble. Few management agencies have attempted even to archive all their data in one place, let alone put it in accessible forms such as computerized databases. Also, it is often impossible to find out exactly how the earlier data were collected, since sampling procedures are typically not documented in detail; this makes assessment of measurement errors and comparison of alternative monitoring techniques very difficult. However, the arduous task of digging through old records and files can be quite rewarding; a management agency can accumulate staggering amounts of information over many years, all stored away by people with the best of intentions for future users.

An accounting system for overall value

A second obvious need is to develop a generally acceptable protocol for measuring the aggregate value from managing all replicates together. This may be simply an additive accounting of yields, or a complex scheme for weighting several benefit–cost measures from each replicate against standards or constraints on average performance.

The development of this protocol should involve some examination of the factors that have determined the historical distribution of harvesting effort among replicates, such as travel costs and proximity to dependent communities. In view of such factors and the need for cooperation (consensus) between harvesters and the management agency, it will typically be wise to adopt a more sophisticated valuation scheme than just the sum of harvests. At least, the replicates of major historical importance will have to be weighted more heavily when the costs of experimentation are assessed.

Model for propagation of uncertainty

An essential feature of adaptive management planning is the deliberate attempt to predict how future state changes and data gathering are likely to affect learning about parameters (or the odds placed on alternative models). To make such predictions, it is necessary (1) to specify beforehand precisely how future data are likely to be analyzed (estimation schemes, etc.), and (2) to develop a stochastic model framework for generating possible data sets and applying the analysis techniques to them.

When there are many replicates that will be analyzed through a complex joint estimation scheme, the safest technique to use initially is "brute force" Monte Carlo simulation. Set up a model for all the replicate dynamics (and interactions) and for the observation process (monitoring scheme). Establish an automated procedure to confront the estimation scheme with many fake data sets from this model, each set representing different dynamic parameter choices and sequences of errors (process and measurement). The confrontation with many data sets will then have to be repeated for each experimental management plan under consideration.

Method for calculating the value of reducing uncertainty

Suppose that methods such as Monte Carlo simulation indicate that a proposed experimental plan will likely result in parameter covariances Σ_{1t}, ..., Σ_{Rt} after t years, starting from initial uncertainties Σ_{10}, ..., Σ_{R0} across the R replicates. This outcome does not really mean much unless it can be translated into an effect on aggregate value. About the simplest reasonable estimate of this effect would be the change in expected value of perfect information (EVPI, see Chapter 7) from time 0 to time t. This change measures the expected improvement due to using the plan up to time t as opposed to a nonadaptive (no-learning) alternative, then switching back to the nonadaptive plan at time t. If EVPI is calculated as suggested in Chapter 7, then

$EVPI_t - EVPI_0$ will be a relatively small, marginal value for each replicate but may be very large in aggregate across replicates.

An alternative to EVPI calculations is a tedious estimation, by Monte Carlo simulations, of expected performance when each experimental plan is accompanied (or followed) by feedback management based on the parameter estimates obtained. In other words, simulate the whole "closed-loop" dynamics (for all replicates) of natural state responses, measurement, estimation, and management action as a function of estimates. This approach will provide a conservative estimate of future management performance since it will not account for future innovations in monitoring technology and experimental design.

Mechanisms for generating alternative plans

Given some perspective on the system from historical data analysis, and machinery for tracking the possible effects of any plan that is devised, the key problem then becomes how to develop some alternative plans that make statistical sense and are practical in terms of considerations like "political salability" that are difficult to quantify in the value accounting system. One possibility is to have a computerized procedure that can generate many plans by systematically or randomly selecting replicates for disturbances, and disturbance patterns to apply to them. But such procedures will only be effective if they can be provided with very tight search criteria beforehand, such as the number of replicates to test each year and the best magnitude and frequency of disturbance for each replicate.

The best plans are likely to come from some process of imaginative synthesis that involves discussion among people with different areas of expertise, ranging from experimental design to administration of sampling programs. Automated procedures for plan generation and testing can be used to focus the discussions and provide quick feedback about options as they are suggested, but it is important that the discussants (planning team) see these procedures as tools to stimulate imagination.

Problems

10.1. Spatial replication or subdivision of renewable resources creates flexibility for experimental management, provided harvesters can move around so as to make up local losses with gains elsewhere. But each replicate may still have a "dependent economic community" (resorts, equipment service facilities, harvesters with limited technology or means to travel) that will oppose experimentation with it, especially if the experiment involves immediate reduction in harvests. Assuming that such communities should not (or politically cannot) be ignored, suggest schemes to avoid making them bear the major burden of experimental losses and risks.

10.2. It has been proposed to experimentally vary exploitation rates on Pacific cod in the Hecate Strait, British Columbia. The Strait can be divided into two areas (replicates) with roughly equal initial stock sizes, and it is thought that there is little dispersal of fish between these areas. The current stock biomass is thought to be around 10 000 tons (5000 in each area), producing an annual trawl catch of 3000 tons with 10 000 hours of trawl effort (5000 hours in each area). Due to exploitation competition [Chapter 4, equation (4.12)], it is thought that the harvest rate u_t (now 0.3) in each replicate area will vary with effort according to the catch equation $u_t = 1 - e^{-qE_t}$, where $q = 7.13 \times 10^{-5}$ and E_t is hours of trawl effort in the replicate. What will happen to total catch this year if all the effort (10 000 hours) is concentrated in one replicate (with a stock size of 5000 tons)? What will the total catch be next year if all the effort is then moved to the other replicate, assuming that the net stock increase in each area is 30% of the stock size left after harvest this year? If it is required that the total catch be kept constant (at 3000 tons), what effort will be required this year if it is all exerted in one replicate? What effort will be required next year if it is all exerted in the other replicate, and the net stock increases are again 30% of the stocks left after harvest this year? Do these results suggest that the evaluation of experimental plans should include economic factors, such as the cost of exerting each unit of fishing effort?

10.3. For replicate systems that may be subject to shared (similar) "environmental effects," it may be possible to estimate the effects as extra dynamic "parameters" when historical data from all the replicates are analyzed together; then, after correcting for the estimated shared effects, each data set may appear much less "noisy." However, this possibility is lost when all replicates have been subject to the same harvest policies over time. Explain this finding in terms of the classical scientific idea of treatment and control measurements in experimental design. When unpredictable environmental effects may be present, is it possible, even in principle, to logically distinguish between these effects and the effects of policy change, unless there is at least one replicate where the policy has not changed? In fitting models to time series data from a single (unique, isolated) replicate, we pretend to avoid this treatment control issue by assuming a stationary model structure (same equations, parameter values over time); what price do we pay for this assumption? (Hint: see Figure 6.4.)

10.4. The sockeye salmon example in Figure 1.1 has been used repeatedly in this book, to make various points about uncertainty. The figure may be misleading, since it represents aggregate data across many biological stocks that spawn in widely separated tributaries within a large watershed. Combined analysis is valid only if the stocks have

Table 10.1(a). Spawning escapement and subsequent recruitments for four hypothetical salmon stocks from a large watershed. Total returns are from eggs spawned in brood years, and are combined for return years 3-5 years after spawning.

Stock #1			Stock #2		
Brood year	Escapement	Total return	Brood year	Escapement	Total return
1948	10 356	232 876	1948	670 622	1 909 842
1949	3593	29 456	1949	58 247	618 785
1950	1 259 381	9 220 024	1950	17 308	203 432
1951	143 428	522 087	1951	100 116	743 398
1952	9317	16 451	1952	485 585	1 837 630
1953	3472	29 548	1953	200 691	615 701
1954	2 651 231	15 072 461	1954	34 296	697 671
1955	63 336	852 458	1955	121 167	1 479 484
1956	3321	7672	1956	646 906	2 421 690
1957	2807	21 365	1957	138 464	138 089
1958	3 287 678	2 013 436	1958	120 104	427 605
1959	134 545	879 895	1959	463 060	2 187 519
1960	1907	2412	1960	426 546	1 046 818
1961	1118	6215	1961	39 101	68 788
1962	1 113 088	2 777 736	1962	77 713	974 905
1963	156 454	3 033 433	1963	998 231	1 164 753
1964	604	17 132	1964	238 272	2 031 487
1965	1795	50 353	1965	35 335	155 375
1966	1 255 893	3 851 506	1966	209 619	861 265
1967	838 945	3 054 910	1967	174 715	1 965 450
1968	3686	20 551	1968	413 862	2 413 817
1969	4986	11 834	1969	70 902	397 863
1970	1 495 504	4 990 517	1970	135 388	667 540
1971	283 791	635 367	1971	157 193	575 904
1972	4153	38 740	1972	562 650	1 900 778
1973	1014	88 234	1973	55 675	203 224
1974	1 061 774	6 264 261	1974	107 563	579 665

Table 10.1(b). Spawning escapement and subsequent recruitments for four hypothetical salmon stocks from a large watershed. Total returns are from eggs spawned in brood years, and are combined for return years 3–5 years after spawning.

Stock #3			Stock #4		
Brood year	Escapement	Total return	Brood year	Escapement	Total return
1948	19 979	198 153	1948	19 431	131 635
1949	582 228	1 030 708	1949	12 725	54 928
1950	59 104	241 087	1950	32 539	182 836
1951	60 423	173 645	1951	12 856	116 935
1952	30 212	88 572	1952	28 050	10 933
1953	154 036	540 597	1953	8989	217 870
1954	35 050	155 482	1954	28 137	232 492
1955	2159	27 456	1955	21 636	72 378
1956	25 280	110 394	1956	8690	21 572
1957	234 850	1 222 183	1957	20 237	8801
1958	38 807	102 352	1958	38 439	30 715
1959	2670	20 835	1959	8363	39 208
1960	14 447	74 127	1960	7033	4623
1961	198 921	255 212	1961	4246	57 472
1962	26 716	75 785	1962	15 824	47 854
1963	4607	92 222	1963	14 469	161 915
1964	2390	41 860	1964	1196	24 962
1965	23 045	416 779	1965	13 521	32 377
1966	10 830	83 040	1966	13 360	12 377
1967	21 044	339 270	1967	19 720	30 940
1968	1522	10 412	1968	2407	43 587
1969	109 655	1 374 870	1969	41 716	46 276
1970	32 578	180 924	1970	6108	63 126
1971	95 940	432 732	1971	2482	20 372
1972	4657	32 401	1972	15 193	72 938
1973	299 892	1 337 312	1973	27 806	34 050
1974	51 374	138 337	1974	40 032	41 873

highly correlated responses, and so act in effect as a single unit. Table 10.1 shows hypothetical time series data for four large contributors to the total production. Plot total recruitments over time; do you see obvious correlations? Then develop a computer program for estimating shared environmental effects as deviations from Ricker stock–recruitment relationships [equations (10.6)–(10.11)], and apply it to the four data sets. Plot stock–recruitment relationships for the four stocks, corrected for the estimates of shared effects. Are your variance estimates for the Ricker b parameter (key determinant of optimum spawning stocks) improved by doing the joint estimation?

10.5. Table 10.2 shows summer population size and fall harvest data for four reindeer herds in northern Finland. Fit each data set to the logistic model $N_{t+1} = R_1 S_t - R_2 S_t^2$, where R_1, R_2 are production parameters, N_{t+1} = summer herd size in year $t + 1$, and S_t = herd size after harvest in the fall of year t. Is the density-dependence parameter R_2 significantly different from zero for any case? From other herds and independent data on reproductive rates, we know that the maximum rate of increase should be around $R_1 = 1.5$. Assuming this value, calculate time series of R_2 values $[R_{2t} = (1.5 S_t - N_{t+1})/S_t^2]$ for the four herds; these time series presumably reflect

Table 10.2. Summer population size (N_t) and fall harvests (H_t) for four reindeer herds in northern Finland. Comparable data are available for 52 other herds. The Paistunturi and Ivalo herds depend mainly on ground lichens for food in late winter. The Jokijarvi and Oivanki herds have access to arboreal lichens and have been given supplemental food (hay) in late winter since the mid-1970s. (Data provided by T. Helle, University of Oulu, Finland.)

Year	Paistunturi		Jokijarvi		Oivanki		Ivalo	
t	N_t	H_t	N_t	H_t	N_t	H_t	N_t	H_t
1959	8750	1716	695	140	1285	231	3268	1161
1960	1211	2689	628	202	1244	276	4341	1632
1961	8831	1988	663	101	1167	123	3208	691
1962	7900	2196	738	211	1537	274	2090	534
1963	10 436	2724	755	197	1623	310	2957	877
1964	10 472	2715	554	142	1575	198	3138	989
1965	7619	2978	662	141	1856	261	1535	589
1966	8215	2789	820	157	1947	248	1450	595
1967	7696	2466	938	235	2380	333	1990	798
1968	7129	2181	947	274	2531	638	2981	1096
1969	7140	1032	804	158	1925	209	1679	518
1970	9565	2257	870	186	2415	232	1918	694
1971	8515	2564	965	186	2439	370	2171	793
1972	8083	2553	1175	262	2711	420	1949	624
1973	4830	2553	986	288	2994	417	1814	534
1974	2895	1301	1000	255	3187	654	1526	213
1975	5280	340	1226	323	3446	544	1831	245
1976	7967	1239	1303	356	2827	389	2379	330
1977	8761	2128	1582	409	2833	689	2583	285
1978	9020	2349	1547	583	3041	750	3344	447
1979	9743	1635	1723	642	3212	730	4075	513
1980	11 283	2653	1695	676	3231	583	4910	901
1981	12 526	3199	1529	386	3598	997	4126	746
1982	14 858	5091	1445	471	2910	720	5416	1033

varying winter "carrying capacities" for the herds. Do the herds appear to be subject to similar winter conditions? Can you develop a formal regression method, like equations (10.6)–(10.11), for estimating shared variations in carrying capacities along with the average R_1 − R_2 response curves for each herd?

10.6. The reindeer herds in Table 10.2 show no convincing evidence of decreasing productivity (reproductive rates, winter survival rates) as herd sizes increase. The current Finnish government policy is to set upper limits on the herds, at near the current herd sizes; there is concern that larger herds would damage the winter ranges (the animals feed on ground lichens that are very unproductive) so as to reduce carrying capacities in the long term. There is also concern about the long term sustainability and hidden costs of supplemental feeding as practiced with many herds like Jokijarvi and Oivanki. The lichen ranges now appear to be at equilibrium, at very low biomasses (50–300 kg/ha compared to 3000 kg/ha on ungrazed ranges), with annual net production rates of around 10% (300 kg produces 30 kg net growth) being taken by reindeer in winter. Develop alternative models for possible changes in range condition and animal productivity if herd sizes are allowed to increase. Assuming that total harvest across the four herds of Table 10.2 is to be kept relatively constant, simulate changes in herd size and range condition for a variety of experimental policies that try to allow one or more herds to increase until biological limits become evident. Note that you will need to consider experiments of very long duration (50–100 years); looking more broadly at Finland as a whole (56 herds), can you suggest plans that might be less risky and bring results more quickly?

Chapter 11

Adaptive Policy Design for
Complex Problems

> *Decision theory is well adapted to coping with*
> *such probability distributions. Unfortunately,*
> *people are not.*

R. Yorque (Holling, 1978)

So far in this book we have concentrated on relatively well defined problems of harvest management, where it has been possible to specify clear alternative models for the managed system and the process of learning about it through adaptive estimation. While, in principle, the various recipes of Chapters 6-10 can be applied to any system, they may fail in practice when (1) the system is of high dimension (many variables); (2) consensus cannot be reached about a small set of alternative hypotheses that capture key uncertainties; and (3) no single objective function will represent the conflicting interests of the various actors involved in management. In time we can expect to see considerable progress in methods for dealing with the technical difficulties of modeling and formal optimization for systems that involve a whole panorama of biophysical and economic variables. But such technical developments will be of little value unless they are accompanied by progress in dealing also with the very human problems of reaching consensus by embracing uncertainty, and of reaching some balance when there is, in fact, no identifiable decision maker and policies proceed from the competitive or cooperative activities of many actors.

In this final chapter I will review some steps that we have found helpful in working with the human problems, then examine two case studies where the problems are vividly clear. These cases (acid rain in Europe, fishery management in the Great Lakes) involve a range of environmental and economic issues that extend far beyond renewable resource management, and each involves many policy design problems that, in practice, cannot be lumped together in any single coordinated framework. However, in each case it is possible to suggest a basic adaptive strategy that somehow cuts across various subsystems to influence management performance in each of them. It is emphasized that the discovery of such strategies is a matter of luck and imaginative synthesis, not of mechanical systems analysis. In closing, I will discuss some of the tactical implications of moving away from traditional methods of problem analysis towards methods more favorable for adaptive policy design.

Moving from Analysis to Synthesis

Looking back over earlier chapters, you will see that I have recommended three essential steps in adaptive policy design: modeling to pinpoint uncertainties, compression for understanding, and optimization to seek best policy options in recognition that uncertainties will change over time. Below, each of these steps is reviewed with a view to its application in complex problems, along with a fourth step that I consider essential: imaginative synthesis, the search for innovative options and strategies. I place this step last for a special reason: the search requires motivation, even desperation, and this is provided by the frustrations that inevitably accompany the first three steps.

Modeling to pinpoint uncertainties

I have emphasized throughout this text that an important first step in policy design is to obtain an honest picture of uncertainties by trying to develop formal models that are consistent with historical experience. For complex systems, the development of such models must be undertaken with the cooperation and involvement of the actors who are expected to understand the results; here the AEA workshop approach discussed in Chapter 3 becomes a valuable tool. It is critical that this model-building step involve the deliberate attempt to construct predictive models, even when it is known in advance that the predictions would have little credibility.

The basic reason for emphasizing modeling as a first step is simple enough: other processes for defining uncertainties will lack a necessary focus on and definition of policy options, and so will be used as a forum by various scientists to promote their own research interests. It is entirely too easy to concoct very convincing verbal arguments about why almost any detailed process research is important to the "understanding" of big problems like acid rain. But many details have a way of paling into insignificance (properly) when included in models that deliberately try to make calculations about the whole problem.

Serious tactical and political difficulties usually arise at this step, because the modeling almost inevitably alienates (appears threatening or superficial to) various members of the scientific community whose advice has traditionally been sought by policymakers. Thus, it is essential that the modeling involve this community, challenge it to see the problem more broadly, and dispassionately embrace and evaluate various alternative hypotheses that have emerged from within the community. To be dispassionate implies not favoring any single hypothesis or previous model, and not openly condemning any of them. Formal methods for assigning odds to various alternatives, as discussed in Chapter 6, may be helpful at this stage, but it is essential to use the statistical tools only to sort out hypotheses that are clearly inconsistent with historical experience. The set left over after this sorting usually leaves considerable room for controversy about the future.

Compression for understanding

The next step is to develop systematically a range of predictions about key policy indicators, using the alternative models and basic policy options identified during the initial modeling work. Again, it is essential at this stage not to seek any best scenario or most likely outcome. Instead, the key goals should be (1) to gain consensus about how large is the range of future outcomes and how deep are the conflicts about which outcomes would be best, and (2) to engender a healthy frustration about the state of affairs. This frustration will help later in the search for imaginative policy options, but at this stage it has the more immediate value of motivating those involved in the analysis to "get down to essentials." That is, it motivates the search for a compressed representation in terms of a few extreme alternative hypotheses, management options, and scenarios of future development.

The quantitative details of alternative predictions are almost always unimportant in large problems. They pale into insignificance in comparison with the qualitative differences that are noticed and highlighted by actors with conflicting objectives. This point is really what makes compression for understanding possible in the first place. To understand it better, I recommend conducting a simple modeling exercise. Construct a little simulation model for some resource, just complicated enough to display dynamic transients of such indicators as catch and industry employment in relation to a key policy variable, such as the harvesting effort allowed. Then run several simulations (scenarios) with various temporal patterns of change in the policy variable, and graphically compare the predicted indicator patterns. Having done this many times with all sorts of actors, let me make a strong prediction: you will never even bother to read the scales on the vertical axes of these graphs. That is, your comparison of the options will not (and generally should not) depend strongly on the quantitative scales (arbitrary units of measurement) that you have chosen for the indicator variables. You will look for ordering in, and basic differences between, the options (effort patterns) treated.

Systems analysts are fond of various formal techniques for "sensitivity analysis" of model predictions to changes in parameters and actions. Calculation of sensitivity to parameters is certainly an integral part of parameter estimation (see Appendix 5A, and Chapters 6 and 7) and approximation of odds to place on alternative hypotheses. Also, it can help to sort out various unnecessary redundancies, such as strings of parameters that multiply together and therefore each have the same effect. But for complex systems, particularly those that can exhibit multiple equilibria and other sudden qualitative changes in behavior as parameters are varied, formal sensitivity testing can be deceptive by lulling the analyst into thinking that a thorough testing has been done when, in fact, the possible behaviors have barely been touched. It is much more important to develop and "play" with various model structures, both simple and complex, so as to gain experience in where to look for key interactions and relationships that affect qualitative prediction. Good compressed models are usually discovered by noticing first that the original complicated model "behaves like" a simpler one with which the analyst has had previous experience.

Seeking the best option

The final section of Chapter 9 highlighted the need to proceed carefully with adaptive optimization, moving back and forth between the results based on formal objective functions and the reactions of actors whose objectives these functions are supposed to represent. The usual result of the steps outlined in that section is not consensus about the best action, but rather a clarification of which objectives are really conflicting in terms of policy choice. Often, apparently conflicting objectives in fact imply the same best policy choice and so lead to coalitions of interest that would not be intuitively obvious. But basic conflicts usually still remain, between short-term versus long-term values and between temporal stability versus informative variability.

Some analysts find it disturbing that modeling exercises intended to bring actors together often result initially in a deepening of conflicts, by highlighting trade-offs that cannot be avoided. It is somehow expected that cooperation in clarifying what the trade-offs are should be accompanied by a commitment to accept some formal calculation of the best compromise policy. There has been much interest in "multiobjective decision analysis" (see

Keeney and Raiffa, 1976), which emphasizes precisely such formal methods. But by seeming to provide a reasonable compromise among options that are all bad in the first place, such formal methods are a lot like sensitivity analysis mentioned above: they may lull the actors into accepting a solution too early. Again, let us recall that there is value in allowing tension and conflict to build, as motivation for seeking innovative policy options.

Consider for a moment that the model building/optimization process is itself a problem in "adaptive management" of the *people* involved. Here the "system" is the analytical machinery for working out optimum feedback policies given a specified input set (hypotheses, objective function, action choices). "Management" consists of varying the inputs, then observing reactions to the formal outputs. It is unnecessary to quantify such output observations as "level of tension among actors;" the key point is that by being even a bit sensitive to them, it is usually possible to provide "rewards" (predictions, policy options) that are sufficiently interesting to keep the actors involved in the process. In my experience there is eventually a "breakthrough" that involves not compromise, but rather some quite different way of seeing the problem (i.e., a whole new input set). The challenge is to keep the process going long enough for this change to take place. My only advice about how to become such an adaptive manager is to try it, while being prepared to make *and learn from* a lot of very embarrassing mistakes. As in any complex management activity, there is in the end no substitute for hard experience; the real trick is to keep firmly in mind at the most difficult moments that mistakes are healthy, and if quickly admitted are unlikely to cause the whole process to break down.

Imaginative synthesis

Perhaps the greatest mistake that an analyst can make is to suppose that better management options will emerge from dispassionate and relaxed discussion about the managed system. There is an excellent warning about this mistake in the old adage "necessity is the mother of invention," and plenty of empirical evidence about how major innovations in technologies and social systems have come mainly in times of crisis (see Chapter 2). The AEA workshop process uses the pressure of short meetings to stimulate thinking about model building, and I have watched people (including myself) accomplish more in a five-day meeting than they normally would in a year of puttering about the office.

There is a strong tendency in resource management to defer hard decisions as long as possible, in the hope that natural events will produce a favorable outcome. In a way we might call this behavior "adaptive policy by default," since it often has the effect of letting major and informative system changes take place before there is any (feedback) reaction. The wait-and-see attitude is a very difficult one to overcome, and I expect that it can only be dealt with in general by trying deliberately to build up among the management actors some of the same emotions (frustration, desperation) that would eventually move them to act. The idea here should not be to try to fool anyone into thinking that there really is a crisis when, in fact, the system is doing reasonably well. Rather, it is simply a matter of recognizing that emotional involvement is a strong prerequisite for creative thinking, a fact that is obvious to artists, but that many scientists (the bad ones) fear to admit.

So far we have discussed only one example of imaginative synthesis, the Hilborn plan (Chapter 2) that provides a way to avoid a very difficult choice in Pacific salmon management (rehabilitation versus enhancement). Two other examples will be discussed in the following case studies, and both also involve finding a way to sidestep some very difficult choices. All three of these examples first appeared as sudden "leaps of imagination" or "intuitive jumps" by participants in AEA workshops, at moments of high frustration in the midst of discussions about uncertainties and policy options. On several occasions we have tried without success to foster such leaps during more relaxed and freewheeling "brainstorming sessions" where participants are urged to think up wild ideas while agreeing not to be critical of one another. Though my sample sizes are obviously very small, I have become convinced that there is no substitute for hammering away systematically at a complex problem until it is clearly and simply expressed in terms of basic strategic options; only when all these options look bad is the magic of imagination likely to be displayed.

Experience is important in the search for imaginative policies. Rather than plucking something entirely fresh out of the air, we usually gain new ideas by seeing analogies or similarities between the problem of current concern and earlier ones where advances have been possible (model building works this way as well). I hope that the Hilborn plan, and the policies suggested for the two cases below, will be useful as analogues for many other situations in renewable resource management.

Acid Rain in Europe

Many people feel that acid rain is the largest environmental problem of this century, in terms of the magnitude of potential ecological impacts and the economic/political difficulties of policy formulation on such large spatial scales. For readers who have been residing on another planet or never look at newspapers, let me briefly review what is understood about the problem. Anthropogenic releases of sulfur and nitrogen into the atmosphere have increased greatly in this century, largely due to increased use of fossil fuels and especially high-sulfur coal for electrical power generation. Through various chemical transformations in the atmosphere, acidic SO_x and NO_x compounds are produced and precipitated as dust and rain. In Europe, as in North America, atmospheric transport from highly industrialized areas (the UK, France, Italy, West and East Germany) carries at least some of the material great distances, and in particular into parts of Scandinavia, where soils and surface waters have very low buffering capacities. The pH has become too low in many lakes to support fish life, and the effects are exacerbated by the seasonal release of very acidic waters during spring snow melts. Effects on terrestrial ecosystems have been less pronounced except in a few areas where trees have been killed by massive concentrations of the pollutants, although there is some evidence of effects on forest growth and there is concern about the mobilization of highly toxic metals (especially aluminum) where soil pH is reduced.

There is little doubt that acid rain is now causing some important ecological damage, and that this will increase over time due to the accumulation of acidic compounds in waters, the erosion of soil buffering capacities, and the continued growth in emissions, if current energy development trends continue without regulation. There has been a great deal of research aimed at trying to understand the chemical and ecological mechanisms involved in the ecosystem response, in the hope that it will be possible to predict long-term changes more precisely. However, direct experiments so far with ecosystems suggest that there are some rather subtle mechanisms at play (such as "buffering" by purple sulfur bacteria in lakes), implying that there is little hope of building accurate predictions (models) for terrestrial systems until more direct experience (with increasing inputs) is obtained. Thus, acid rain is a classic example of the issue raised in Chapter 1: we can build as many quite credible impact models as you would like, but in the end only experience or deliberate large-scale experimentation will reveal the correct alternative.

The emission–transport–impact system has three basic control points: (1) emission can be reduced by changing fuel mixes (coal types, nuclear power, etc.) and/or installing emission control devices, such as scrubbers, that remove compounds from emission gases; (2) long-range transport can be reduced by going back to lower stacks on power plants (admitting that dilution is not always the solution), thereby accepting much greater deposition near the large emission sources; and (3) impacts can be ameliorated by direct application of buffering compounds (lime, etc.) to the affected ecosystems. All these suggestions have a variety of direct and hidden (not obvious) costs, that are potentially staggering on a scale such as that of Europe. One OECD study proposed spending US\$3 billion on scrubber retrofits, which it was estimated would reduce sulfur emissions by about 50%. Economic analysis of possible control strategies is greatly complicated by the fact that many nations are involved, and none of them wants to be placed at a competitive disadvantage in exports/industrial production by bearing differentially high control costs. However, these costs are directly paid, they will inevitably touch whole economies through taxes and/or energy prices.

Critical uncertainties

We have already noted two major uncertainties about the acid rain problem: ecological impacts on terrestrial systems have not yet become clearly evident, and the total economic impact of various control strategies may extend far beyond what any direct analysis would indicate. A third major uncertainty concerns the importance of long-range transport. When acidic deposition is measured at some sampling station, such as in central Norway, it is as yet impossible to say what proportion of the compounds originated from different source locations. Within the Scandinavian countries that have been most impacted so far, there are local emission sources that could account for at least half of the current estimated deposition rates in these countries. This is a critical policy issue, since the impacted regions are hardly justified in calling for control measures by other countries if, in fact, they are creating the problem for themselves.

In an effort to estimate who is to blame for the existing deposition patterns, detailed atmospheric transport models have been used to simulate annual deposition patterns from various point source locations. The basic outputs of these models for policy purposes are emission-deposition tables, showing how much of the simulated deposition in each nation is due to emissions from each other nation. Unfortunately, alternative simulations using reasonable parameter values do not agree in their predictions about the key issue of transport from continental Europe into Scandinavia. The basic difficulty is that Scandinavia sits on the tails of the emission plumes from southern areas like the UK, and receives total inputs that are only a small part of the southern emissions. The amount of material reaching the tails of the simulated plumes is very sensitive to various model parameters, such as wind speeds (reduce wind speeds a little and more will be deposited near emission sources, leaving less available for deposition further away ...). In the end, it is just not possible with current modeling technology to resolve the uncertainty about what fraction of Scandinavian deposition comes from local as opposed to distant sources.

Hard options

In early 1983, the United Nations Economic Commission for Europe (ECE) ratified the Geneva Convention on Transboundary Air Pollution. Under this convention, the nations of the ECE in effect recognized the *potential* importance of acid rain, and agreed that they "shall endeavor to limit and, as far as possible, gradually reduce and prevent air pollution including long-range transboundary air pollution." No purse strings were attached to this agreement, and there are key clauses about the need for more research and monitoring to determine the efficacy of whatever control measures are taken. It was agreed that control measures should be "compatible with balanced development," meaning that all parties should share the economic burden of control costs and should *move together* in reducing emissions and/or impacts.

So a strategic option has in effect been chosen, though it is not at all clear how to proceed tactically (scrubbers versus liming, etc.). Also, the massive investments required may well prove to be too much for many ECE nations to stomach. No matter how far the actual expenditures finally go, a glaring feature of this strategy is its *presumption of certainty* about the sharing of blame for deposition patterns. If all nations move together to reduce emissions, it will never be determined whether the long-range transport estimates were correct (if all national emissions fall by roughly the same percentage, so will all national depositions, no matter what the spatial emission/deposition pattern actually is).

Most of the debate has been about the tactics and economic feasibility of emissions reduction, and the common default option of "let's wait and see" has been pushed aside through a great deal of diligent publicity, mostly by scientists, about the imminence of the crisis. There is just enough visible damage to make a credible case that more is soon to follow. At the heart of arguments against waiting is the very long recovery time for terrestrial ecosystems should damage over large areas suddenly start to appear. This is a very compelling argument: a large fraction of the public takes pride and solace in the forests of central and northern Europe; the possibility that some of these might be lost and not regenerate within their lifetimes is a risk that Europeans are simply not willing to take.

An actively adaptive option

It is clear that something must be done soon, at least to reduce the risks associated with current deposition rates in the Scandinavian countries. It is equally clear that whatever actions are taken must be shared in cost by the member nations of the ECE. However, I think it is a terrible mistake to adopt without serious question the "engineering mentality" of a strategy based on widespread and simultaneous application of technologies for emission reduction. There is still that nasty little uncertainty about where the depositions are coming from in the first place. Luckily, it is simply unnecessary to proceed as though the long-range transport models were correct.

A rather radical policy suggestion emerged from discussions during an AEA workshop on acid rain held at the International Institute for Applied

Systems Analysis (IIASA), Laxenburg, Vienna, in early 1983. No consensus was reached about this suggestion, and it is not supported by or seen as a topic for further research by IIASA. The suggestion was simply for the ECE nations to share the costs of subsidizing all fossil fuel power plants in Scandinavia to use only low-sulfur fuels for a few years. This would mainly mean paying the price and transportation cost differentials for such fuels, which are not readily available in Europe, along with the costs of improving deposition monitoring networks to ensure that any effects of the local emission reduction would be measured accurately.

To invest in this very large-scale "experiment" would certainly be a gamble for the ECE. If sharp reductions in deposition rates were detected, this would be evidence against the importance of long-range transport. Other European nations might then opt not to invest so much in emission control, letting the Scandinavian nations find their own solution to what had been demonstrated to be their own problem. So, on this side of the gamble, the other nations would gain and Scandinavia would lose. If only small changes in deposition rates were detected, then two possibilities would have to be explored. The first is that monitoring systems were just not good enough to detect the changes. The second is that long-range transport is, indeed, important. Either possibility would favor a return to the original strategy of shared emission reduction, with all nations the losers. Notice that the gamble is asymmetrical: Scandinavia would lose either way (unless the ECE agreed to continue fuel subsidies indefinitely, which is quite unlikely). So you should not be surprised that Scandinavians who heard about the suggestion were not at all pleased.

On closer inspection, the original suggestion turns out not to be very good. Based on the most extreme estimates from transport models and emission source distribution data, it is very unlikely that the reduction in deposition rates would exceed 25%. Monitoring systems capable of detecting this change against background natural variation would currently be very costly to install and operate. To avoid confounding of effects, other nations would have to forgo or delay emission reduction programs aimed at reducing local as well as distant impacts (this is especially a concern for West Germany).

But the suggestion triggered a whole collection of other ideas based on the concept of large-scale, shared-cost experimentation, aimed at defining actual emission/deposition patterns. One suggestion was to reverse the fuel subsidies, to temporarily reduce sulfur emissions from the Germanies and the UK rather than Scandinavia. This would be much more costly, but would be immediately beneficial in terms of environmental impacts outside Scandinavia. Another suggestion was to design a continent-wide plan for sequentially installing permanent emission control devices on the largest point sources (mainly power plants), with the sequence chosen so as to be as informative as possible about the long-range transport pattern.

Perhaps the most interesting thing about these options is that they are viewed as radical in the first place. It is quite natural for most people to think about other large investment programs in terms of a careful sequence of tests using devices such as market surveys and pilot studies. Somehow it is viewed as unscientific or threatening to talk about experimentation on large spatial scales, as though experiments were things to be done only in boxes or on benches in university laboratories. Worse, some scientists involved in our discussions were worried about the very notion of publicly admitting uncertainty, and felt that it was important to maintain at least the appearance of consensus within the scientific community. The arrogance of this attitude is shocking; in effect the scientists are saying that policymakers cannot be trusted to deal directly with uncertainty (or perhaps more to the point, that the policymakers might not continue to support "valuable" research if it does not appear to be giving consistent answers).

Lake Trout Rehabilitations

The Laurentian Great Lakes of North America have had a fascinating management history. For reviews see the symposium proceedings of SCOL (1971), PERCIS (1976), and SLIS (1980). Fishery development began in the 1800s, with the exploitation of large species like lake sturgeon and lake trout. As the larger species and individuals were depleted, commercial harvesting turned to progressively smaller forms and more inaccessible stocks (deep water, far offshore, etc.). By about the 1930s, it appears that a rough bionomic equilibrium had been reached. But then there came a series of rapid and disastrous changes, ranging from the invasion of the parasitic sea lamprey to the introduction of much more efficient monofilament gillnets. Many species, including the prized lake trout, virtually disappeared from the upper lakes (Superior, Huron, Michigan), and a number of complete extinctions occurred. In Lake Erie, pollution effects became obvious with the depletion of oxygen from large areas in the lake's central basin. Dramatic changes continued into the 1960s and 1970s, with such events as the explosive growth of alewife populations, the sudden reduction in lamprey abundance through control efforts with the chemical pesticide TFM, and the development of an extremely valuable sport fishery for introduced salmonids and hatchery-produced lake trout.

In the following discussion, let us concentrate just on the rehabilitation of lake trout, which is seen as a key indicator of success in efforts to restore the "health" of the Great Lakes ecosystems. Further, let us restrict the analysis to three key policy variables: harvesting, stocking, and lamprey control. It must be kept in mind that responses of lake trout to these variables will take place in a complex and changing environment. Effective fishing efforts and mortality rates will be influenced by changes in the abundance of other species, such as the introduced Pacific salmon. Stocking policies are evolving with the introduction of new hatchery techniques, the diversification of hatchery gene pools, and the discovery of more effective tactics for putting the stocked fish where they are more likely to survive. The abundance and attack patterns of lamprey will continue to change due to pollution control in their spawning rivers and changes in the availability of alternative prey. There will be changes in the community of "forage" fishes (alewife, small coregonids, sculpins) upon which the lake trout feed. There may be substantial improvements in some aspects of water quality (reduced eutrophication), but deterioration in others (accumulation of heavy metals and toxic organic residues). Exotic species, such as the pink salmon, will continue to effect unpredictable changes in the community trophic structure. In short, it is not technically feasible (and there will probably never be enough data) to build any comprehensive and credible model that captures all of the dynamic variables and assorted policy variables associated with these changes. Through a series of AEA workshops that tried to look at some of the possible changes, we sorted out the three key policy variables mentioned above as dominant factors in all of the models that were produced.

Considering how many management agencies and extragovernmental interest groups are involved, it is a wonder that some reasonably well coordinated management policies have been designed. There is a coordinating body, the Great Lakes Fishery Commission, which promotes cooperation through various committees and meetings, and which exerts direct authority over the lamprey control program. The federal governments of Canada and the US have taken responsibility for the implementation of lamprey control and for most of the lake trout stocking program. State and provincial agencies are responsible for regulation of harvesting through tactics such as bag limits and closed seasons. At every level, the government agencies are responsive to pressures from special interest groups, such as sport fishing and tackle manufacturing associations.

Critical uncertainties

There is little doubt that a harvestable population of lake trout can be maintained by continued stocking and lamprey control, at least for the next decade or two. The biggest uncertainties are about what will happen in the long term. All stocked lake trout are marked with fin clips, so the development of natural reproduction can be monitored through juvenile and catch sampling. So far, the development of an unmarked stock has proceeded roughly as expected only in Lake Superior, where there were residual wild stocks after the initial collapse. In Lakes Michigan and Huron, where wild stocks apparently became extinct, the stocked fish have apparently not reproduced successfully. Thus, a first major uncertainty is about whether wild stocks can be reestablished at all, even if fishing mortality rates can be effectively regulated.

A second major uncertainty concerns the future of lamprey impacts. Chemical pesticides usually do not work in the long term (genetic resistance develops), and in any case the attack/mortality rates depend on the relative abundance of prey and predator (see Chapter 5). Now that prey abundances are higher, it may be possible to reduce control efforts considerably (and delay the development of resistance) without changing such visible indicators as the number of fish marked by lamprey wounds. The fact that fish do survive attacks (and show up much later in catches with old scars) has led to controversy about whether the lamprey has ever been a major mortality agent except in laboratory experiments. The stock collapses were associated with high fishing rates as well as high wounding and scarring rates, and it is possible that the lamprey caused high mortality rates only after fishing had reduced stocks so far that the number of attacks per fish remaining became intolerable.

The third major uncertainty concerns regulation of fishing mortality rates. As noted in earlier discussions on Pacific salmon management (Hilborn plan, Chapter 2), sport fishing efforts are notoriously difficult to regulate when the abundance of fish is attractively high. Native Indian fisheries may cause local depletions, as old treaty rights are reaffirmed through court tests. There is pressure to allow the reestablishment of commercial net fishing.

Finally, there is great uncertainty about future changes in the forage fish base for trout production. There are signs that native species, which are more difficult for the trout to capture, are becoming reestablished. Alewife and smelt (both exotics) populations have declined in some areas. There may be continued growth in stocking rates and natural reproduction of introduced Pacific salmon, which use the same forage species.

Difficult policy options

In an effort to promote understanding and consensus among the various actors, there has been a strong move to define simple targets for management action, such as the establishment of self-reproducing wild stocks of lake trout. There is general consensus about only one policy requirement, namely that lamprey abundance must be kept as low as possible through continued chemical treatment and development of less environmentally dangerous control measures (weirs to block lamprey spawning, biological control techniques, etc.). A basic conflict has arisen about stocking and harvesting regulations, and it is essentially one of short-term versus long-term objectives.

According to the long-term view, fishing rates and stocking policies should be adjusted to restore a naturally reproducing system as quickly as possible. The basic argument is that nature can do the best (and cheapest) job at producing fish in the long term, and that anyway there is direct value in having a "balanced" natural ecosystem with a diverse (and statistically stable) collection of natural stocks. It is not yet clear what stocking policies are needed (since there are risks like the possibility that stocked fish may act like sterile male releases with respect to self-producing stocks that have already been established), but it does seem obvious that fishing rates should be kept very low. Let us call this policy of much restrained fishing the "rehabilitation option."

According to the short-term view, the most important thing is that a valuable fishery has been reestablished and the sportsmen engaged in this fishery form a powerful political lobby. Thousands of them have invested in expensive boats and gear, resort development is booming, and it is rapidly being forgotten why the lake trout stocking programs were initiated in the first place. Hatchery managers and lamprey control agents use the sport catch and economic values as measures of their own performance, and join many sportsmen in opposition to restrictive fishing regulations. It is argued that while it would be better to have naturally reproducing fish stocks, there is just too much now invested in the fishery to make severe cutbacks politically feasible. Let us call the policy of letting fishery development continue the "put-and-take" option.

A matter of some mystery to me is that some biologists do not seem to understand that there is a fundamental conflict between the rehabilitation and put-and-take options. I have seen several reports emphasizing the need to increase stocking rates so as to speed wild stock rehabilitations, without any mention at all of harvest regulation or the likelihood that increased trout abundance due to more stocking would in turn attract higher fishing efforts. This is not just lack of a systems view; it is plain stupidity. A variation on the theme is an option we might call "balanced development" (it is not!), which proposes that catches and efforts be allowed to grow at a moderate pace set by incoming data about how fast the stocks are recovering. The flaw in this argument is that fishing rates in some areas may already be high enough to greatly delay or preclude wild stock recovery, so it is essentially just a tactical variation of the put-and-take option.

An adaptive option: Surfing in the Great Lakes

The above argument about two extreme options should be starting to sound quite familiar. It is much like the situation we faced with British Columbia salmon before the Hilborn plan was discovered. Indeed, the Hilborn plan might well be a good way to proceed. But let me now discuss another option, which confronts the existence in the Great Lakes situation of a "cliff edge" associated with depensatory lamprey mortality. Chapter 5 presented a quite simple model for how equilibrium stock size in lake trout is likely to vary as a function of harvest rate, stocking, and lamprey abundance (Figure 5.3). The essential prediction of this model is that the stock may have two equilibria, with movement from the upper to the lower involving a catastrophic collapse (falling off a cliff edge) as fishing rates increase. This prediction is certainly in line with the history of Great Lakes stocks. The upper equilibrium consists of a mixture of stocked and wild fish, with the large stock size resulting in relatively low mortality rates due to the lamprey. The lower equilibrium consists mainly of stocked fish, with high enough mortality rates due to lamprey predation to prevent wild stock recovery. At very low fishing rates the lower equilibrium may vanish entirely (if stocking rates are high enough and lamprey abundance low enough), permitting movement toward the high (mixed) equilibrium. At high fishing rates, the high equilibrium vanishes entirely (wild stocks are not sustainable), even if lamprey abundance is negligible. It was noted in Chapters 5 and 7 that the location of the cliff edge is likely to vary over time, so that there may exist no completely safe harvest rate low enough to guarantee maintaining the mixed (or pure wild) stock equilibrium if ever it is achieved.

For the sake of argument, let us take it as given that the rehabilitation option (or a Hilborn plan) has won out, and that harvest rates have been low enough to allow a mixed (or pure wild) stock to develop and reach temporary equilibrium. This situation will likely be reached by the mid-1980s in Lake Superior, although reinvasions of old spawning areas may keep the equilibrium moving slowly upward over time. It is still unclear about what will happen in the other lakes.

Now we face an interesting question: given reasonable recovery, how should the harvest rates then be allowed to vary, especially considering that there may be no completely safe rate and there will continue to be strong demand to allow growth in sport catches? One option would be to try and hold the harvest rate low and steady, then monitor abundance and lamprey wounding rates closely so as to respond quickly if the cliff edge moves toward lower harvest rates. A basic difficulty with this option is that if it succeeds for a long time (10–20 years), expectations and commitment will develop so as to make effective response more and more difficult. When the collapse does come, it may go a long way before there is the political will to take action (as has happened repeatedly in the history of Great Lakes management).

An alternative option I call "surfing on the cliff edge," and its development was motivated by two observations. First is Mehra's (1981) discovery (see Chapter 7) that the optimum policy for estimating parameters in fold catastrophe models (the model in Figure 5.3 is like a fold catastrophe) involves staying in the region of instability (around the cliff edge) as much of the time as possible. Second is the fact that equilibrium yield would be maximized by harvest rates near the cliff edge, if the cliff edge were stable in position and if harvest rates could be regulated exactly.

Equilibrium stock size and yield are shown as functions of harvest rate in Figure 11.1, for a combination of parameter estimates and fixed stocking rate/lamprey abundance levels that are "reasonable" for Lake Superior. Also shown are regulated trajectories of abundance and harvest rate for three policy options. In option A (associated with stock size N_A, yield Y_A, and harvest rate u_A), the harvest rate and yield are held "safely" low and constant. In options B and C there is a cycle (surfing) around the cliff edge. In these options, harvesting effort (u) is allowed to grow (catch increases, stock decreases) until a catastrophic decline begins. Then, in option B, the decline is allowed to continue until most of the natural stock is lost, then fishing efforts are reduced and held down until the stock increases. Notice that the average yield under this policy would be lower than under policy A, since the recovery part of the cycle would probably take many years. Under option C, response to the decline occurs much earlier and less effort reduction is necessary to bring about a stock recovery (since the stock declines less, mortality rates due to lamprey are also less). If option C were practical, it would result in higher average yields than option A and would allow regular monitoring of the location of the cliff edge.

Notice that surfing policies B and C involve two key variables: how far the collapse goes before harvest rate restriction is initiated, and how rapidly the harvest is reduced once the decision is made to do so. For simplicity, I am assuming here that the unregulated (bionomic) equilibrium would be at a quite low stock size, so that effort would not begin to decrease naturally before the collapse is well under way. The first of these variables we might call a "detection threshold;" if it were too small (small stock decrease), then random fluctuations in apparent abundance would be confused with collapses, and harvest rates would be restricted unnecessarily often. If it were too large, the stock and yield fluctuations would become intolerably large. Thus, the optimum detection threshold is a function of the quality of monitoring statistics available (i.e., investment in monitoring) and the expected amount of natural random variation in stock sizes. The second of the variables we might call the "power to respond," since it is determined by how drastic an effort reduction (how severe the regulations) the management authorities dare to impose without inviting political decisions to override their recommendations and allow high harvests to continue. Such overriding decisions happen with discouraging frequency in the management of North American sport fisheries.

Given a "kernel of understanding" about responses to varying fishing efforts, a next step would be to examine how stocking and lamprey control

Option A: (stable, conservative)

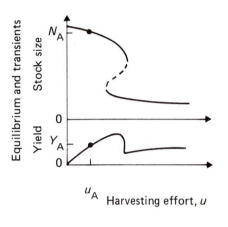

Option B: pathological surfing   Option C: productive surfing

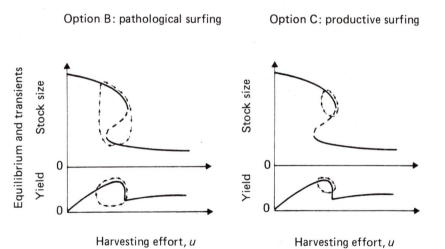

Figure 11.1. Three policy options for regulation of harvesting effort on lake trout in the Great Lakes (see Figure 5.3). In option A, effort is kept low and steady. In option B, effort is allowed to increase until a major collapse occurs, and then there is a long recovery period. In option C, effort also increases until collapse starts, but detection and response to the collapse is much faster. B and C are "surfing" policies.

might also be manipulated to improve the adaptive surfing policy. Time delays in response to these actions become a critical concern: the effects of altered lamprey control will not be seen for at least one year, and effects on stocking do not become apparent for at least 4-6 years. This means that stocking cannot be used to effect a quick reduction in mortality rates during declines, although increases in it can (1) reduce the power to respond (reduce fishing effort) that is required for natural recovery, and (2) hasten the recoveries. Changes in lamprey control are expected to have little effect unless the stock size is driven very low before management responses are effected.

A side benefit of the surfing option might be to prevent the development of unrealistically high expectations about sustainable harvests. This in turn might lead to mandates for investment in better monitoring systems (reduced detection thresholds) and larger powers to respond. So far as I know, no management agency has ever knowingly substituted a policy of regularly allowing small, manageable crises for the typical policy of pretending that none will occur. Thus, I can offer no solid evidence about whether expectations would, in fact, remain reasonable. The users (and many managers) of natural resources often seem to have very short memories, and we are talking here about cycles with periods of 10-30 years. Perhaps it is unreasonable to expect any mandates to persist for so long.

I suspect that in the long term we will have no choice but to develop surfing policies for many natural resources. The pressures to allow growth away from cautious equilibria (like u_A in Figure 11.1) are always great, and will continue to grow if demands keep increasing as they have in this century. Also, we can expect improvement in monitoring systems, which will eliminate one of the biggest excuses for caution. Rather than fighting for a temporal stability that cannot be maintained for long anyway, perhaps we should be concentrating on how to manage the cycles most effectively and informatively.

Tactics for Adaptive Managers

At various points in this book I have stressed that effective resource analysis takes more than good biology or good economics or good mathematical modeling. Management is done by and for people; even the best ideas will be cast aside in favor of easy courses of action like pretending certainty or waiting for problems to take care of themselves. It is just too easy for people to hide behind platitudes like the need for caution, or the importance of detailed understanding before action, or the need to apply methods and models that have stood the test of time (usually without any real test, of course). Adaptive policy design stresses the use of methods and concepts that are often not simple to explain, demand the explicit admission of ignorance, and place a premium on imagination rather than on precision of thinking. Anyone who is convinced that it is important to design and use adaptive policies should be prepared for an uphill battle: he implicitly places high importance on long-term objectives and will have to act as an active advocate of these objectives while trying to be dispassionate about the available scientific evidence. Table 11.1 reviews some of the strategic changes in attitude that an advocate of adaptive management might try to promote. Hardly anyone would argue with items (6)-(8); the real tactical difficulties come in trying to be convincing about items (1)-(5), where conventional attitudes have become deeply rooted through the educational system within which most people with scientific training have spent many years.

Table 11.1. Conventional versus adaptive attitudes about the objectives of formal policy analysis.

Conventional		Adaptive	
(1)	Seek precise predictions	(1a)	Uncover range of possibilities
(2)	Build prediction from detailed understanding	(2a)	Predict from experience with aggregate responses
(3)	Promote scientific consensus	(3a)	Embrace alternatives
(4)	Minimize conflict among actors	(4a)	Highlight difficult trade-offs
(5)	Emphasize short-term objectives	(5a)	Promote long-term objectives
(6)	Presume certainty in seeking best action	(6a)	Evaluate future feedback and learning
(7)	Define best action from a set of obvious alternatives	(7a)	Seek imaginative new options
(8)	Seek productive equilibrium	(8a)	Expect and profit from change

There is, perhaps, only one really essential principle to keep in mind when searching for tactics that will convince people to change conventional attitudes and engage willingly in the process of adaptive policy design. I have heard educators call this *the principle of self-discovery*, and it says that people only change their basic attitudes when they devise the arguments to do so for themselves; people generally do not respond strongly to the arguments made by other people, no matter how tight the reasoning behind them. We can see this principle at work in most matters of everyday life, but, like motivation for creativity, it is something that many scientists are loath to admit. Even in matters that do not touch closely on personal pride or self-justification for past courses of action, most people act as though they simply do not hear very well. Let me mention again an example of how this principle can be used. We had virtually no luck in the mid-1970s in convincing government managers and fishermen to use probing escapement experiments for Pacific salmon management, no matter how carefully we explained the decision theoretic basis for such policies. What finally did the trick was an almost trivial computer simulation game that let people try to manage a randomly varying population over time; most individuals who play this game start intuitively making probing decisions almost immediately, and before long are asking for guidance about how to quantitatively optimize the probing pattern.

Working from the principle of self-discovery, it is simple to construct a list of tactical changes for more effective communication, as shown in Table 11.2. None of these recommendations is new, and you can find items (2)-(4) repeated over and over in textbooks on organizational management, instructions to contributors for practically every major scientific conference, and posted as admonitions to employees on bulletin boards of government offices everywhere. The amazing thing is how seldom they are heeded. Government managers still waste day after day in rambling meetings, scientists try to hide their ignorance behind vast tables of statistics, and professional status is wielded like a club at very intelligent people who happen not to carry the standard credentials.

Table 11.2. Conventional versus adaptive tactics for policy development and presentation.

Conventional		Adaptive	
(1)	Committee meetings and hearings	(1a)	Structured workshops
(2)	Technical reports and papers	(2a)	Slide shows and computer games
(3)	Detailed facts and figures to back arguments	(3a)	Compressed verbal and visual arguments
(4)	Exhaustive presentation of quantitative options	(4a)	Definition of few strategic alternatives
(5)	Dispassionate view	(5a)	Personal enthusiasm
(6)	Pretense of superior knowledge or insight	(6a)	Invitation to and assistance with alternative assessments

Tactics (3a) and (4a) in Table 11.2 represent the greatest intellectual challenge. It is not difficult to provide simplistic representations of renewable resource problems; arguments based on extreme hypotheses (predictions) and policy options are regularly used by government and industry to promote objectives like increased funding. But these simplistic arguments are usually based on weak analogies with frightening events of everyday life, rather than careful consideration aimed at finding interactions and quantitative details that can be ignored without grossly misrepresenting the basic dynamic structure of the problem. The challenge involves more than the technical steps of laying out a problem systematically, then doing things like sensitivity analysis to obtain simpler models; it is also a matter of being willing to step out from behind a barrier of professional details and stand exposed to criticism for possibly leaving out some important factors. Perhaps there is some comfort in the fact that *all* analyses leave out important factors (see Problem Bounding, Chapters 2 and 3), and that the most stinging criticisms are valuable prods to learning.

Cutting through the list of adaptive tactics are the suppositions of enthusiasm to communicate, and willingness to accept fresh viewpoints. Possibly no human being can proceed with much balance in these dimensions. To be enthusiastic, one must develop a certain affection for a set of models and policy options. Then this affection makes it difficult to be open about criticism and new ideas. This brings me to a final tactical suggestion, the implications and implementation of which I will leave you to ponder: seek laughter at every opportunity.

Problems

11.1. Explain why a uniform (all nations equal) reduction in sulfur and nitrogen emissions across Europe (or North America) would not help resolve uncertainties about the importance of long-distance transport of air pollutants. Then consider two basic groups of actors, "southern Europe" and "Scandinavia." Develop a series of alternative experimental plans for emission reduction by one or the other group, and for sharing costs of these reductions. For each plan, construct a table listing the immediate and long-term costs and benefits to each of the groups, for both extreme hypotheses about the importance of long-range transport to Scandinavian deposition rates (insignificant versus dominant). Does this table help you to see why Scandinavia is likely to oppose any experimental plan?

11.2. Compressed models, such as Figure 11.1, are intended to show general response patterns and impacts of strategic policy choices, and are not the best tool for detailed temporal prediction or analysis of specific questions, such as how to regulate fishing effort. Explain why the reverse is also true, i.e., detailed models are not the best tool for strategic analyses. Would examination of a few detailed scenarios likely reveal the patterns predicted in Figure 11.1? What concerns will tend to be the focus of discussion and debate, if management actors work together to develop overall (i.e., strategic) management plans using a detailed model?

11.3. Tables 11.1 and 11.2 define the adaptive analyst as basically an inventor, salesman, and broker of ideas. This view is hardly in keeping with the common view that analysis should be dispassionate and objective. Using examples from your own experience, criticize my assumption that it is possible to be both enthusiastic (inventive, convincing) and objective in the development of models and the recognition of uncertainties about them.

11.4. The business of developing compressed, understandable representations of complex problems involves at least two types of logical operation (or creative activity): *selection* of a subset of variables (such as lake trout population size in the Great Lakes) that are somehow representative of many other variables or factors, and *aggregation* of variables and processes (age classes into total population, harvesting activities and regulations into overall harvest rates, etc.) so that only broad actions and outcomes are considered. Can you think of other logical operations that might be productively applied to big problems? What is implicitly assumed about other variables when a few are selected for further study? What mathematical operations (addition, multiplication, etc.) can be used in the process of aggregation?

11.5. For complex systems, there will be a variety of compressed representations as in Figure 11.1, representing different major variables and policy instruments. One can "connect" these representations intuitively, for example, by moving the stock equilibrium curve in Figure 11.1 to represent changing impacts of water pollutants. Identify some basic guidelines for avoiding deceptive conclusions from such a qualitative analysis. For example, one must be careful about using a dynamic response variable from one model as a control variable in another model, as would happen in the Great Lakes if lake trout abundance were treated as a control variable in a compressed model of population changes in small fish species (like alewife) that are lake trout prey.

11.6. One might argue that for really large and complex systems, all compressed representations may be misleading; further, in such cases the best adaptive policy might be to allow only slow, incremental changes in variables that are subject to direct control, while monitoring carefully so as to react in the event of sudden catastrophes should they arise. Criticize this argument, paying particular attention to (1) hidden assumptions about the "smoothness" (linearity, simplicity) of most system responses, and (2) the assumption that flexibility to respond when necessary can be maintained during periods of relative stability.

References

Allen, K.R. (1973) The influence of random fluctuations in the stock–recruitment relationships on the economic return from salmon fisheries, in B.B. Parrish (Ed) *Fish Stocks and Recruitment. Rapports et Procès-Verbaux Reun. Cons. Int. Explor. Mer* 164:350–9.

Allison, G.T. (1971) *The Essence of Decision: Explaining the Cuban Missile Crisis* (Boston, MA: Little, Brown, & Co.).

Anderson, K.P. and Ursin, E. (1977) A multispecies extension to the Beverton and Holt theory of fishing, with accounts of phosphorus circulation and primary production. *Meddr. Dansk. M. Fiskeri-og-Havunders.* 7:319–435.

Argue, A.W., Hilborn, R., Peterman, R.M., Staley, M.J., and Walters, C.J. (1983) Strait of Georgia chinook and coho fishery. *Can. Bull. Fish. Aquatic Sci.* 211:91.

Bard, Y. (1974) *Nonlinear Parameter Estimation* (New York: Academic Press).

Bar-Shalom, Y. (1981) Stochastic dynamic programming: Caution and probing. *IEEE Trans. Automatic Control* 26(5):1184–95.

Bar-Shalom, Y. and Tse, E. (1976a) Caution, probing, and the value of information in the control of uncertain systems. *Ann. Econ. Social Measurement* 5:323–37.

Bar-Shalom, Y. and Tse, E. (1976b) Concepts and methods in stochastic control, in C.T. Leondes (Ed) *Control and Dynamic Systems*, vol. 12 (New York: Academic Press).

Bazykin, A.D. (1976) *Structural and Dynamic Stability of Model Predator-Prey Systems*. Research Memorandum RM-76-8 (Laxenburg, Austria: International Institute for Applied Systems Analysis).

Bazykin, A.D., Berezovskaya, F.S., Denisov, G.A., and Kuznetzov, Yu.A. (1981) The influence of predator saturation effect and competition among predators on predator-prey system dynamics. *Ecological Modelling* 14:39–57.

Beddington, J.R. and Cooke, J.G. (1982) Harvesting from a prey–predator complex. *Ecological Modelling* 14:155–77.

Beer, S. (1975) *Platform for Change* (London: Wiley).

Behn, R.D. and Vaupel, J.W. (1982) *Quick Analysis for Busy Decision Makers* (New York: Basic Books).

Bellman, R. (1957) *Dynamic Programming* (Princeton, NJ: Princeton University Press).

Bellman, R. (1961) *Adaptive Control Processes* (Princeton, NJ: Princeton University Press).

Beverton, R.J. and Holt, S. (1957) On the dynamics of exploited fish populations. *Fish. Invest., London*, Ser.102, 19.

Bilton, H.T., Alderdice, D.F., and Schnute, J.T. (1982) Influence of time and size at release of juvenile coho salmon (*Oncorhynchus kisutch*) on returns at maturity. *Can. J. Fish. Aquatic Sci.* 39:426–47.

Botsford, L.W. (1979) Population cycles caused by inter-age, density-dependent mortality in young fish and crustaceans, in E. Naybr and R.G. Hartnoll (Eds) *Cyclic Phenomena in Marine Plants and Animals, Proc. 13th European Marine Biology Symp.* (New York: Pergamon) pp 73–82.

Botsford, L.W. (1981) The effects of increased individual growth rates on depressed population size. *Am. Naturalist* 117:38–63.

Brauer, F. and Soudack, A.C. (1979) Stability regions in predator–prey systems with constant-rate prey harvesting. *J. Math. Biol.* 8:55–71.

Buckingham, S. (1979) *Adaptive Environmental Management: Your Role in the Workshop*. User's guide and slide show (Vancouver, BC: Clover Productions).

Buckingham, S. and Walters, C. (1975) *A Control System for Intraseason Salmon Management, Proc. Workshop on Salmon Management.* Collaborative Paper CP-75-2 (Laxenburg, Austria: International Institute for Applied Systems Analysis).

Burgoyne, G.E. (1981) Observations on a heavily exploited deer population, in C.W. Fowler and T.D. Smith (Eds) *Dynamics of Large Mammal Populations* (New York: Wiley) pp 403–13.

Caddy, J.F. (1983) *An Alternative to Equilibrium Theory for Management of Fisheries.* FAO Series WCFMD/prep 1. Sess. 2./Panel 2.2 (Rome: FAO), mimeo.

Caughley, G. (1976) Wildlife management and the dynamics of ungulate populations, in T.H. Croaker (Ed) *Applied Biology*, Vol. 1 (London: Academic Press) pp 183–246.

Caughley, G. (1981) What we do not know about the dynamics of large mammals, in C.N. Fowler and T. Smith (Eds) *Dynamics of Large Mammal Populations* (New York: Wiley) pp 361–72.

Clark, C.W. (1973) The economics of overexploitation. *Science* 181:630–4.

Clark, C.W. (1976) *Mathematical Bioeconomics: The Optimal Management of Renewable Resources* (New York: Wiley).

Clark, W.C., Jones, D.D., and Holling, C.S. (1979) Lessons for ecological policy design: A case study of ecosystem management. *Ecological Modelling* 7:1–53.

Cochran, W.G. (1963) *Sampling Techniques*, 2nd edn (New York: Wiley).

Cody, M.L. (1976) Optimization in ecology. *Science* 183:1156–64.

Cooper and Smith (1981) unpublished.

Costanza, R. and Sklar, F.H. (1983) *Mathematical Models of Freshwater Wetland and Shallow Water Ecosystems: An Articulated Review.* Paper prepared for SCOPE Int. Conf. on Freshwater Wetlands and Shallow Water Bodies, Tallinn, USSR, August 1983.

Cushing, D.H. (1968) *Fisheries Biology, A Study in Population Dynamics* (Madison, WI: University of Wisconsin Press).

Deriso, R.B. (1980) Harvesting strategies and parameter estimation for an age-structured model. *Can. J. Fish. Aquatic Sci.* 37:268-82.

Deriso, R.B. (1983) *Sources of Variability in a Delay-Difference Model: Application to Pacific Halibut.* IPWC Working Paper (Seattle, WA: Int. Pacific Halibut Commission), mimeo.

Detchmendy, D. and Sridhar, R. (1966) Sequential estimation of states and parameters in noisy nonlinear dynamical systems. *Trans. ASME, J. Basic Eng.* 88:362-8.

Eberhardt, L. (1981) Population dynamics of the Pribolof fur seals, in C.W. Fowler and T.D. Smith (Eds) *Dynamics of Large Mammal Populations* (New York: Wiley).

Ermoliev, Yu.M. (1976) *Stochastic Programming Methods* (Moscow: Nauka) (in Russian).

Ermoliev, Y. and Gaivoronski, A.A. (1984) *Stochastic Quasigradient Methods and Their Implementation.* Working Paper WP-84-55 (Laxenburg, Austria: International Institute for Applied Systems Analysis).

ESSA Ltd (1982) *Review and Evaluation of Adaptive Environmental Assessment and Management* (Vancouver, BC: Environment Canada).

Fel'dbaum, A.A. (1960, 1961) Theory of dual control I-IV. *Automatic Remote Control USSR* 21:1240-9, 1453-65; 22:3-16, 129-43 (in Russian).

Fowler, C.W. and Smith, T.D. (Eds) (1981) *Dynamics of Large Mammal Populations* (New York: Wiley).

Gaivoronski, A.A. and Ermoliev, Y. (1979a) Stochastic optimization problems with simultaneous parameter estimation. *Izv. Akad. Nauk. SSSR* 4:29-38 (in Russian).

Gaivoronski, A.A. and Ermoliev, Y. (1979b) Problems of stochastic programming with consideration of parameter consistency. *Technical Cybernetics* 4:29-38 (in Russian).

Gallant, A.R. (1975) Seemingly unrelated nonlinear regressions. *J. Econometrics* 3:35-50.

Gatto, M. and Rinaldi, S. (1976) Mean value and variability of fish catches in fluctuating environments. *J. Fish. Res. Board Canada* 33:189-93.

Gelb, A. (Ed) (1974) *Applied Optimal Estimation* (Cambridge, MA: MIT Press).

Glantz, M. (1983) *Man, State, and Fisheries: An Inquiry into some Societal Constraints that affect Fisheries Management.* Rep. FAO Tech. Commission to Examine Changes in Abundance and Species Composition of Neritic Stocks, San Jose, Costa Rica (mimeo).

Glantz, M.H. and Thompson, J.D. (Eds) (1981) *Resource Management and Environmental Uncertainty: Lessons from Coastal Upwelling Fisheries* (New York: Wiley).

Goodman, D. (1981) Life history analysis of large mammals, in C.W. Fowler and T.D. Smith (Eds) *Dynamics of Large Mammal Populations* (New York: Wiley).

Graybill, F.A. (1961) *An Introduction to Linear Statistical Models,* Vol. I (New York: McGraw-Hill).

Gross, J.E., Roelle, J.E., and Williams, G.L. (1973) *Program ONEPOP an Information Processor: A System Modelling and Communication Project.* Progress Report, Colorado Coop. Wildlife Research Unit (Fort Collins, CO: Colorado State University).

Gulland, J.A. (1961) Fishing and the stocks of fish at Iceland *UK Ministry of Agriculture, Fisheries and Food, Fish Invest,* Ser. 2, 23(4).

Gulland, J.A. (1981) An overview of applications of quatious research in fisheries management, in K.B. Haley (Ed) *Applied Operations Research in Fishing* (New York: Plenum) pp 125-35.

Haber, G.C., Walters, C.J., and Cowan, I.M. (1976) *Stability Properties of a Wolf-Ungulate System in Alaska and Management Implications.* Research Report R-5-R, Institute of Animal Resource Ecology, University of British Columbia, Vancouver, Canada.

Hilborn, R. (1973) A control system for FORTRAN simulation programming. *Simulation* 20:172-3.

Hilborn, R. (1979) A comparison of fisheries control systems that utilize catch and effort data. *J. Fish. Res. Board Canada* 36:1477-89.

Hilborn, R. (1985) Apparent stock-recruitment relationships in mixed stock fisheries. *Can. J. Fish. Aquatic Sci.* 42:718-23.

Holling, C.S. (1965) The functional response of predators to prey density and its role in mimicry and population regulations. *Mem. Entomol. Soc. Canada* 45:1-60.

Holling, C.S. (1973) Resilience and stability of ecosystems. *Ann. Rev. Ecol. Systems* 4:1-23.

Holling, C.S. (Ed) (1978) *Adaptive Environmental Assessment and Management.* Wiley International Series on Applied Systems Analysis, Vol. 3 (Chichester, UK: Wiley).

Holling, C.S. (1980) Forest insects, forest fires, and resilience, in H. Mooney *et al.* (Eds) *Fire Regimes and Ecosystem Properties.* Gen. Tech. Rep. WO-26 (USDA Forest Service) pp 445-64.

Hourston, A.S. (1981) Summary tables for annual assessments of the status of British Columbia herring stocks in the 1970s. *Can. Data Rep. Fish. Aquatic Sci.* 250.

Hourston, A.S. and Nash, F.W. (1972) Millions of adult fish (catch + spawners) at age for British Columbia herring areas, 1950-51 to 1969-70. *Fish. Res. Board Canada MS Report* 1228.

Jazwinski, A.H. (1970) *Stochastic Processes and Filtering Theory* (New York: Academic Press).

Jones, D.D. and Walters, C.J. (1976) Catastrophe theory and fisheries regulation. *J. Fish. Res. Board Canada* 33:2829-33.

Jones, R. (1981) *The Use of Length Composition Data in Fish Stock Assessments (With Notes on VPA and Cohort Analysis).* FAO Fisheries Circular No. 734 (Rome: FAO).

Kalman, R.E., Falb, P.R., and Arbib, M.A. (1969) *Topics in Mathematical System Theory* (New York: McGraw-Hill).

Keeney, R. (1977) A utility function for examining policy affecting salmon in the Skeena River. *J. Fish. Res. Board Canada* 34:49-63.

Keeney, R.L. and Raiffa, H. (1976) *Decisions with Multiple Objectives* (New York: Wiley).

Krutilla, J.V. and Fisher, A.C. (1975) *The Economics of Natural Environments* (Baltimore, MD: Johns Hopkins University Press).

Kuhn, T.S. (1962) *The Structure of Scientific Revolutions* (Chicago: University of Chicago Press).

Laponce, J.A. (1972) Experimenting: A two-person game between man and nature, in J.A. LaPonce and P. Smoker (Eds) *Experimentation and Simulation in Political Science* (Toronto: University of Toronto Press).

Larson, R.E. (1968) *State Increment Dynamic Programming* (New York: Elsevier).

Larson, R.E. and Casti, J. (1978) *Principles of Dynamic Programming. Part I: Basic Analytic and Computational Methods* (New York: Dekker).

Larson, R.E. and Casti, J. (1982) *Principles of Dynamic Programming. Part II: Advanced Theory and Applications* (New York: Dekker).

Leopold, A. (1933) The conservation ethic. *Journal of Forestry* 31:634-43.

Lett, P.F. and Benjaminson, T. (1977) A stochastic model for the management of the northwestern Atlantic harp seal (*Pagophilus groenlandicus*) populations. *J. Fish. Res. Board Canada* 34:1155-87.

Levastu, T. and Favorite, F. (1977) *Preliminary Report on Dynamical Numerical Marine Ecosystem Model (DYNUMES II) for the Eastern Bering Sea* (Seattle, WA: US Dept Commerce, NOAA, Northwest Fisheries Center).

Lindblom, C.E. (1959) The science of muddling through. *Public Administration Review* 19:79-88.

Ludwig, D. and Walters, C. (1981) Measurement errors and uncertainty in parameter estimates for stock and recruitment. *Can. J. Fish. Aquatic Sci.* 38:711-20.

Ludwig, D.W. and Walters, C.J. (1985) Are age-structured models appropriate for catch-effort data? *Can. J. Fish. Aquatic Sci.* 42:1066-72.

Ludwig, D., Jones, D., and Holling, C.S. (1978) Qualitative analysis of insect outbreak systems: The spruce budworm and the forest. *J. Animal Ecol.* 47:315-32.

Mace, P. (1983) *Predator-Prey Functional Responses and Predation by Staghorn Sculpins (Leptocottus armatus) on Chum Salmon Fry (Oncorhynchus keta).* PhD Dissertation, University of British Columbia, Vancouver, BC, Canada.

Mangel, M. (1982) Search effort and catch rates in fisheries. *European J. Operations Research* 11:361-6.

Marchetti, C. (1980) Society as a learning system: Discovery, invention, and innovation cycles revisited. *Technological Forecasting and Social Change* 18:267-82.

May, R.M., Beddington, J.R., Clark, C.W., Holt, S.J., and Laws, R.M. (1979) Management of multispecies fisheries. *Science* 205:267-77.

McCullough, D. (1979) *The George Reserve Deer Herd: Population Ecology of a K-Selected Species* (Ann Arbor, MI: University of Michigan Press).

Mehra, R.K. (1974) Optimal input signals for parameter estimation in dynamic systems—survey and new results. *IEEE Trans. Automatic Control* AC-10(6):753-68.

Mehra, R.K. (1981) Choice of input signals, in P. Eykhoff (Ed), *Trends and Progress in System Identification* (New York: Pergamon) pp 305-66.

Meinhold, R.J. and Singpurwalla, N.D. (1983) Understanding the Kalman filter. *Am. Statistician* 37:123-7.

Mendelssohn, R. (1980) Using Markov decision models and related techniques for purposes other than simple optimization: Analysing the consequences of policy alternatives on the management of salmon runs. *Fish. Bull.* 78:35-50.

Moskowitz, B.A. (1978) The acquisition of language. *Scientific American* 239(5):92-108.

Murray, D.M. and Yakowitz, S.J. (1979) Constrained differential dynamic programming. *Water Resources Research* 15:1017-27.

Overton, S. (1978) A strategy of model construction, in C.A.S. Hall and J. Day (Eds) *Ecosystem Modelling in Theory and Practice: An Introduction with Case Histories* (New York: Wiley).

Paloheimo, J.E. (1980) Estimation of mortality rates in fish populations. *Trans. Am. Fish. Soc.* 109:378-86.

Paulik, G.J. (1973) Studies of the possible forms of the stock-recruitment curve, in B.B. Parrish (Ed) *Fish Stocks and Recruitment. Rapports et Procès-Verbaux Reun. Cons. Int. Explor. Mer.* 164:302-15.

Paulik, G.J. and Greenough, J.W. (1967) Management analysis for a salmon resource system, in K.E.F. Watt (Ed) *Systems Analysis in Ecology* (New York: Academic Press) pp 215-52.

Pauly, D. (1979) On the interrelationships between natural mortality, growth parameters, and mean environmental temperature in 175 fish stocks. *J. Cons. Int. Explor. Mer.* 39:175-92.

Pearse, P.H. (1982) *Turning the Tide: A New Policy for Canadian Pacific Fisheries.* Final Report, Commission on Pacific Fisheries (Vancouver, BC: Canadian Dept of Fisheries and Oceans).

Pekelman, D. and Tse, E. (1980) Experimentation and budgeting in advertising: An adaptive control approach. *Operations Research* 28:321-47.

PERCIS (1976) The PERCIS International Symposium. *J. Fish. Res. Board Canada* 34(10).

Peterman, R.M. (1975) New techniques for policy evaluation in ecological systems: Methodology for a case study of Pacific salmon fisheries. *J. Fish. Res. Board Canada* 32:2179-88.

Peterman, R.M. (1977) A simple mechanism that causes collapsing stability regions in exploited salmonid populations. *J. Fish. Res. Board Canada* 34:1130-42.

Peterman, R.M. (1981) Form of random variation in salmon smolt-to-adult relations and its influence on production estimates. *Can. J. Fish. Aquatic Sci.* 38:1113-19.

Poljak, B.T. and Tsypkin, Ya.Z. (1980) Robust identification. *Automatica* 16:53-69.

Pope, J.G. (1972) An investigation of the accuracy of virtual population analysis using cohort analysis. *Res. Bull. ICNAF* 9:65-74.

Priestly, M.B. (1982) On the fitting of general non-linear time series models, in O.D. Anderson (Ed), *Time Series Analysis: Theory and Practice* Vol. 1 (Amsterdam: North-Holland) pp 717-31.

Pycha, R.L. (1980) Changes in mortality of lake trout (*Salvelinus namaycush*) in Michigan waters of Lake Superior in relation to sea lamprey (*Petromyzon marinus*) predation, 1968-78. *Can. J. Fish. Aquatic Sci.* 37:2063-73.

Raiffa, H. (1968) *Decision Analysis* (Reading, MA: Addison-Wesley).

Regier, H.A. and Henderson, H.F. (1973) Towards a broad ecological model of fish communities and fisheries. *Trans. Am. Fish. Soc.* 102:56-72.

Ricker, W.E. (1954) Stock and recruitment. *J. Fish. Res. Board Canada* 11:559-623.

Ricker, W.E. (1963) Big effects from small causes: Two examples from fish population dynamics. *J. Fish. Res. Board Canada* 20:257-64.

Ricker, W.E. (1973a) Linear regressions in fishery research. *J. Fish. Res. Board Canada* 30:409-34.

Ricker, W.E. (1973b) Two mechanisms that make it impossible to maintain peak-period yields from stocks of Pacific salmon and other fishes. *J. Fish. Res. Board Canada* 30:1275-86.

Ricker, W.E. and Manzer, J. (1974) Recent information on salmon stocks in British Columbia. *Int. N. Pacific Fish. Comm. Bull.* 29:1-24.

Rinaldi, S. and Gatto, M. (1978) On the determination of a commercial fishery production model. *Ecological Modelling* 8:165-72.

Robbins, H. and Monro, S. (1951) A stochastic approximation method. *Ann. Math. Statistics* 23:400-7.

Ruppert, D., Reish, R.L., Deriso, R.B., and Carroll, R.J. (1983) *Development of a Biological Simulation Model for Atlantic Menhaden. Volume II: The Stochastic Model*, NMFS Beaufort Laboratory, Report No. 2, 135 pp.

Schaefer, M.B. (1977) A study of the dynamics of the fishery for yellowfin tuna in the eastern tropical Pacific Ocean. *Inter.-Am. Trop. Tuna Comm. Bull.* 2:247-68.

Schnute, J. (1977) Improved estimates from the Schaefer production model: Theoretical considerations. *J. Fish. Res. Board Canada* 34:583-603.

Schnute, J. (1985) A general theory for the analysis of catch and effort data. *Can. J. Fish Aq. Sci.* 42:414-29.

SCOL (1971) The International Symposium on Salmonid Communities in Oligotrophic Lakes. *J. Fish. Res. Board Canada* 29(6).

Sharp, G. (Rapp.) (1980) *Workshop on the Effects of Environmental Variation on the Survival of Larval Pelagic Fishes.* Intergovernmental Ocean. Commission Workshop Rep. No. 28 (Rome: FAO).

Shepard, M.P., Withler, F.C., McDonald, J., and Avo, K.V. (1964) Further information on spawning stock size and resultant production for Skeena sockeye. *J. Fish. Res. Board Canada* 21:1329-31.

Shepherd, J.G. (1982) A versatile new stock-recruitment relationship for fisheries, and the construction of sustainable yield curves. *J. Cons. Int. Explor. Mer.* 40:67-75.

Skud, B.E. (1979) Revised estimates of halibut abundance and the Thompson-Burkenroad debate. *Int. Pac. Halibut Comm., Sci. Rept.* 56, 35pp.

SLIS (1980) Proceedings of the Sea Lamprey International Symposium. *Can. J. Fish. Aquatic Sci.* 37(11):1585-2214.

Smith, A.D.M. (1979) *Adaptive Management of Renewable Resources with Uncertain Dynamics.* PhD Dissertation, University of British Columbia, Vancouver, BC, Canada.

Smith, A.D.M. and Walters, C.J. (1981) Adaptive management of stock-recruitment systems. *Can. J. Fish. Aquatic Sci.* 38:690-703.

Smith, T. and Polacheck, T. (1981) Reexamination of the life table for northern fur seals with implications about population regulatory mechanisms, in C.W. Fowler and T.D. Smith (Eds) *Dynamics of Large Mammal Populations* (New York: Wiley) pp 99-120.

Soeda, T. and Yoshimura, Y. (1973) A practical filter for systems with unknown parameters. *Trans. ASME, J. Dynamic Systems, Measurement, and Control* 74:396-401.

Soutar, A. and Isaacs, J.D. (1974) History of fish populations inferred from fish scales in anaerobic sediments off California. *NOAA/NMFS Fish. Bull.* 72:257.

Southwood, T.R.E. and Comins, H.N. (1976) A synoptic population model. *J. Animal Ecol.* 45:949-65.

STOCS (1981) Proceedings of the 1980 Stock Concept International Symposium. *Can. J. Fish. Aquatic Sci.* 38:1457-921.

Tse, E., Bar-Shalom, Y., and Meier, L. (1973) Wide-sense adaptive dual control of stochastic nonlinear systems. *IEEE Trans. Automatic Control* AC-18:98-108.

Tsypkin, Ya.Z. (1971) *Adaptation and Learning in Automatic Systems* (New York: Academic Press).

Tukey, J. (1977) *Exploratory Data Analysis* (Reading, MA: Addison-Wesley).

Uhler, R.S. (1979) Least squares regression estimates of the Schaefer model: Some Monte Carlo simulation results. *Can. J. Fish. Aquatic Sci.* 37:1284-94.

Van Hyning, J.M. (1973) Stock-recruitment relationships for Columbia River Chinook salmon, in B.B. Parrish (Ed) *Fish Stocks and Recruitment. Rapports et Procès-Verbaux Reun. Cons. Int. Perm. Explor. Mer.* 164.

Walters, C.J. (1969) A generalized computer simulation model for fish population studies. *Trans. Am. Fish. Soc.* 98:505-12.

Walters, C.J. (1974) An interdisciplinary approach to development of watershed models. *Technological Forecasting and Social Change* 6:299-323.

Walters, C.J. (1975) Optimal harvest strategies for salmon in relation to environmental variability and uncertain production parameters. *J. Fish. Res. Board Canada* 32:1777-84.

Walters, C.J. (1977) Confronting uncertainty, in D.V. Ellis (Ed) *Pacific Salmon: Management for People* (Victoria, BC: University of Victoria Press) pp 261-97.

Walters, C.J. (1980) Systems principles in fisheries management, in R.T. Lackey and L.A. Nielsen (Eds) *Fishery Management* (New York: Wiley) Ch. 7, pp 167-95.

Walters, C.J. (1981) Optimum escapements in the face of alternative recruitment hypotheses. *Can. J. Fish. Aquatic Sci.* 38:678-89.

Walters, C.J. and Hilborn, R. (1976) Adaptive control of fishing systems. *J. Fish. Res. Board Canada* 33:145-59.

Walters, C.J. and Ludwig, D. (1981) Effects of measurement errors on the assessment of stock-recruitment relationships. *Can. J. Fish. Aquatic Sci.* 38:704-10.

Walters, C.J., Steer, G., and Spangler, G. (1980) Responses of lake trout (*Salvelinus namaycush*) to harvesting, stocking, and lamprey reduction. *Can. J. Fish. Aquatic Sci.* 37:2133-45.

Walters, C.J., Hilborn, R., Staley, M., and Wong, F. (1982) *An Assessment of Major Commercial Fish and Invertebrate Stocks on the West Coast of Canada, with Recommendations for Management.* Commission on Pacific Fisheries Policy Report (mimeo, available from ESSA Ltd, Vancouver, BC).

Ware, D.M. (1980) Bioenergetics of stock and recruitment. *Can. J. Fish. Aquatic Sci.* 37:1012-24.

Watt, K.E.E. (1968) *Ecology and Resource Management: A Quantitative Approach* (New York: McGraw-Hill).

Weinburg, G.M. (1975) *An Introduction to General Systems Thinking* (New York: Wiley).

Wells, L. (1980) Lake trout (*Salvelinus namaycush*) and sea lamprey (*Petromyzon marinus*) populations in Lake Michigan, 1971-78. *Can. J. Fish. Aquatic Sci.* 37:2047-51.

Wenk, C.J. and Bar-Shalom, Y. (1980) A multiple model adaptive dual control algorithm for stochastic systems with unknown parameters. *IEEE Trans. Automatic Control* AC-25:703-10.

Wierzbicki, A. (1977) *Models and Sensitivity of Control Systems* (Warsaw: Wydawnictwa Naukowo-Techniczne) (in Polish; English translation forthcoming).

Yoshimura, T., Watanabe, K., Konishi, K., and Soeda, T. (1979) A sequential failure detection approach and the identification of failure parameters. *Int. J. Systems Sci.* 10:827-36.

Young, P. (1974) Recursive approaches to time series analysis. *Inst. Math. Appl. Bull.* 10(5/6):209-24.

Index